Remapping Ethiopia
Socialism & After

EDITED BY

WENDY JAMES
DONALD L. DONHAM
EISEI KURIMOTO
ALESSANDRO TRIULZI

James Currey
OXFORD

Ohio University Press
ATHENS

Addis Ababa University Press
ADDIS ABABA

James Currey Ltd
73 Botley Road
Oxford
OX2 0BS

Addis Ababa University Press
P.O. Box 1176
Addis Ababa

Ohio University Press
Scott Quadrangle
Athens, Ohio 45701

© James Currey Ltd 2002
First published 2002

1 2 3 4 5 06 05 04 03 02

British Library Cataloguing in Publication Data
Remapping Ethiopia : socialism and after. - (Eastern African studies)
I. Ethiopia - History - 1974 -
1. James, Wendy, 1940-
963'.07

ISBN 0-85255-456-7 (James Currey Cloth)
0-85255-455-9 (James Currey Paper)

**Library of Congress Cataloging-in-Publication Data
available on request**

ISBN 0-8214-1447-X (Ohio University Press Cloth)
0-8214-1448-8 (Ohio University Press Paper)

EASTERN AFRICAN STUDIES

Revealing Prophets
Edited by David M. Anderson &
Douglas H. Johnson

*East African Expressions
of Christianity*
Edited by Thomas Spear &
Isaria N. Kimambo

The Poor Are Not Us
Edited by David M. Anderson &
Vigdis Broch-Due

*Potent Brews**
Justin Willis

Swahili Origins
James de Vere Allen

Being Maasai
Edited by Thomas Spear &
Richard Waller

Jua Kali Kenya
Kenneth King

Control & Crisis in Colonial Kenya
Bruce Berman

Unhappy Valley
Book One: State & Class
Book Two: Violence &
Ethnicity
Bruce Berman &
John Lonsdale

Mau Mau from Below
Greet Kershaw

*The Mau Mau War
in Perspective*
Frank Furedi

*Squatters & the Roots
of Mau Mau 1905–63*
Tabitha Kanogo

*Economic & Social Origins
of Mau Mau 1945–53*
David W. Throup

Multi-Party Politics in Kenya
David W. Throup
& Charles Hornsby

Empire State-Building
Joanna Lewis

*Decolonization & Independence
in Kenya 1940–93*
Edited by B.A. Ogot
& William R. Ochieng'

*Eroding the Commons**
David Anderson

Penetration & Protest in Tanzania
Isaria N. Kimambo

Custodians of the Land
Edited by Gregory Maddox,
James L. Giblin &
Isaria N. Kimambo

*Education in the Development
of Tanzania 1919–1990*
Lene Buchert

The Second Economy in Tanzania
T.L. Maliyamkono
& M.S.D. Bagachwa

*Ecology Control &
Economic Development in
East African History*
Helge Kjekshus

Siaya
David William Cohen
& E.S. Atieno Odhiambo

*Uganda Now
Changing Uganda
Developing Uganda
From Chaos to Order
Religion & Politics in East Africa*
Edited by Holger Bernt Hansen &
Michael Twaddle

*Kakungulu & the Creation
of Uganda 1868–1928*
Michael Twaddle

Controlling Anger
Suzette Heald

Kampala Women Getting By
Sandra Wallman

*Political Power in Pre-Colonial
Buganda**
Richard Reid

Alice Lakwena & the Holy Spirits
Heike Behrend

Slaves, Spices & Ivory in Zanzibar
Abdul Sheriff

Zanzibar Under Colonial Rule
Edited by Abdul Sheriff &
Ed Ferguson

*The History & Conservation of
Zanzibar Stone Town*
Edited by Abdul Sheriff

Pastimes & Politics
Laura Fair

*Ethnicity & Conflict
in the Horn of Africa*
Edited by Katsuyoshi Fukui
& John Markakis

*Conflict, Age & Power
in North East Africa*
Edited by Eisei Kurimoto
& Simon Simonse

*Property Rights & Political
Development in Ethiopia & Eritrea*
Sandra Fullerton Joireman

*Revolution & Religion
in Ethiopia*
Øyvind M. Eide

Brothers at War
Tekeste Negash &
Kjetil Tronvoll

From Guerrillas to Government
David Pool

*A History of Modern Ethiopia
1855–1991*
Second edition
Bahru Zewde

*Pioneers of Change
in Ethiopia**
Bahru Zewde

Remapping Ethiopia
Edited by W. James, D. Donham,
E. Kurimoto & A. Triulzi

*Southern Marches
of Imperial Ethiopia**
Edited by Donald L. Donham &
Wendy James

* forthcoming

Contents

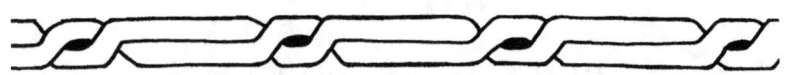

List of Illustrations	vii
Preface *by Wendy James*	ix
List of Contributors	xi
Introduction *Donald L. Donham*	1

I The Political Framework   9

1 Controlling Space in Ethiopia   9
Christopher Clapham

II Looking Back on Projects of the Socialist State, 1974–91   33

Introduction   33
Donald L. Donham

2 Evading the Revolutionary State: 37
The Hor under the Derg
Tadesse Wolde

3 Memory & the Humiliation of Men: 59
The Revolution in Aari
Alexander Naty

4 Close yet Far: 74
Northern Shewa under the Derg
Ahmed Hassan Omer

5 Garrison Towns & the Control of Space   90
in Revolutionary Tigray
Jenny Hammond

6 Modernist Dreams & Human Suffering: 116
Villagization among the Guji Oromo
Taddesse Berisso

7 Surviving Resettlement in Wellegga: The Qeto Experience *Alula Pankhurst*	133

III The Promise of 1991: Re-shaping the Future & the Past — 151

Introduction *Donald L. Donham*	151
8 Paradoxes of Power & Culture in an Old Periphery: Surma, 1974–98 *Jon G. Abbink*	155
9 Political Visibility & Automatic Rifles: The Muguji in the 1990s *Hiroshi Matsuda*	173
10 Evangelical Christianity & Ethnic Consciousness in Majangir *Ren'ya Sato*	185
11 Capturing a Local Elite: The Konso Honeymoon *Elizabeth Watson*	198
12 Fear & Anger: Female Versus Male Narratives among the Anywaa *Eisei Kurimoto*	219
13 Imperial Nostalgia: Christian Restoration & Civic Decay in Gondar *Cressida Marcus*	239

IV 'Ethiopia' from the Outside — 257

Introduction *Donald L. Donham*	257
14 No Place to Hide: Flag-waving on the Western Frontier *Wendy James*	259
15 Battling with the Past: New Frameworks for Ethiopian Historiography *Alessandro Triulzi*	276
Bibliography	289
Index	300

List of Illustrations

Maps
1	Ethiopia: provincial boundaries, 1974	xii
2	Ethiopia: regional boundaries, 1995	xiii
1.1	The central highlands (to illustrate Chapter 4)	8
1.2	The northern highlands (to illustrate Chapters 5, 13)	8
1.3	The southern plateaux & valleys (to illustrate Chapters 2, 3, 6, 8, 9, 11)	31
1.4	The western plateaux & valleys (to illustrate Chapters 7, 12, 14)	32

Plates

Frontispiece Taking the palace in Addis Ababa, May 1991

1.1	Mengistu's effigy awaiting burning at the land reform celebrations, Woldia, Wello	5
2.1	Cattle gates of the Hor *qawot*	43
2.2	Rufo Ali, *qawit*	44
2.3	Lemma, *fund'o* leader and elders	47
5.1	New roles for women: ploughing near Awhie	94
5.2	Roman, Aregash & Mebrat, three of the first women fighters	95
6.1	Villagization (i): Guji Oromo	120
6.2	Villagization (ii): Guji Oromo	121
8.1	The *komoru* of Chai-Suri, 1996 (son of Dollote IV)	160
8.2	EPRDF troops presenting themselves to the Suri, November 1991	162
8.3	Missionary teacher Mike Bryant with Daniel Kibo (a Christian convert) and a non-converted Suri November 1999, Tulgi	169
11.1	Konso: listening to the new message	203
11.2	Flag waving at the Konso rally	203
11.3	Acclaim for the new policies, Konso	204
12.1	Dance of an Anywaa Youth Association on May Day, 1990, at Abwobo	220
12.2	Anwyaa petty traders at Pinyudo, 1989	221
12.3	'Linking arms.' Painting on the wall of the administrative office in Gambela town, 1993	231
13.1	Celebrating a church festival at Gondar	251
14.1	The UN flag flies over a damaged Nasir, 1991	271
14.2	Six hundred miles on foot and several years from home: a transit camp in Ethiopia, 1994	273

List of Illustrations

Figures
1.1	The spatial patterns of resettlement, Assosa, early 1980s	18
1.2	Plan of an ideal village layout	18
9.1	Origins of Muguji rifles	181
13.1	The sacred geography of Gondar	241

Tables
1.1	Populations of Ethiopia & Eritrea by ethnic identification, 1984	22
4.1	Landowners in Northern Shewa, 1976	80
6.1	Planned villages in Guji Oromo	126
9.1	Types of Kalashnikov held by the Muguji, 1999	180
9.2	Inflow routes	181
9.3	Automatic rifles purchased from 1990–March, 1999	182

Note on Transliteration

The system of writing Amharic words in this volume is based on that used by Allan Hoben in his *Land Tenure among the Amhara of Ethiopia* (1973). We are grateful to Dr. Girma Getahun for his advice and assistance in preparing transcriptions and translations from Amharic. In representing other languages of Ethiopia, we have accepted the judgement of individual contributors. For well-known place and personal names we use forms already recognized internationally (Addis Ababa, Gondar, etc.).

Preface

The Ethiopian revolution of 1974, which toppled the imperial regime of Haile Selassie I, ushered in a decade and a half of centrally-directed rule by the socialist regime known as 'the Derg'. This was a period of state control penetrating many areas of life for the Ethiopian people, causing many difficulties and suffering. Academic life managed to survive at the University of Addis Ababa, partly by keeping a low profile, but active researchers, especially field researchers in the social sciences, faced many frustrations. Little was published during this period about the conditions of ordinary life across the country. With the change of government in 1991, it once again became possible for researchers, both Ethiopians and scholars from elsewhere, to pursue field projects. A new generation of writing about the country has begun, and the present volume is a contribution to this development. The studies collected here are by scholars from a range of disciplines, including anthropology, political science, history and geography, all of whom introduce first-hand knowledge of Ethiopia. A few represent an older generation who have been able to return to regions they knew before the revolution and to reflect on changes that have happened since, while the majority are younger scholars who embarked on their research in the 1990s.

An earlier collection of field-based studies devoted to examining the relations between centre and periphery in twentieth-century imperial Ethiopia, *The Southern Marches of Imperial Ethiopia: Essays in Social Anthropology and History*, was edited by Donham and James in 1986 and was well received. However, the book only dealt with the period up to 1974. After the change of government in Ethiopia in 1991, and the general resumption of research activity, the idea of a new volume took shape in discussions among the present editors. We felt it would be worth attempting a sequel to *Southern Marches* in order to trace the way in which the centralizing policies of the socialist government, and then the regional devolution policies of the new government, actually affected and are still affecting the lives of people in specific localities across the country.

The project was first launched at a special panel on 'Rethinking Centres and Peripheries in Ethiopia' at the XIIIth International Conference of Ethiopian Studies, held in Kyoto, Japan, in December 1997. We are much indebted to Professor Masao Kawai, chair of the organizing committee for the Conference, and Professor Katsuyoshi Fukui, chair of the executive committee and long-established pioneer of Ethiopian studies in Japan. Our panel was supported by the Murata Science and Technology Foundation,

Preface

to whom we express our sincere gratitude. The important place now occupied by Japanese scholars in the field of Ethiopian studies was marked by the Kyoto Conference as a whole. All the field-based chapters in the present volume are developed versions of papers presented in Kyoto or studies closely related to them (for detailed references and further information, see *Ethiopia in Broader Perspective*, the Kyoto Conference papers edited in three volumes by Fukui, Kurimoto, and Shigeta 1997). In order to take our project forward, two workshops were held in Oxford, in January and in September, 1999. At this stage we invited Professor Christopher Clapham to write the opening chapter about Ethiopia's changing political relations with its regions from 1974 to the present, as a context for the local studies. We are indebted to all those who offered papers in our panel at Kyoto, to those who presented papers with similar topics in other panels on that occasion, and to those who took part in the Oxford workshops and contributed to the clarification of important issues. In constructing this book, we had a hard task in having to select from a wide range of possible contributions those case studies which would fit together best and collectively illustrate our main themes.

We are most grateful to the organizers of the Kyoto conference for providing the opportunity for this project to be launched in a fertile context. We would also like to express our gratitude to other bodies which provided financial assistance for the Oxford workshops, in particular the British Academy, the Institute of Social and Cultural Anthropology at Oxford, and participants' own institutions where these were able to provide fares. Both Harris Manchester College, and St. Anne's College, went out of their way to provide convenient facilities and hospitality. Editorial help in fashioning our chapters into a coherent book was provided by Robert Parkin, Eric Henry, and Girma Getahun, to whom we express our thanks. Jonathan Rae and Ronan Foley drew the main maps. This has been very much an international and collective venture, made possible not only by the series of formal meetings we have held but by individual discussions between members of the network as they have met each other informally in half a dozen countries since the Kyoto conference.

Assistance with publication costs was given by Emory University, Atlanta, and by the Istituto Universitario Orientale of Naples. We would like to express our gratitude for this help, which will be applied to the production costs of this volume and in support of the locally-available edition being brought out by the Addis Ababa University Press. The assistance is also being used to support the re-issue of *Southern Marches* in a paperback edition, as a companion to the present volume.

Wendy James

Notes on Contributors

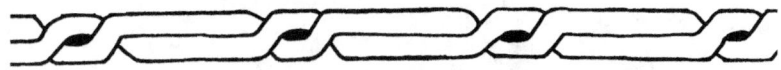

JON ABBINK is a senior researcher at the African Studies Centre, Leiden, and Professor of African Ethnic Studies at the Vrije Universiteit, Amsterdam. His most recent book (ed. with G. Aijmer) is *Meanings of Violence* (2000).

AHMED HASSAN OMER is completing his Ph.D. at the Institut d'Etudes Africaines, Université de Provence. He has been a lecturer at the Institute of Ethiopian Studies and the Department of History, Addis Ababa University.

CHRISTOPHER CLAPHAM is Professor of Politics & International Relations at Lancaster University, and has written extensively on government in Ethiopia, including *Transformation and Continuity in Revolutionary Ethiopia* (1988).

DONALD L. DONHAM is Professor of Anthropology at Emory University, Atlanta. His most recent book is *Marxist Modern: An Ethnographic History of the Ethiopian Revolution* (1999).

JENNY HAMMOND completed her doctorate at Oxford on the social history of the revolution in Tigray. Her most recent book is *Fire From the Ashes: A Chronicle of the Revolution in Tigray, Ethiopia, 1975-1991* (1999).

WENDY JAMES, FBA is Professor of Social Anthropology in the University of Oxford and a Fellow of St. Cross College. Her most recent book (ed. with P. Dresch & D. Parkin) is *Anthropologists in a Wider World: Essays on Field Research* (2000).

EISEI KURIMOTO is Professor of Anthropology at the National Museum of Ethnology at Osaka. He has carried out sustained research both in the Sudan and Ethiopia, with particular reference to the Pari and the Anywaa.

CRESSIDA MARCUS is completing a doctorate at Oxford, based on her field research in Gondar, and has edited an issue of the *Journal of Ethiopian Studies*.

HIROSHI MATSUDA is Associate Professor in the Department of Cultural Anthropology, Kyoto Bunkyo University. He has carried out research on inter-ethnic relations in the Lower Omo Valley of south-western Ethiopia.

ALEXANDER NATY is Professor in the Department of Anthropology in Asmara University. He has published several articles on the languages and ethnography of N.E. Africa.

ALULA PANKHURST is Associate Professor of Anthropology in the Department of Sociology at Addis Ababa University. His publications include *Resettlement and Famine in Ethiopia: The Villagers' Experience* (1992), and with Dena Freeman he is editing *Peripheral People: the Excluded Minorities of Ethiopia* (London: Hurst).

REN'YA SATO is Associate Professor at the Graduate School of Social and Cultural Studies, Kyushu University. He has been conducting research on historical ecology and subsistence economy among the Majangir.

TADESSE WOLDE obtained his Ph.D. from the London School of Economics and is a post-doctoral researcher at the Max Planck Institute for Social Anthropology at Halle in Germany. He has published several articles on the Hor (Arbore) of the far south of Ethiopia.

TADDESSE BERISSO is currently Associate Director of the Institute of Ethiopian Studies. He teaches anthropology at Addis Ababa University.

ALESSANDRO TRIULZI is Professor of African History at the Istituto Universitario Orientale of Naples. Recent publications include *Being and Becoming Oromo: Historical and Anthropological Approaches* (ed. with P. Baxter & I. Hultin, 1996).

ELIZABETH WATSON carried out research with the Konso of Ethiopia, and in Kenya. She is currently Assistant Lecturer in the Department of Geography, University of Cambridge and Fellow of Newnham College.

Map 1   *Ethiopia: provincial boundaries, 1974*

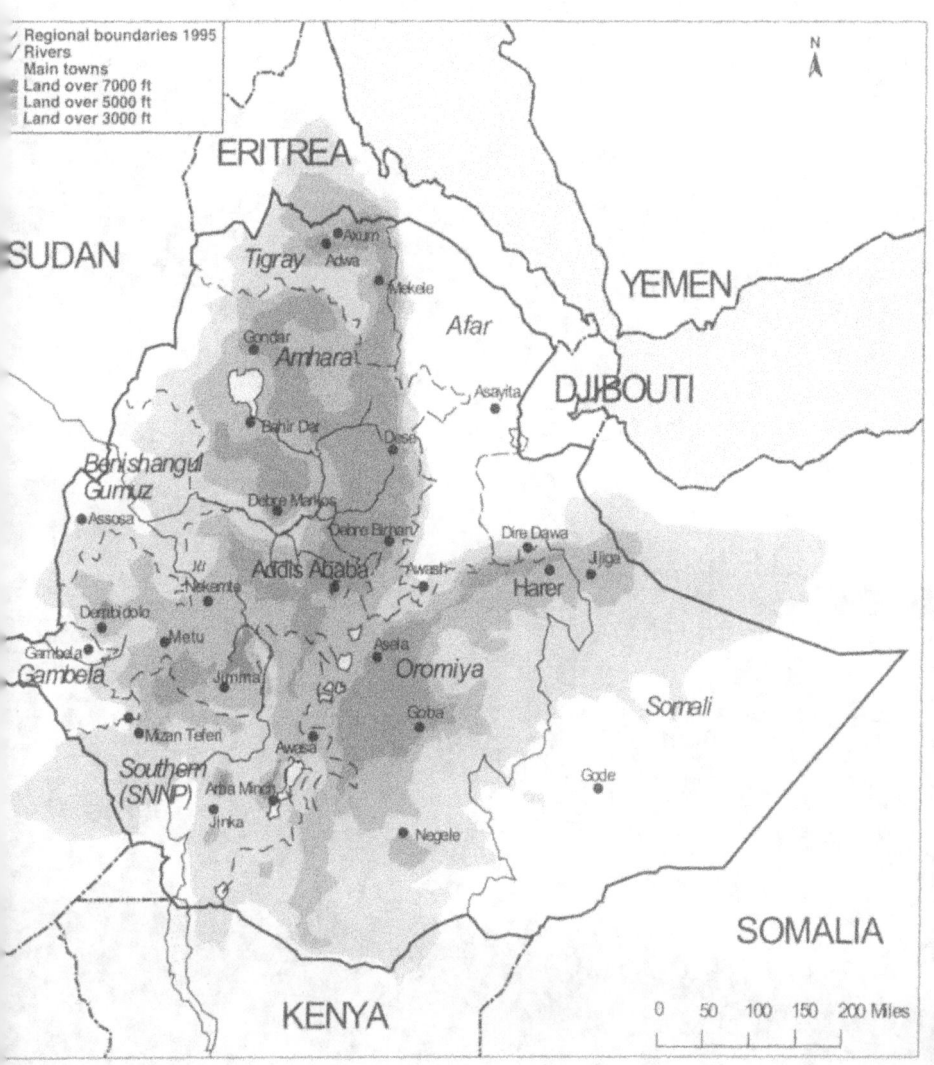

Map 2   Ethiopia: regional boundaries, 1995

Introduction

DONALD L. DONHAM

When *The Southern Marches of Imperial Ethiopia: Essays in History and Social Anthropology* was published in 1986, it marked a number of departures. To anthropologists, it demonstrated the ways in which the interpretation of local case materials must proceed in relation to an understanding of wider historical contexts. For Ethiopia – unlike most of the rest of Africa – that context was provided by the re-consolidation and expansion of an indigenous empire. From the mid-nineteenth century until the revolution of 1974, the Ethiopian state was dominated by and associated with a cultural core – Orthodox Christian and Amharic- or Tigrinya-speaking. From the turn of the century onward, power was exerted from a hierarchy of administrative centres dominated by the newly established city of Addis Ababa. Analysed against this background, the social and cultural patterns of particular Ethiopian peoples can be seen, in many ways, as responses to this wider sequence of events.

To Ethiopian historians, on the other hand, *Southern Marches* marked other turns. Ethiopian history, like a good deal of historical writing, had been focused on metanarratives of the nation – in the case of Ethiopia, the preservation of its independence against all colonial odds. Emperors' lives, their battles and foreign intrigues, from Tewodros to Yohannes to Menilek to Haile Selassie, furnished a seemingly natural way to shape stories and to distinguish periods. *Southern Marches*, in contrast, looked at the project of state-making from the point of view of the southern peripheries. There, the darker side of history came into focus – the assertion of cultural superiority by the Orthodox Christian core, the serfdom and slavery, and the extraction of resources in demand by the world market. Like much other work at the time, *Southern Marches* demonstrated the value of a history 'from below' expanded by new sources such as local memory, oral history, and the indirect evidence contained in sedimented custom.

As the first edition of *Southern Marches* went out of print, the editors

Introduction

began to consider the possibility of bringing out a second edition. The question of how to 'update' the book naturally posed itself. The more this issue was considered, the more problematic it appeared, for what had transpired in Ethiopia after 1974 – the point at which *Southern Marches* left off – had, more than once, revised and transformed fundamental patterns. In light of the significance of these changes, we have decided to assemble the present volume as a companion to a reissued paperback edition of *Southern Marches*. Here we present a set of fresh local studies from 1974 to the present.

The centre–periphery dynamic that *Southern Marches* documented remained stable for roughly three-quarters of a century. From the end of the nineteenth century until the revolution of 1974, the superiority of core cultural values was little questioned, and forms of reactive local identity were little developed in the peripheries (Eritrea, because of its colonial heritage, was perhaps the major exception, but even there until the 1970s, the feeling of difference and opposition seems to have been concentrated among Muslims rather than Orthodox Christians). The hierarchy of administrative centres, the demarcations of provincial boundaries, the lines of communication – all these changed in detail of course, but by and large, the overall pattern of how Ethiopia 'fit together' was not different in 1970 compared to 1900 (Map 1 shows the pre-1974 international and provincial boundaries).

The same assertion can hardly be made for 2000. The very shape of the country – the iconic outline that symbolizes the nation – has changed, as Eritrea has become its own country. Not only has the shape of Ethiopian space changed, but sizeable and influential populations of both Ethiopians and Eritreans – vitally identified as such – now reside in diasporas outside that space. One visible indicator of this transformation is the presence of 'Ethiopian' restaurants in most European and North American cities of any size. But more than this, a younger generation of Ethiopians and Eritreans has grown up outside its borders, steeped in other cultures, mainly North American. Given the increased pace of travel and communication, new ideas, as well as remittances, have found their way back to north-east Africa. In this more interactive and 'globalized' setting, the very image of Ethiopian centres and peripheries – which depended, after all, upon the relative closure of former times – no longer captures the most critical of current dynamics.

To simplify, one could say that Ethiopia has gone from a hierarchical arrangement of cores and peripheries, apparent to all and inscribed upon geographical surfaces, to a more open series of interactions drawing upon partially shared and intersecting 'ethnoscapes' of the imagination (Appadurai 1996). How and why did this shift occur? What are its local consequences? And, most importantly, how do local peoples themselves understand and experience these changes?

The approach taken to these questions in this volume can be best introduced with the concept of 'mapping'. James C. Scott has recently called

Introduction

attention to the way that all modern states 'map' local communities, and in doing so, make them legible for the purposes of taxation, conscription, and the prevention of rebellion:

> These state simplifications, the basic givens of modern statecraft, were rather like abridged maps. They did not successfully represent the actual activity of the society they depicted, nor were they intended to; they represented only that slice of it that interested the official observer. They were, moreover, not just maps. Rather, they were maps that, when allied with state power, would enable much of the reality they depicted to be remade.
>
> (Scott 1998: 3)

Since the 1970s, three political organizations have played major roles in remapping Ethiopia: the Provisional Military Administrative Council or 'the Derg',[1] later giving rise to the Workers' Party of Ethiopia; the Tigrayan People's Liberation Front (TPLF), later to take the lead in forming the Ethiopian People's Revolutionary Democratic Front (EPRDF); and the Eritrean People's Liberation Front (EPLF). Each of these was steeped in Marxist ideology, hierarchically organized in Leninist fashion, and formed in war. The two key moments during which Ethiopia was remapped were 1974 and 1991. According to Theda Skocpol's definition, 1974 constituted a social revolution in the classic sense. That is, both class relations and the form of the state were interactively transformed (Skocpol 1979; Clapham 1988). In contrast, 1991 remade only the state, although as I shall make clear below, the way the Ethiopian state related to its international environment was substantially changed.

Each of these two moments of rupture concealed long-term continuities of significance. After 1974, the Derg took over the old imperial and Orthodox Christian project of incorporating and controlling the peripheries – now secularized and Leninized – and turned it to its logical and self-defeating end. Indeed, the Derg's single-minded reliance on coercion in the northern periphery was an essential ingredient in the sequence of events that followed. By alienating Tigrayan and Eritrean peasants so totally, and by creating the conditions for a series of diasporas that began to interact with local politics, the Derg itself set in train the rise of the TPLF and EPLF.

At the climactic moment of the second change, 1991, the Workers' Party disintegrated, and the TPLF and EPLF assumed the levers of state power. Once again, however, much remained the same, and in fact, the events of the 1990s, rather than a reversal of the revolution of 1974, represent in some respects its continuation and working out. For example, both

---

[1] Formally, Ethiopia was ruled by the Provisional Military Administrative Council from September 1974 until September 1984, when it gave way to the Workers' Party of Ethiopia. The PMAC was quickly dubbed 'the Derg' (Amharic, 'committee'). In popular usage, this term stuck as the name of the government for the whole period up to the fall of the socialist regime in May 1991, and we have followed this usage in the present book.

Introduction

the TPLF and the EPLF maintained class relationships within the broad framework established by the Derg. All peasant lands remained in state hands, even as markets were allowed to operate much more freely (a process that had already begun during the last days of the Workers' Party). And perhaps more significantly, Stalin's proposed solution to the so-called national question – the right of each nationality to 'self-determination', up to and including the right to secede – grew in acceptance in Ethiopia (if not in Eritrea where the struggle was understood in terms of national liberation from colonial rule).

Why the notion of a federation of nationalities appealed to the TPLF in the 1990s has often puzzled observers. Part of the answer, as Christopher Clapham shows in Chapter 1, is contained in the observation that Tigray was and is the most culturally homogenous province in Ethiopia. There, the TPLF's appeal to a common identity united people. It did so particularly in light of the fact that Tigray had been increasingly impoverished over the twentieth century as a result of overpopulation and the lack of production of any commodity in demand by the world economy. In this context, intellectuals within the TPLF used appeals to a sense of regional and linguistic difference to mobilize peasants. Tigrayans, they were told, were poor because of 'Amhara domination' (see Alemseged Abbay 1998; Young 1998).

By doing so, the TPLF helped to create a fundamentally new kind of consciousness, a Tigrayan 'ethnic' identity. Before, Tigrayan peasants' sense of difference had related primarily to their identification with Tigrayan elites who had competed with other lords in Ethiopia for the office of 'king of kings'. This hardly opposed Tigrayans to the cultural core – just the opposite. By the late 1980s, however, many Tigrayans had apparently begun to see themselves in a different light – as another 'kind' of people, a people called forth by narratives of group injustice and suffering. That this transformation took place in the context of the Derg's brutal attempt to stamp out any opposition in Tigray is significant. The experience of war, trauma, and death seems always to hold the potential for transforming people in the most fundamental ways.

At the very moment, however, that a new reactive sense of Tigrayan identity was being created, this change was obscured by the claim that things had always been so. TPLF intellectuals presented their insurgency as a continuation of a regional rebellion of the 1940s called Weyane (even though the political motives and personnel of the two were substantially different – Gebru Tareke 1991). Remapping Tigray (and later Ethiopia) required, then, a rewriting of history – a process that has occurred from the 1980s onward in the Horn of Africa in an ever-widening series of conflicting circles.

Once the EPLF had defeated the Derg in Eritrea, and the TPLF (in alliance with other groups in what would become the EPRDF) had begun to expand beyond Tigray itself, Stalin's model of an ethnic federation, Clapham observes, was virtually the only option the TPLF leadership had

Introduction

Plate 1.1 *Mengistu's effigy awaiting burning at the land reform celebrations, Woldia, Wello [J. Hammond]*

with which to occupy central power. The fact that it was indeed Stalin's model was, however, forgotten as the TPLF made an ideological about-face in the early 1990s. As the world entered the post-Cold War era, the TPLF gave up Marxism and Leninism for a celebration of democracy and human rights. In this way, curiously, North American notions of human rights would be grafted onto Soviet ideas of nationality, and this layering was apparently crucial to the degree that it helped elicit United States support for the TPLF as it took power in Ethiopia.

No one has yet been able to explain in full just how these ideological changes occurred in the TPLF / EPRDF. Perhaps it is too early to attempt to do so. One hypothesis that may need to be explored is the role of the diaspora in educating Ethiopian politicians in how to mirror Western assumptions (while pursuing local projects that often contradicted them). In any case, Meles Zenawi and Isaias Afewerki, heads of Ethiopia and Eritrea by the 1990s, were hailed by United States policy makers as examples of Africa's 'new leaders' – leaders who would usher in a new progressive era (Ottaway 1999).

As the global environment of Ethiopia underwent a sea change after the end of the Cold War, the EPRDF remapped Ethiopia in the most literal sense. Previously, as *Southern Marches* set out, the exigencies of transport and communication across a mountainous terrain (rather than continuities of language or culture) had determined the lines of administration – first in imperial and then into Derg times. In the 1990s, the EPRDF used materials from the Derg's old institute of nationalities to redraw administrative

5

Introduction

boundaries in order to conform to lines of supposed cultural difference. Just as in the Soviet Union, the cultural autonomy of the new units was officially celebrated, while the federation itself was firmly controlled by the centre. Not the least of ironies created is that EPRDF intellectuals emulated the Soviet model just as the Union fell apart – along the lines that, decades before, Marxist intellectuals had decided were 'national' boundaries.

In some sense, what the EPRDF created in the 1990s was a form of 'indirect rule' based on official definitions of ethnicity. Local peoples were told that they had the right to rule themselves at long last – as long as they constituted an 'ethnic group' associated with an appropriately defined territory. Outside Tigray, this condition held in few areas of Ethiopia, particularly in the south. Günther Schlee and A. A. Shongolo (1995) have shown just how problematic the equation of language with ethnic identity with territory was among agropastoral groups in southern Ethiopia. The EPRDF programme – far from providing a solution for so-called ethnic differences – actually produced ethnic conflict in this area as various groups competed to be accepted as representatives of 'Oromo' or 'Somali' peoples in the early 1990s. A recent thesis by Dena Freeman (1999) on the Gamo highlands further to the north demonstrates that identity is just as complexly structured among many settled farming groups. There, the local notion of community, *dere*, operated like a set of Chinese boxes, contextually variable and expandable. At its most extensive, *dere* identity could comprehend the whole Gamo highlands, but this was not where most peasants lived most of their lives.

As James Scott (1998: 3) observes, new maps allied with state power 'enable much of the reality they depict to be remade'. In other words, they tend to call forth the reality assumed. The case studies in this volume present some of the complexities of this process with regard to 'ethnicity' during the 1990s. To generalize, one could say that educated elites in the old peripheries have often responded positively to the new state – at least in the beginning. They, rather than ordinary peasants, appear to be the primary advocates of 'identity politics' at the local level. In some few areas, whole subaltern communities have occasionally re-negotiated their local relationships and turned themselves into 'ethnic groups', in order to take advantage of the new order (see for example Chapter 9). Finally, in the old core, the response has been different once again. There, as *Southern Marches* pointed out, Amhara identity had previously implied not just a commitment to an 'ethnic group' in a horizontal array of others. Rather, being Amhara, well into Derg times, was assumed to set the (vertical) standard of what it meant to be Ethiopian. This being so, 'the Amhara' as discussed in newspapers and seminars today have not simply been 'an ethnic group', of the kind that many commentators imagine, at least not before the early 1990s. If the experience of colonial Africa is any guide (Fardon 1996; Ranger 1999), the new state itself may well be playing a crucial role in creating 'the Amhara' as understood in today's ethnic discourse.

Introduction

Behind all these local changes in Ethiopia (which sometimes point in contradictory directions) lies the fundamentally new world order created in the post-Cold War era. With the disintegration of the Soviet Union, there is now no ideological alternative to 'democracy', 'human rights', and the 'principles' of the market. Political elites outside the cores of the world system may resist by telling hegemonic powers what they want to hear (and then proceeding on their own), but unlike the world context for the 1974 revolution, no principled opposition appears now possible.

This uni-polar world has created a much more open context for political, cultural, and material flows in Ethiopia. Christopher Clapham notes the example of United States humanitarian aid at the end of the 1980s (see Duffield and Prendergast 1994). While the Workers' Party was still the internationally recognized power in Ethiopia, the United States, in contradiction with international law, began to funnel food aid into Tigray from across the Sudanese border. USAID granted funds to a Norwegian Christian church-supported NGO based in the Sudan, Emergency Relief Desk, who then transported aid across the border to the Relief Society of Tigray, another NGO created by the TPLF. This flow linked thus a bureaucratic and conservative USAID, a left liberal Norwegian church group, and the then-Marxist TPLF – links that no doubt played some role in bringing the TPLF to power within Tigray and thence within wider Ethiopia. Politics has always created strange bedfellows, but these juxtapositions illustrate well what some observers have called the 'postmodern' aspects of our current world order. What seems especially notable in this case is the savvy with which Ethiopian political actors have begun to exploit its possibilities.

Not only is the context in which Ethiopian elites act more open, it is also more unstable. As we write, Ethiopia and Eritrea, so recently the closest of allies, are themselves at war. Movements towards cleansing have taken place on both sides, with 'Eritreans' being expelled from Ethiopia, and 'Ethiopians' fleeing Eritrea. What had been mixed and hybrid – often the result of intentional strategies by ordinary people to expand their options and their lives – has now been rendered simple and oppositional. At war, moreover, both Ethiopia and Eritrea have begun to support opposition movements within the other. But the matter does not end here. A whole band of conflicts – religiously, regionally, and 'ethnically' based – now interrelate, intertwine, and amplify one another, all the way from the Horn of Africa through Sudan, Uganda, and Rwanda to the Congo. In some areas, the modern state has collapsed, in others, massive genocide has occurred, with war – as always – playing a critical role in the formation of ever new identities of suffering.

This, then, is the larger context for the series of case studies that follow. As will become clear, the way people respond to the 'same' change can be quite different depending on local context and living memory. Certainly, the ironies of change appear clearer closer to the ground – as well as, we believe, the poetry, endurance, and perhaps even hope.

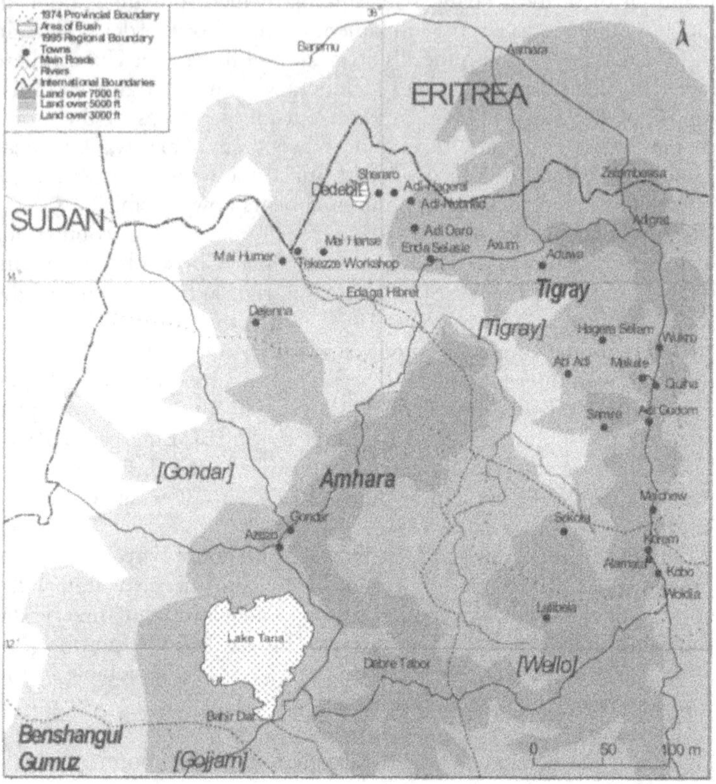

Map 1.2 The northern highlands (to illustrate Chapters 5, 13)

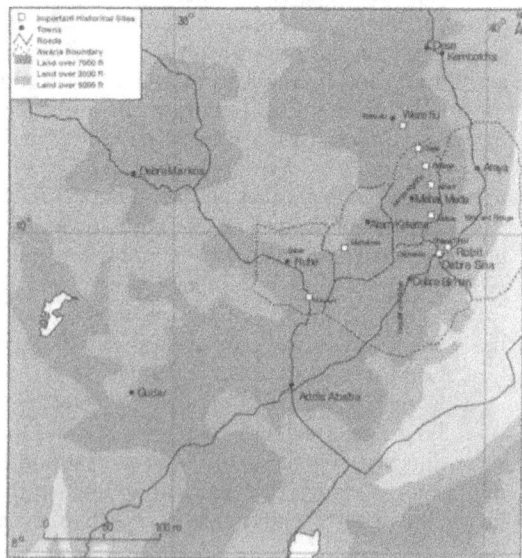

Map 1.1 The central highlands (to illustrate Chapter 4)

I

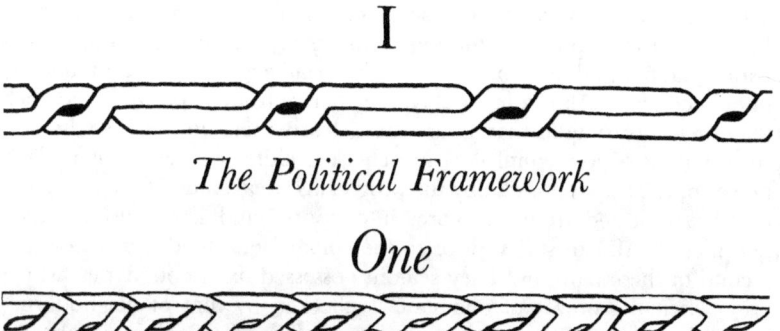

The Political Framework

One

Controlling Space in Ethiopia
CHRISTOPHER CLAPHAM

Governance everywhere is to an appreciable extent concerned with spatial relationships. States require mechanisms – conceptual and ideological, as well as coercive and economic – through which to relate the social and territorial units assigned to them by the essentially spatial conventions of boundary and sovereignty. These mechanisms are all the more important in states whose own construction is problematic. The upheavals that have shaken Ethiopia since 1974 have readily been ascribed to a combination of competition for the control of central state power and insertion into the international system maintained by the Cold War. With time, however, the relationships between the extremely diverse elements of which Ethiopia is composed – and indeed the definition of the very idea of 'Ethiopia' itself – have emerged as the critical issues at stake. This chapter seeks to explore the development of these relationships at an Ethiopia-wide level, in order to provide a framework within which to place the specific and often highly localized struggles with which the remaining chapters in this book are concerned. It traces the shift from an initial conceptualization in very largely 'centre–periphery' terms – as a relationship between a 'national' state and the subordinate communities of which it is composed – to the much more fluid and ambivalent interactions that have characterized the final years of the twentieth century. Although it will largely be concerned with the period since 1974, it must start by going back to examine the spatial construction of Ethiopian statehood, which in turn provides the essential foundation for understanding all of its subsequent developments.

The spatial construction of Ethiopian statehood

The process of state creation in Ethiopia has followed a pattern which, though familiar in many parts of the world, was pre-empted in most of

Looking Back on the Projects of the Socialist State

sub-Saharan Africa by the imposition of colonial rule. Historically, states have arisen in response to the requirements of territorial control, surplus expropriation and external defence in relatively wealthy and densely populated areas. Almost invariably, they derive from a core zone in which the criteria for state formation are most clearly met and then expand into surrounding regions, until they are checked either by the countervailing power of rival states, or else by the progressive weakening of the force that can be projected from the core, into poorer and less densely settled peripheries. African states, prior to the colonial era, tended to follow the second of these patterns. They seldom possessed fixed boundaries (which arise when the power of one state is checked by that of neighbouring states), but spread out from the core into hinterlands of tributary rule and mere raiding, which expanded or contracted with the strength and ambition of the ruler. Like other states, but to a greater extent than most, African states relied on revenues from external trade, which were generally much easier to collect and control than direct exactions from their populations. This pattern, which long preceded colonial rule, can be traced in the Ethiopian case to the trading networks of the Axumite kingdom, as revealed by *The Periplus of the Erythrean Sea* (Huntingford 1980; compare also Phillipson 1998). From the era of the slave trade onwards, African state formation was closely associated with the continent's incorporation into the global economy.[1]

Ethiopian state formation followed its own distinctive variant of this common pattern, being characterized by a peculiar combination of the underlying ecological and social structure of the region, with its precolonial political experience and the way in which the modern process of state creation coincided with (and to some extent replicated) that of colonial and post-colonial states elsewhere in the continent.[2] At its base, obviously enough, lay the ox-plough agriculture of the northern plateau and the hierarchical social formations which this sustained, coupled with the pronounced localism induced by the remarkable topography of the northern highlands. The possession of a long-established and politically dominant state also promoted a set of attitudes or ideologies, compounded of Orthodox Christianity, a set of historical mythologies and a written language, which defined its members in their own eyes as being more civilized than their neighbours and in turn fostered a sense of manifest destiny in their claims to govern surrounding territories. These attitudes, intensified by the conflict between Christianity and Islam, have long marked Ethiopian regional administration – though never more so than

---

[1] The best succinct discussion of African state formation that I have come across is in Herbst 2000; see also Iliffe 1995, and for the dependence of African states on externally generated resources (which he terms 'extraversion'), Bayart 1993. For a comparable discussion of the development of European statehood, see Tilly 1990.

[2] Though writers such as Holcomb and Sisai Ibssa (1990) have characterized Ethiopia as a 'colonial state', it is actually a state of a very different kind: paradoxically, colonial states in Africa have generally proved to be rather more stable and effective than those resulting from indigenous processes of state formation.

Controlling Space in Ethiopia

under the 'Derg', the socialist administration from the mid-1970s[3] – and negated any idea of equality between the peoples and cultures of the core state tradition and those of other regions. Although individuals from these regions could fairly readily associate themselves with the state tradition through the adoption of Orthodox Christianity, the Amharic language and Ge'ez/Amharic names – and a great many did so – their cultures and societies were condemned to subordination by the nature of the state itself.

This formula was readily conceptualized in terms of 'centre' and 'peripheries'. The centre, represented at its most extensive by the highland Christian 'Great Tradition', and more specifically by a centralized source of political power with the emperor at its apex, tended to move southwards over the two millennia of 'national' history that this tradition espouses, reaching its current location with the foundation of Addis Ababa in the late nineteenth century. Symbolically located close to the geographic centre of the state, Addis Ababa formed the nodal point of a communications system spreading out to the furthest parts of the empire. Specific regions and communities could be defined as more or less peripheral, in accordance with their physical distance from the capital, their level of incorporation into the coercive and economic structures of government, and their degree of association with the legitimizing myths of nationhood. In its turn, this conception of the state readily lent itself to a teleology of 'nation-building', in which the manifest destiny of peripheral areas was to be drawn ever closer to the national core. This teleology received powerful support from the historiography of 'modern' Ethiopia from the accession of Emperor Tewodros in 1855.[4]

Ethiopia's association with the international system during the colonial period reinforced this idea of state formation by enabling its government to benefit from the norm of sovereignty, and to participate in the global economy and political system on terms very different from those available to other African rulers. This in turn had critical consequences for that government's relations with its own peoples, notably those at the periphery. The key requirement was international recognition, which had been tacitly accorded during Ethiopia's longstanding (if fragmented) relations with European states from the early sixteenth century onwards, but was formally ratified after the defeat of the Italian invading force at Adwa in 1896. The most obvious expression of recognized statehood was then the ability to gain access to imported arms with which (aided by greatly enhanced internal leadership and organization) to impose control over what then became the national territory and the peoples within it. The need to impose this control arose not simply from entrenched ideologies of manifest destiny but from the need to raise state revenues from inter-

---

[3] Ethiopia was ruled by a Provisional Military Administrative Council from 1974, referred to as the 'Derg'. In 1984 the PMAC was replaced by the Workers' Party of Ethiopia, but the name 'Derg' went on being used up to the fall of the government in 1991.
[4] See Bahru Zewde 1991, Marcus 1994, for examples of works constructed essentially along these lines.

national trade, which the northern highlands were in no position to provide. Their dense populations were already hard-pressed to meet their own subsistence needs, they were ecologically unsuited to serve the international economy's demand for tropical products, and their people were to some extent protected against the grosser forms of surplus expropriation by a measure of political autonomy and continued rule by local elites. This very exclusion from the processes of state-formation and especially economic incorporation that characterized much of twentieth-century Ethiopian history meant that, in some respects, historically 'core' areas of Ethiopia became a peculiar kind of periphery of their own, in ways explored later in this book by Ahmed Hassan Omer (Chapter 4), Jenny Hammond (Chapter 5) and Cressida Marcus (Chapter 13). This helps to explain the paradox that, in recent times, the most serious armed challenges to the central Ethiopian state have arisen disproportionately from the core zone.[5]

The conquered peoples to the west, south and east of the core varied in their capacity to produce internationally marketable goods, but enjoyed no such political protection. As Donham has noted (1986a: 24), 'there was never a single homogeneous periphery'. Instead, different parts of the territory that, from the early twentieth century, was internationally recognized as constituting Ethiopia were incorporated in different ways, and to very different degrees. Though Donham's broad distinction between semi-independent enclaves, *gebbar* areas and fringe peripheries provides a useful starting point, this in its turn may be supplemented and subdivided by additional forms of differentiation. One of these was religion, in that Islamic societies posed a challenge to the Christian Orthodox state in a way that peripheries not associated with major world religions did not. Another was the structure of indigenous subsistence production, broadly distinguished between grain, *ensete* and pastoralist zones – *ensete* cultivation, which sustained the densest populations in south-western Ethiopia, appears to have enabled the people dependent on it to maintain a greater level of autonomy than grain cultivators, whereas pastoralists (most of whom were also Muslim) were generally treated with the contempt reserved by representatives of a historic agricultural state for persons of no fixed abode.[6] A third was the ability or otherwise of different parts of peripheral Ethiopia to produce internationally tradable goods, of which coffee was by some way the most important. Finally, different areas were of greater or lesser strategic importance to the imperial state. After the period of the 'scramble' of the late nineteenth century, when the Ethiopian state, like its colonial rivals, sought to incorporate the maximum amount of territory into its own domain, cross-border relations in the west and south-west became largely a matter of policing. The south-east, on the other hand, which controlled the major artery for Ethiopian foreign trade, presented the main Islamic threat and faced potential routes for colonial invasion, remained politically and economically critical.

[5] I have examined this paradox in Clapham 1992.
[6] For a recent set of studies of pastoralism in Ethiopia, see Hogg 1997.

Controlling Space in Ethiopia

After Haile Selassie's restoration in 1941, the level of autonomy open to the formerly semi-independent enclaves was greatly reduced, and although some deference was paid, notably in Wellegga, to former traditions of indirect rule, the periphery as a whole was incorporated into a uniform structure of provincial government through *teklay gizat* (region) and *awraja* (province). The overwhelming majority of regional and provincial governors were central officials, especially from Shewa, many of whom (most notoriously Ras Mesfin Seleshi in Kefa) ruled with notable rapacity. Any regional politics that acknowledged the distinct identities of different societies within the empire was limited to interests that could be mediated through the patrimonial mechanisms of the court. In both Wellegga and Tigray, for example, indigenous ruling houses were associated through marriage with the imperial family. That this mechanism was open to the Christian Oromo rulers of Wellegga but not the Muslim ones of Jimma probably does as much as anything to explain the difference in treatment between the two regions. At the same time, the incorporation of the southern periphery into a particularly exploitative relationship with the international and domestic monetary economy was accelerated: in the Awash valley through concessions to foreign multinationals, HVA (N.V. Handelsvereeniging, involved in sugar production) and Mitchell Cotts; in Arsi through agricultural development schemes funded by foreign aid; but elsewhere largely through the commodification of landlord–*gebbar* relationships.

Paradoxically, the thirty years preceding 1974 were probably the most peaceful, in the consistently violent history of Ethiopian regional politics, that the country has ever known. Moreover, of the three major revolts analysed by Gebru Tareke, two were in historically core regions, Tigray and Gojjam, and were prompted by a threat to cherished autonomy; the third, in Bale, was greatly assisted by its proximity to Somalia (see Gebru Tareke 1991). Nor did the relative tranquillity of the Haile Selassie years rest on massive and overt military force, though the underlying social and political structure was unquestionably coercive: compared with the later socialist state, the military capacity of the imperial regime was small. This impression of stability, however, rested on unsustainable political, social and economic foundations. Politically, an unmobilized population was bound eventually to find a voice. By comparison with colonial Africa, the pre-revolutionary decades represented an equivalent to the hiatus between the primary resistance to colonial conquest and the nationalist upheavals of the post-war period. Moreover, the exclusion of any acceptable process of political mobilization analogous to that of the nationalist parties helped (as in Africa's other non-colonial state, Liberia) to prompt an ultimately much more violent reaction. Socially, as already noted, the structure of the Ethiopian state made the incorporation of peripheral peoples on any terms approaching equality impossible: the inevitable process of politicization could only intensify the divisions between Christian and Muslim, settler and indigene, central state and local autonomy. Economically, the structures of land ownership and surplus expropriation on which the imperial regime rested were likewise incompatible with any process of national integration. For all their apparent tranquillity, the imperial regime's relations with its peripheries were at an impasse.

This was nowhere clearer than in Eritrea, which despite (or because of) its subsequent independence, is critical to any general consideration of centre–periphery relations in Ethiopia up until 1991. Eritrea is the exception that proves the rule, because by the time of federation in 1952, its population had already been politically mobilized by the experience of Italian colonialism and the subsequent formation of political parties under the aegis of the interim British administration and the UN-orchestrated consultations on its political future. Even though the dominant party after 1952, the Unionist Party, had been formed explicitly in order to promote union with Ethiopia, it was quite unable to develop any relationship with the imperial government that was compatible with the maintenance of its popular base. Its degeneration into a mere instrument of imperial overrule thus prompted armed opposition to develop in Eritrea both earlier and more effectively than in other parts of Ethiopia.

Revolutionary Ethiopia: the project of *encadrement*

Unlike its imperial predecessor, the military government that seized power in 1974 was very much aware of the spatial contradictions of Ethiopian statehood and devised a reasonably coherent policy to deal with them. This policy may plausibly be designated Jacobin, in emulation of the French revolutionaries of 1791 to 1794. It amounted to a project of *encadrement*, or incorporation into structures of control, which was pursued with remarkable speed and ruthlessness. It sought to intensify the longstanding trajectory of centralized state formation by removing the perceived sources of peripheral discontent and espousing an ideal of nation-statehood in which citizens would equally be associated with, and subjected to, an omnipotent state. In keeping with the Marxist preconceptions that were universally shared by Ethiopian radicals at that time, it conceived the sources of discontent in terms of class interests and economic exploitation, an approach for which – given the levels of exploitation especially in the *gebbar* zone – a very plausible case could be made. The removal of certain cultural symbols of the great tradition, such as the emperor and the protected position of the Orthodox Church, was by contrast largely incidental. The Derg, as the regime and all its works came to be known, represented the centre–periphery conceptualization of Ethiopia in its most intense form.

The central elements of this policy were put in place over a period and called for the brutal suppression of alternative conceptions of Ethiopian statehood. Its first and most important constituent was the land reform of 1975, under which all land was nationalized, use rights then being allocated among local inhabitants by newly created peasant associations. This measure had a dramatic impact on central–local relations in several respects. First, it destroyed the power of the landlords, as a result provoking armed opposition in those parts of the country, especially in the northern highlands, in which locally based notables still retained an appreciable following. Second, it prompted an often violent reaction against immigrant landlords and *neftennya* settlers in much of southern Ethiopia, in which the combination of indigenous numbers and central government backing proved decisive. Third, it resulted in the creation – at least in the agricultural areas of the country, though far less so among pastoralists and fringe populations – of a structure of local government, built on the peasant associations, which incorporated these areas far more intensively into a national administrative structure than had ever been the case in the past. Land reform thus involved not the distribution of land among a peasantry which could thus gain control over its own means of production, but rather the 'capture' of the peasantry in a way that subjected them increasingly to state control. This process may be regarded as the single most important feature of the revolution, and the success or failure of the revolutionary regime in incorporating the peasantry would ultimately determine not only its own fate, but the future of Ethiopia. Fourth, and temporarily, land reform initially left control of their production in the hands of peasants themselves, in the process breaking the links through which, under the *gebbar* system and its equivalents, surplus production had been appropriated in order to maintain the state and to link Ethiopia to the global economy. In some way or other, these links had to be restored.

Although land reform provided the legal basis for the creation of a new relationship between the centre and peripheries, revolutionary change was initially implemented to a very large extent through the Development Through Cooperation Campaign of 1975 to 1976, popularly known as the *zemecha*. Under this campaign, students from urban centres were sent out to revolutionize the countryside, partly in an effort to remove them from the cities, where they posed a threat to the new regime, but also to compensate for the lack of any administrative machinery in the countryside through which the Derg could implement policies which struck at the power base of the rural ruling class. Since these students were very heavily drawn from urban backgrounds and had become accustomed to think of the countryside as a zone of backwardness and squalor, the encounter between peasants and students was one of discovery on both sides: it is scarcely too much to say that the creation of a politics of urban–rural relations in Ethiopia dates from that moment. On one side, this encounter helped empower the peasantry, both by providing them with tangible central support in contests with readily identified local adversaries, and more intangibly by helping to spread hitherto undreamed of ideas. On the other side, students brought to it a stereotyped conception of rural class relations as a Marxian contradiction between exploitative landlords and subjugated peasants. However, they had difficulty in adapting to situations in which this stereotype failed to work, either because the peasants already controlled their own means of production, or because they maintained solidary relations with their supposed exploiters, or indeed because land simply did not have that centrality to many of the peripheral peoples that it did in the economies and value systems of the arable areas of the country. The *zemecha* also encapsulated and reinforced an attitude to 'development' that, deeply entrenched in longstanding Ethiopian ideas of governance, was to reach its apogee under the Derg – the idea that change had necessarily to descend from the top. The student *zemach*, like the old provincial governor, was the representative of a higher wisdom, here encapsulated in 'modern' learning, and the role of his peasant subjects was to be developed. However much the peasantry might be 'empowered' against their local class adversaries, the idea that they might be empowered against the central government ran counter not just to the centralizing instincts of a military Marxist dictatorship, but to much more basic attitudes to government.

Almost immediately, therefore, the revolution raised the critical issue of the relationship between central power and local autonomy. The initial challenge raised by dispossessed former landowners, described in Ahmed Hassan Omer's chapter on Northern Shewa, could readily be dismissed as 'feudal', but other demands for autonomy were more problematic. The most important was in Eritrea, where, as early as November 1974, General Aman Andom's willingness to deviate from a rigidly centralist policy prompted his overthrow and death at the hands of the Jacobins within the Derg, led by Mengistu Haile-Mariam. Thereafter, a solution to the Eritrean issue could be obtained only by military means, as eventually happened when the Derg's forces were defeated by the EPLF in 1991. But equally, the upheavals of 1974 to 1975 raised the question of whether, and to what extent, 'ethnicity' should be acknowledged in the government of other areas of Ethiopia. The year 1974 symptomatically marked the formation of the Oromo Liberation Front (OLF); and amongst the numerous urban-based political organizations that emerged in the wake of the revolution, one of the most important, MEISON,[7] received appreciable support from Oromo intellectuals and entered into a tactical alliance with the Derg. During the 1975 to 1977 period, several governors (most of them MEISON supporters) were appointed to the regions from which they originated.

This form of tacit local representation was unable to survive the Somali war of 1977–78 and the breach between the Derg and MEISON that coincided with it. MEISON supporters, like those of the EPRP (the Ethiopian People's Revolutionary Party), its main rival, were ruthlessly hunted down and killed. This period of blood-letting, commonly described as the 'red terror', was by no means restricted to Addis Ababa and was most intense in provincial towns, such as Gondar, where it had a lasting

---

[7] The Amharic acronym for the All Ethiopia Socialist Movement.

impact in alienating local elites from political life. The Derg emerged from the terror, the Somali war and a subsequent offensive in Eritrea, which substantially reduced the territory controlled by the EPLF, with its confidence massively enhanced. Not only had it seen off three major threats, but it now also had a firm alliance with the Soviet Union, which guaranteed it an apparently limitless flow of weapons through which to impose its control. The Somali war also prompted another development, next only to the land reform and the *zemecha* in reshaping central–local relations, which was the formation of massive and largely conscript armed forces. The forces which (with Soviet arms and Cuban assistance) defeated the Somalis appear to have been recruited very largely from the southern areas of the country, in which land reform had had the greatest impact and in which (initially at least) the Derg could call on the support of a peasantry which had every interest in defending the revolution. The militarization of rural Ethiopia, through conscription and training in arms, and the eventual spread through the countryside of the ubiquitous AK-47, can be traced back to this period, ultimately proving the undoing of the Derg itself.

The Soviet alliance represented one manifestation of the close relationship between access to external resources and the control of local space that went back to the foundation of the Ethiopian state. Moreover, as that state came under increasing challenge, so the salience of external resources increased. Strategically, Ethiopia was incorporated into the global bipolar security structure established by the Cold War, which appeared to offer its rulers a guarantee of ultimate triumph over their local opponents. Ideologically – and it is important to remember that the late 1970s represented the apogee of communism in the developing world – the regime and its supporters could draw on a set of ideas that appeared to represent modernity in its most advanced form and to tie it to the consolidation of centralized state power.[8] Diplomatically, it was protected by the international conventions of state sovereignty, which were espoused in their most extreme form by the Organization of African Unity. Economically, the regime was able to control and profit from the flow of goods across state frontiers, including even most of the famine relief supplied by Western donors. International backing appeared to represent not so much the subordination of Ethiopia to a position of dependence in the global system, but rather the key role of the state in implementing the inevitable and unstoppable project of *encadrement*.

After the victory of 1978, therefore, the triumphant Derg felt able to consolidate its power through the creation of centralized and hierarchical institutional mechanisms intended to bind the peripheries into a permanent and subordinate relationship with the state. Among the most important of these were the mechanisms designed to recreate the structures of surplus expropriation that had been destroyed by land reform. The key

---

[8] See Donham 1999 for a fascinating exploration of this idea of modernity in the Ethiopian context.

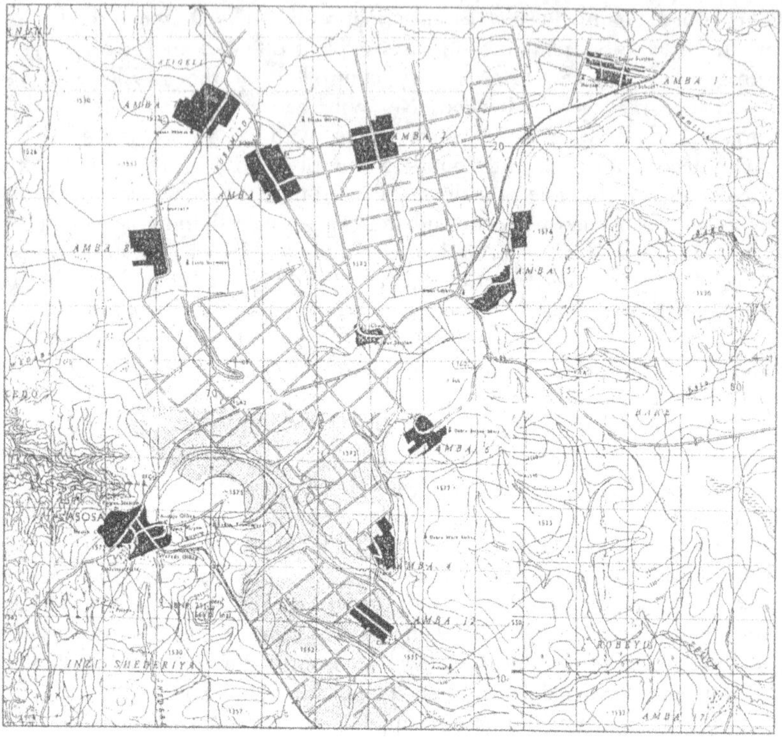

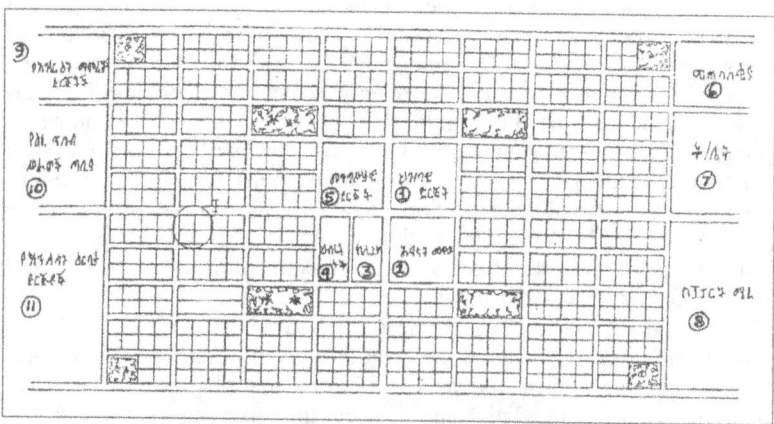

Figure 1.1 The spatial patterns of resettlement. Assosa, early 1980s
[Ethiopian Mapping Agency]

Figure 1.2 Plan of an ideal village layout, according to Ethiopian Government Guidelines of the mid-1980s
[reproduced in John M. Cohen & Nils-Ivar Isaksson, Villagization in the Arsi Region of Ethiopia (Uppsala, Swedish University of Agricultural Sciences, 1987), p. 129.]

institution here was the Agricultural Marketing Corporation (AMC), which laid down grain quotas to be delivered (at highly disadvantageous prices) by each region, and in turn by each province and district, down to the individual peasant association and farmer. The extraordinary hierarchical efficiency of the revolutionary Ethiopian state is demonstrated by the ability of the AMC to exact grain from peasants at a time of famine, which in turn those peasants could only acquire by selling livestock and buying the grain on the open market. Unsurprisingly, this not only alienated the peasantry from the regime but also resulted in their progressive withdrawal from the market into subsistence production, a process which Dessalegn Rahmato (1985: 61–2) has characterized as 'agrarian involution'.

Other mechanisms for surplus appropriation included state farms, which took over large commercial farms previously operated by landlords and multinational companies and also attempted to establish large 'virgin lands' projects for grain and export crops, with uniformly disastrous effects; and producers' cooperatives, through which peasant farmers were induced to pool their resources in order to create units which the Derg (like all revolutionary socialist governments that equated large-scale mechanized production with progress and modernity) regarded as economically efficient. The most intrusive, however, were resettlement and villagization. Both of these, despite earlier experiments dating back in some cases to the imperial regime, were essentially a product of the period after the great famine of 1984 and 1985, when the Derg's most intensive efforts at social engineering were launched. Resettlement involved the long-distance movement of people from 'overpopulated' areas into planned modern villages elsewhere (see Figure 1.1). The movement was largely from the north-eastern famine zone (Tigray, Wello, and northern Shewa), but also from areas in southern Ethiopia affected by a disease of the *ensete* plant, into 'underpopulated' regions in the west and south-west (primarily in Gojjam, Wellegga, and the Gambela district). Some 600,000 people were moved in this way, sometimes by force or by the arbitrary designation of peasants for resettlement by peasant association leaders who wanted to get rid of them. Alula Pankhurst (Chapter 7) examines the process in much more detail than is possible here.

Villagization, on the other hand, involved the concentration of scattered homesteads into designated villages by peasants who continued to farm the same land as before. Initiated in Bale region as a security measure during the 1977–8 war, this was extended to most of the country in 1985 and 1986, involving a massive process of house destruction and rebuilding that encompassed probably some forty per cent of Ethiopia's rural population. Apart from the war zones and the extreme peripheries, the most important regions to be spared were those of concentrated *ensete* production, as in Gurageland, where the houses were so large, and the need for these to remain in immediate proximity to the *ensete* gardens so vital, that even the Derg regime relented. Like the exactions of the AMC, villagization illustrated the extraordinary (and heavily counterproductive) regulatory efficiency of the regime. It was a remarkable experience at that time to fly over south-western Shewa, with a copy of the government's official villagization guidelines in my hands, and see the newly constructed villages below me correspond precisely to the plan of a model village set out in the guidelines (see Fig. 1.2). Nowhere, of course, do developments on the ground precisely correspond to leadership intentions, and in Ethiopia as in other 'socialist' states, local level bosses pursued their own agendas with greater or lesser levels of brutality and corruption. Villagization was nonetheless the most visible expression of the 'capture' of the peasantry, within residential perimeters accessible to wheeled transport, where they could be taxed, conscripted, and prevented from smuggling their produce to illegal open markets. Since this was achieved at considerable cost in agricultural production, through the loss of efficiency caused by the increased distance between homestead and fields, as well as by the degradation of the area immediately surrounding the village, it could readily be regarded as a deliberate economic cost incurred by the regime in order to enhance its control. This, however, would be to misconstrue the mentalities of the officials concerned, for whom 'development' was virtually coterminous with control. I was shown around both villages and settlements, with unaffected pride, by local administrators who were clearly convinced that only in this way could the entrenched backwardness of the peasantry yield to the benefits of planned and centrally directed modernity. From the peasant viewpoint, of course, it was a very different matter: in Chapter 3, Alexander Naty records the sullen acquiescence that characterized much of southern Ethiopia, while in Chapter 6 Taddesse Berisso explores the very damaging impact of villagization on some of the communities on which it was imposed.

The project of *encadrement* also had a significant political element. During the early years of the revolution, much of this took a coercive form, through mechanisms such as the creation of peasant association defence squads, effectively armed gangs at the disposition of the peasant association leadership, which on the one hand had an interest in supporting the regime that had empowered them, and on the other could impose this power (by selecting individuals for conscription, for example) on rivals within their area. After the end of the Somali war, the regime set about constructing the formal apparatus of a Marxist-Leninist state, central to which was the creation of a vanguard single party, which was eventually launched as the Workers' Party of Ethiopia (WPE) in September 1984. Given the collapse of this organization less than seven years later, it is easy to regard it as no more than a formal shell. It enjoyed the dubious distinction of becoming the last ruling communist party to be created anywhere in the world, before the debacle of 1989; and the attempt to impose over the whole country a disciplined Leninist structure through which, in a phrase much used at the time, to 'translate into deeds the decisions of the party leadership' proved (like so many of the Derg's actions) to be counterproductive. At the time, however, it did provide a remarkably effective instrument for the implementation of the top-down approach to government so characteristic of the Derg, and activities such as villagization and resettlement were greatly accelerated by the eagerness of local party officials to vie for central approval. Major resettlement schemes, for example, had their own party leadership, distinct from that of the ordinary local administration; and there was no doubt at all that, in the Leninist duality between party and state institutions, the party held the upper hand. Regional and provincial administrators became auxiliaries of the local WPE First Secretary.

The creation of the party, and the formal establishment in 1987 of the People's Democratic Republic of Ethiopia (PDRE), with its short-lived constitution, also provided the first glimmerings of a representative structure in which various localized 'nationalities' were accorded a distinct identity. The use of the term 'nationalities', widely shared by Ethiopian radicals (including the subsequent EPRDF) from 1974 onwards, and the development of the belief in a measure of self-government for each nationality, both derive from Stalin's writings on the 'national question' within the USSR. In order to cope with the ethnic diversity of the former Russian empire within the framework of a formally democratic and socialist state, Stalin (who himself came from a peripheral region, Georgia) developed a system in which nationalities would enjoy cultural rights and a limited amount of administrative autonomy within their own home areas, subject to the overarching control of a communist party, which (since the class interests of the workers, in Marxist ideology, necessarily overrode the superstructural manifestations of ethnicity) would necessarily be paramount. In the eyes of virtually all of Ethiopia's Marxist intellectuals, most explicitly those in the TPLF, this ideology provided an appropriate model for dealing with the national question in Ethiopia. Even under the ruthlessly centralized Derg regime, it led to the creation in 1983 of the Institute for the Study of Ethiopian Nationalities (ISEN), which conducted studies of the distribution of the different nationalities throughout Ethiopia, and led to the introduction of a strictly controlled element of representation for nationalities under the constitution of the PDRE. Eventually and ironically, ISEN studies provided the basis for the division of Ethiopia into regions based on nationality under the EPRDF. Questions on nationality identification were included in the 1984 census, providing the first figures beyond crude estimates as to the relative numerical strength of different nationalities within Ethiopia. Apart from regions such as Tigray and Eritrea, where a large proportion of the population was not under government control and was therefore unavailable to be counted, these figures were not evidently biased or falsified; they were not published at the time, and eventually appeared under the EPRDF. Elections to the National Shengo (as the parliament or Supreme Soviet of the PDRE was called) were usually based on the selection of local nationality representatives: all the members from the Gambela region, for example, were Anuak (sometimes written Anywaa) or Nuer.

Looking Back on the Projects of the Socialist State

Table 1.1 Populations of Ethiopia and Eritrea by ethnic identification, 1984

Group	Total	%	Group	Total	%
Oromo	12,387,664	29.07	Afar*	583,120	1.37
Amhara	12,055,250	28.29	Agew	489,834	1.15
Tigrayan*	4,149,697	9.74	Guji	481,442	1.13
Gurage	1,855,905	4.36	Gamo	463,933	1.09
Somali	1,613,394	3.79	Gedeo	455,408	1.07
Sidama	1,261,721	2.96	Kaffa	443,209	1.04
Welayta	1,092,958	2.57	Kambatta	432,819	1.02
Tigre**	683,085	1.60	Others	3,523,925	8.27
Hadiya	643,512	1.51	TOTAL	42,616,876	

Source: Transitional Government of Ethiopia, *The 1984 Population and Housing Census of Ethiopia: Analytical Report at National Level* (Addis Ababa, December 1991), Table 1.15.

Notes
* = substantial part of population in Eritrea.
** = whole population in Eritrea.
'Others' includes 40 named groups with populations between 20,000 and 275,552, total 2,787,514 (6.54%); further unspecified groups with populations under 20,000, total 538,107 (1.26%); and foreigners, naturalized citizens, and those without ethnic identities, total 198,304 (0.47%).

Even under the Derg, the project of *encadrement* was not all-embracing. Especially in the extreme south-west of the country, with which several of the case studies in this book are concerned, small groups, such as the Hor described by Tadesse Wolde in Chapter 2, remained largely untouched – though even they, bizarrely, had two representatives sent to the Ideological School in Addis Ababa to learn about Marxism-Leninism. Ethiopia's pastoralist peoples were also only partially incorporated into the new state. For one thing, they presented almost insuperable obstacles to the idea of socialist development that loomed so large in the Derg's approach to its peripheries – despite some efforts at sedentarization, they could not plausibly be settled, collectivized, or forced into benevolent modernity. For another, since they inhabited Ethiopia's lowland peripheries and most of them were Muslim, they presented a security threat. They were accordingly dealt with in two main ways. First, agriculturally productive areas (which, as dry season refuges, were critical to the pastoralist way of life) were taken over for resettlement and cash crop cultivation, continuing a process initiated under the imperial regime. Second, they were managed politically by exploiting the internal divisions characteristic of pastoralist societies, with only the thinnest veneer of socialist statehood. Both among the Afar (whose quiescence was important both to the Eritrean war and to maintenance of the supply route from Assab) and among the Somalis (where they were aided by the collapse of the Somali Republic in the 1980s), these tactics were at least reasonably successful. At the same time, however, the pastoralists were the most transnational of all Ethiopia's peoples. The Afar, Somali and Boran not only extended across Ethiopia's frontiers (additionally so for the Afar, once Eritrea became independent), they were relatively unconcerned with the new politics of territoriality, because they had been so weakly incorporated into the old.

Contested space & the idea of 'nationality'

Needless to say, all of this activity proved entirely incapable of furnishing the regime with a powerful and nationwide political structure that would achieve development and crush opposition. Rather, the Derg pushed the project of *encadrement* to its self-destructive limits, rapidly resulting in the reversal of an apparently ineluctable process of centralized state-formation that stretched back to the accession of Emperor Tewodros in 1855. This reversal was already under way before the debacle of 1991. The failure of the top-down model of development on which the Derg's own conception of its mission relied was brutally exposed by the 1984 famine and reinforced by the worldwide crumbling of socialist approaches to development throughout the 1980s. Though capitalist states had supplied most of the food aid during the famine, they were willing to provide long-term agricultural assistance only in accordance with their own model of market-led development. Economic reforms announced in 1990 led to the rapid abandonment of villages and agricultural cooperatives and to the disappearance of the produce marketing system. The collapse of the USSR in 1989 to 1990 coincided with, but did not cause, the regime's military defeat. Much more significant for subsequent developments than the Derg's internal failure and loss of external support, however, was the emergence of the forces of resistance that were to displace it.

This resistance amounted to far more than simple opposition to the idea of *encadrement* within the long-established parameters of the centre–periphery model of Ethiopian statehood. It involved, rather, the recasting of spatial relationships in terms that enabled previously peripheral zones to capture direct control over sources of power that had previously been monopolized by the state, and to reconstitute these in ways that challenged the idea of statehood and threatened to displace it. These sources of power extended to the full range of political resources available to the Derg and notably included alternative ideological models, structures of internal organization, and forms of access to the outside world. Once crowned with military success, they resulted in a reconfiguration of political space under which the ideas of centre and periphery were at most only partially relevant.

For a start, the leading opposition movements, and notably the EPLF and TPLF, were able to contest the power of the Derg so effectively because they succeeded in replicating (and indeed surpassing) the process of *encadrement* in which the Derg had itself engaged. These movements had their own land reforms and peasant associations, their own surplus expropriation and conscription mechanisms, and their own (largely covert) party systems and structures of political control. The *baito*s formed by the

Looking Back on the Projects of the Socialist State

TPLF, for example, served essentially the same function as the peasant associations formed by the Derg in providing a local-level structure through which to organize the countryside. Under the pressure of fighting an extraordinarily successful insurgent war, they developed hierarchies and mentalities that often reinforced the conspiratorial attitudes and top-down approaches characteristic not only of Marxist intelligentsias, but of highland Ethiopian society more generally. Those movements, such as the OLF, which were unable to create such effective organizational structures were noticeably less successful in presenting any viable challenge to the regime. Although the defeat of the Derg marked a major blow to the project of a centralized Ethiopian state, it therefore did not mark the end of the political processes set in train by the revolution, but instead harnessed them in rather different ways.

Second, opposition movements were able to undercut the privileged access to external resources which the state had previously controlled and to build often more effective linkages themselves. One way in which this was done was through the forging of alliances between opposition insurgencies and non-governmental organizations in the developed industrial world, which in turn had significant leverage over their own domestic governments and civil societies. The Emergency Relief Desk, through which food aid was channelled into areas controlled by the opposition fronts during the 1980s, provides a classic example (see Hendrie 1994). It not only directly challenged the ability of the Derg to use food aid supplied through Addis Ababa to maintain its own control over subject populations, but also provided a useful source of cash and built diplomatic and publicity linkages to aid-providing states. Similar linkages were built with human rights organizations (see DeMars 1994). Most fundamentally of all, such operations transformed the spatial significance of the border. Rather than enclosing political communities within territorial states, in which power was allocated by international convention to the state government and domestic actors were required to look inwards to the national capital, the border became a zone of uncertainty and therefore of opportunity, across which local actors could move the resources needed to sustain resistance and could also take refuge themselves when circumstances made this necessary.[9] Wendy James's account in Chapter 14 provides a case study of the ways in which borders can thus be utilized. The concepts of centre and periphery are radically transformed once previous peripheries become points of access to power, requiring sharp symbolic definition and in the process draining resources away from central control.

Moreover, insurgent movements were able to extend their operations well beyond border zones, into arenas that had previously been restricted to internationally recognized states. The EPLF and TPLF established what were, in effect, their own foreign ministries and diplomatic missions (see Clapham 1996: ch. 9). They enjoyed regular relationships with major

---

[9] For a set of studies on the changing significance of boundaries in a broader African context, see Nugent and Asiwaju 1996.

external powers, and despite their own Marxist origins, they were far better placed than the Derg to take advantage of the transformation of global politics that occurred in 1989. One of their most significant contributions to the development of a globalized transnational politics that subverted the privileged role of the state was their ability to organize the diaspora of their exiled supporters in other parts of the world. The EPLF in particular, but also the TPLF to a significant degree, were able to use these diasporas as a source of publicity, expertise and cash. The EPLF was able to tax Eritreans in Italy and North America, far more effectively than most African states were able to tax citizens living within their own territory. The subversion of a centre–periphery model of statehood in Ethiopia forms part of the upheavals that are commonly subsumed under the catchword of 'globalization'.

Opposition movements, finally, achieved considerable success in transforming the politics of identity, the significance of which in defining the difference between the Ethiopian 'Great Tradition' and subordinate peoples has already been emphasized. Only in Eritrea, moreover, with its distinct pattern of social and political development and its peculiar status in international law, did the restructuring of identities take an essentially statist form. There, the EPLF promoted a territorially based nationalism and adopted a nationalities policy that was very similar to the Derg's, its goal being the establishment of a sovereign state, which was effectively achieved in 1991 and ratified following the referendum two years later. It therefore strictly subordinated nationality to the central power structure of the insurgent movement, and subsequently of the independent state. Elsewhere, opposition to the Derg was based on the mobilization of nationality, which in Ethiopia as in the Soviet Union provided by far the most effective mechanism for challenging the power of the Marxist-Leninist state, while at the same time drawing on identities that had been accorded at least some legitimacy by that state itself. It was said in the former USSR that the Stalinist theory of nationalities had life only after death: the complex structure of union republics, autonomous republics and so forth, through which the Soviet nationalities were recognized, was meaningless so long as control by the Communist Party of the Soviet Union survived, but it provided a ready-made formula for the break-up of the state once it collapsed. The aftermath of the Derg's collapse was rather similar.

The EPRDF & the new politics of space

The EPRDF, which seized power with its capture of Addis Ababa in May 1991, therefore conceived Ethiopia in terms very different from those of previous regimes. The first Ethiopian regime in over a century to be drawn from outside Shewa, it replicated in some respects the pattern of the half-century up to Menilek's accession in 1889, when the throne was in effect a prize claimed by the most successful regional warlord. The new government was dominated by the TPLF and closely reflected the distinctive attitudes which the movement had developed in the course of some fifteen years of insurgent warfare in Tigray, for most of which it had been effectively isolated not only from the outside world, but from the rest of Ethiopia. Whereas the Derg saw itself as protecting 'national unity' against divisive forces, the TPLF saw itself as representing one of a diverse group of peoples who had been more or less arbitrarily incorporated into a single political unit and for whom 'national unity' was no more than a pretext for repression. For them, Ethiopia needed to be taken apart and put together again – if it could be put together at all – in accordance with a formula that respected the identities and autonomy of the peoples of which it was composed.

This was, moreover, a programme that offered the prospect of finding allies in other parts of the country. As already noted, opposition to the Derg outside Eritrea (for which the TPLF sensibly recognized that separate independence was the only solution) had by 1991 come to be based almost entirely on ethnicity. By upholding the rights of 'nationalities', the TPLF could therefore make common cause with movements and people in other parts of the country that had likewise been alienated by the heavy-handed and often violent policies of the Derg. On this basis, the EPRDF was established under the leadership of the TPLF as a grouping that also included the Ethiopian People's Democratic Movement (EPDM, subsequently renamed the Amhara National Democratic Movement or ANDM), an Amhara organization that operated with substantial TPLF support in the area immediately to the south of Tigray, and the Oromo People's Democratic Organization (OPDO). Other partners were added after May 1991.

Within two months of seizing power, the EPRDF convened a national conference which led to the establishment of an 87-member Council of Representatives, which in turn formed the basis for a transitional government, pending the preparation of a definitive constitution. The most remarkable feature of the Council was that its members were designated almost entirely (save for representatives of a few professional organizations) on the basis of ethnicity, being drawn either from existing movements, or in some cases, notably in the south-west, from organizations rapidly formed in order to take advantage of the new dispensation. Representation of Ethiopia's largest nationality, the Oromo, was divided between the TPLF-affiliated OPDO, the Oromo Liberation Front and a number of smaller movements. Whether the OLF could have been incorporated into an EPRDF government had it been treated as the sole representative of the Oromo remains open to question. The relationship between OPDO, which was strongly favoured by the government, and the OLF, which regarded itself as the sole authentic representative of the Oromo, was inherently conflictual: in 1992 the OLF withdrew from government and attempted (with only limited success) to relaunch itself as a guerrilla insurgency. Other Oromo insurgent movements notably included the Islamic Front for the Liberation of Oromiya (IFLO), which operated in the south-east of the country, while the OLF was based largely in the west.

Under the constitution of the newly formed Federal Democratic Republic of Ethiopia (FDRE), ratified in 1994, nationalities became the constitutive basis of the state. The document itself opened with the phrase, 'We, the Nations, Nationalities and Peoples of Ethiopia' and went on to state that, 'All sovereign power resides in the Nations, Nationalities and Peoples of Ethiopia', which explicitly retained rights of self-determination, up to and including secession. Accordingly, the country was divided, on the basis of nationality, into nine regions: Tigray, Afar, Amhara, Oromiya, Somali, Beni Shangul-Gumuz, the Southern Nations, Nationalities and Peoples (SNNP, itself a composite of various nationalities, which had agreed to merge), Gambela, and Harar, with Addis Ababa retaining a special status as the national capital (see Map 2). Assemblies were elected for each region, all of them coming under the control of ethnic parties belonging or affiliated to the EPRDF.[10] Within regions, the *wereda* or district became the main subsidiary unit and included about half a dozen 'special *weredas*', which enjoyed enhanced levels of autonomy, especially over issues such as the use of nationality languages. In Chapter 11, Elizabeth Watson explores the implications of this status for one such special *wereda*, Konso. Other smaller nationalities, such as the Suri described by Jon Abbink in Chapter 8, were able to establish a designated *wereda* which, without achieving special *wereda* status, nonetheless assured them a measure of self-government. The process extended down even further to the grass-roots level: in Chapter 9, Hiroshi Matsuda traces the events which led to the granting of *qebele* status to the tiny Muguji population in the Omo valley.

Ethiopia has thus gone far further than any other African state (and almost any state worldwide) in reconstituting itself in ethnic terms. It has adopted what some might see as a peculiarly anthropological approach to state-building, recalling Conti Rossini's famous aphorism (1929), 'L'Abissinia è un museo di popoli'. This volume, however, while presenting among its chapters a series of anthropological studies of local and culturally distinctive communities, takes a more critical line; rather than seeing these groups as the essential building blocks of the wider state, it sees them as partly created by the processes of shared political history. The EPRDF regime's image of Ethiopia as an assemblage of distinct ethnicities draws in part on the specific experience of Tigray, which occupies an anomalous position among the larger Ethiopian nationalities in being

---

[10] Parties representing the central and sedentary nationalities (Oromo, Amhara, Tigray and Southern Peoples) belong to the EPRDF, while those representing peripheral nationalities (Afar, Somali, Gambela and Beni Shangul) are only affiliated with it. This division reflects the difference in EPRDF ideology between core zones at a higher level of socio-political 'development', capable of sustaining democratic parties, and less developed nationalities, which are as yet incapable of sustaining such parties, also indicating in its turn the EPRDF's continuing indebtedness to Stalin's theory of nationalities.

socially far more homogeneous than most areas of the country, despite a longstanding history of factionalism within its internal politics, and with by far the greater number of its people sharing a single language and religion. An economically marginalized area, producing little if anything for export, it is subject to periodic famine and is surplus-consuming rather than producing. As a result, it has a history of emigration rather than of immigration and has therefore been spared the complex ethnic politics induced by population movements in many other areas of the country. While culturally part of the core, and indeed the homeland of the former Axumite empire, it is distinguished from the Amharic-speaking area to the south by a separate language and political identity and has retained, more than any other part of Ethiopia, a tradition of political autonomy, accompanied by the resentments induced by economic dependence and the loss of power to distant Shewa. The separation of Eritrea in 1890, with its substantial Tigrinya-speaking population, divided the region and left the economically most dynamic part of it on the other side of the Mareb. While the Eritreans, with their history of Italian colonization and the distinctive political status conferred upon them by the United Nations' post-1945 disposal of the Italian colonies, could make a convincing case for independence, in Tigray (despite some flirtation with the idea in the early years of the movement) this was not a viable option. A Tigrayan independence movement would have none of the legal basis available to Eritrea and had no plausible source of external support. The region was landlocked and impoverished, dependent on external resources in one form or another. Nor did it even have the history of exploitation and alienation imposed by central conquest on southern and western Ethiopia. For a Tigray-based movement fighting against a centralizing regime in Addis Ababa, some form of regional autonomy within Ethiopia was the only political programme that made any sense.

Regional autonomy likewise had much to offer to many of the smaller nationalities, especially those which needed protection against more powerful neighbours and which for the first time could acquire some stake in a political structure in which they had hitherto been mere subject peoples. Indeed, as several of the chapters in this book indicate, the new dispensation of devolved power provided a more congenial setting in which to pursue the project of *encadrement* than the mechanisms imposed by the Derg, since incorporation into the system offered considerable advantages, especially to local-level elites. Any educated Konso, for example, could find a government job. With these opportunities, certainly, went all the tensions associated with the incorporation of political factionalism into communities in which internal divisions had previously been suppressed. The new structures were subject to a very high level of leadership instability as a result of both factionalism and the unpredictable impact of events at a higher level. There was nonetheless a sense of local political ownership that had not previously existed.

But the 'museum of peoples' approach to regional governance was itself problematical. For one thing, it imposed a need for a clear territorial demarcation on peoples who – as is very commonly the case throughout Africa and elsewhere – had become interspersed through diverse and longstanding patterns of settlement. Under the new system, territory that was 'owned' by one group could not be owned by another, and pressures to assert ownership were intense – Ren'ya Sato shows in Chapter 10 how one tiny and previously scattered group, the Majangir, moved to consolidate their own territoriality. A particularly violent confrontation occurred in mid-1998 between the Gedeo, a settled agricultural people living in the SNNP region of southern Ethiopia, and the agro-pastoralist Guji Oromo. Historically, Guji had lived within Gedeo territory and vice-versa, and disputes between them had been settled with little difficulty. The new spatial politics, however, insisted that one group had to have priority, resulting in a conflict which led to many deaths and the flight of some tens of thousands of Gedeo across what had now become a fixed frontier between the Oromiya and SNNP regions (see Tronvoll 2000).

It soon became clear, too, that there were severe political limitations to the exercise of autonomy. This is most evidently illustrated by the case of the Oromo, who, as the largest and most centrally placed of all the nationalities, were in a position to make or break not only the new political structure, but the continued existence of Ethiopia itself. The EPRDF regime used every means at its disposal to support its own coalition partner, OPDO, against rival claimants to represent the Oromo, including notably the OLF and IFLO. Many thousands of Oromos suspected of sympathy with opposition movements were arrested, and expressions of Oromo identity outside OPDO were suppressed. In practice, the promised 'self-determination, up to and including secession' amounted only to the exercise of local self-government in alliance with the EPRDF. This certainly included some significant changes, such as the displacement of Amharic as the language of public life by Oromifa, written in the roman rather than the ethiopic script; but it raised considerable doubts as to whether the politics of nationality could be contained within the limited sphere of action that the government was prepared to provide.

Amharas faced the converse problem: whereas they had been accustomed to regard being Amhara as virtually coterminous with being Ethiopian, it was now no more than a constituent identity within a larger state, in which they enjoyed neither political power (which rested largely with Tigrayans) nor numerical predominance (which fell to the Oromo) and in addition inhabited one of the poorest, least developed and most environmentally degraded parts of the national territory. The sense of alienation that this induced, together with a resort to Orthodox Christianity as a core expression of cultural identity, are graphically described in Cressida Marcus's account of Gondar in Chapter 13.

The problems in the borderlands are different again. While the new political order provided a place for nationalities within Ethiopia, it did nothing to address the problems of peoples who straddled state frontiers, and in some ways it exacerbated them. There was indeed a substantial devolution of power to the individual region and to subordinate national units within it. Even within the two smallest and most peripheral regions, Gambela and Beni Shangul-Gumuz, indigenes took over the leading administrative positions and a lively local politics displaced the imposition of central rule (see Young 1999). The peoples of the western frontier nonetheless continued to be affected, as Wendy James describes in Chapter 14, by the shifting relations between the governments in Addis Ababa and Khartoum and corresponding levels of support for, or distance from, Sudanese opposition movements. The politics of the Somali zone operate, as before, in accordance with the manipulation of clan identities and alliances with external sponsors, intensified after 1998 by Eritrean support for the factions most opposed to Ethiopia. On the Eritrean frontier, Kunamas and Afars especially were caught between the two sides after the outbreak of the Ethiopian–Eritrean war in May 1998.

The new system of ethnic confederalism also raises all the old questions of political power and surplus appropriation. A workable and acceptable balance between the demands of central management and regional representation within a confederal and nominally democratic political structure would be difficult to work out at the best of times. In a state such as Ethiopia, with its deeply entrenched legacies of domination and complete lack of any tradition of oppositional politics, tensions are inevitable. The EPRDF has attempted to address these by ensuring that all the regional administrations remain under its control and by using any necessary force to suppress threatening opposition movements, both centrally and at the regional level. When the war with Eritrea broke out in May 1998, taking it completely by surprise, it was nonetheless forced by an unexpected level of Ethiopian national sentiment to abandon its previous emphasis on ethnic diversity and resort to much the same nationalist rhetoric as the Derg.[11] The demands of war likewise intensified centralizing pressures over issues such as conscription, while the Eritrean government sought to destabilize its adversary by offering support to opposition insurgencies, notably the OLF. The relationship between 'Ethiopia' and the units of which it is composed remains fragile and uncertain.

---

[11] The causes of conflict lie beyond the scope of this chapter and are in any case obscured by the fierce partisanship that characterizes most writing on the subject. See Abbink 1998a and also Tekeste Negash and Tronvoll 2000.

Controlling Space in Ethiopia

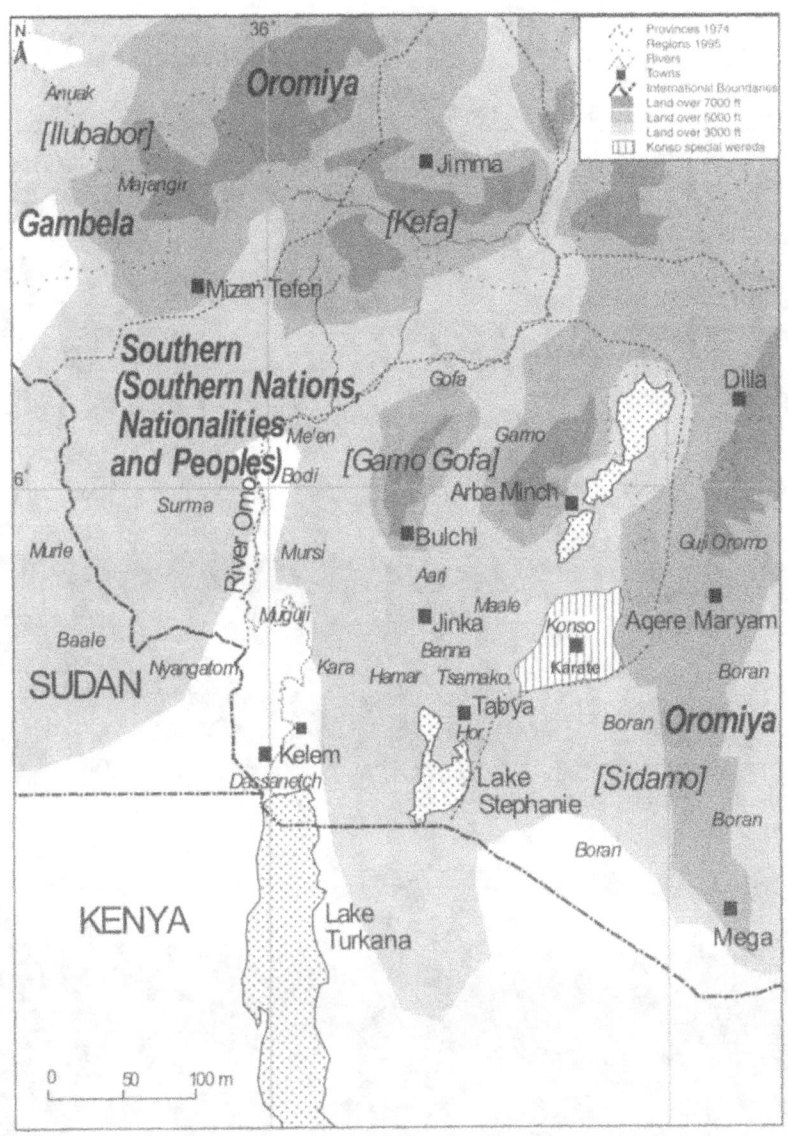

*Map 1.3 The southern plateaux & valleys
(to illustrate Chapters 2, 3, 6, 8, 9, 11)*

Looking Back on the Projects of the Socialist State

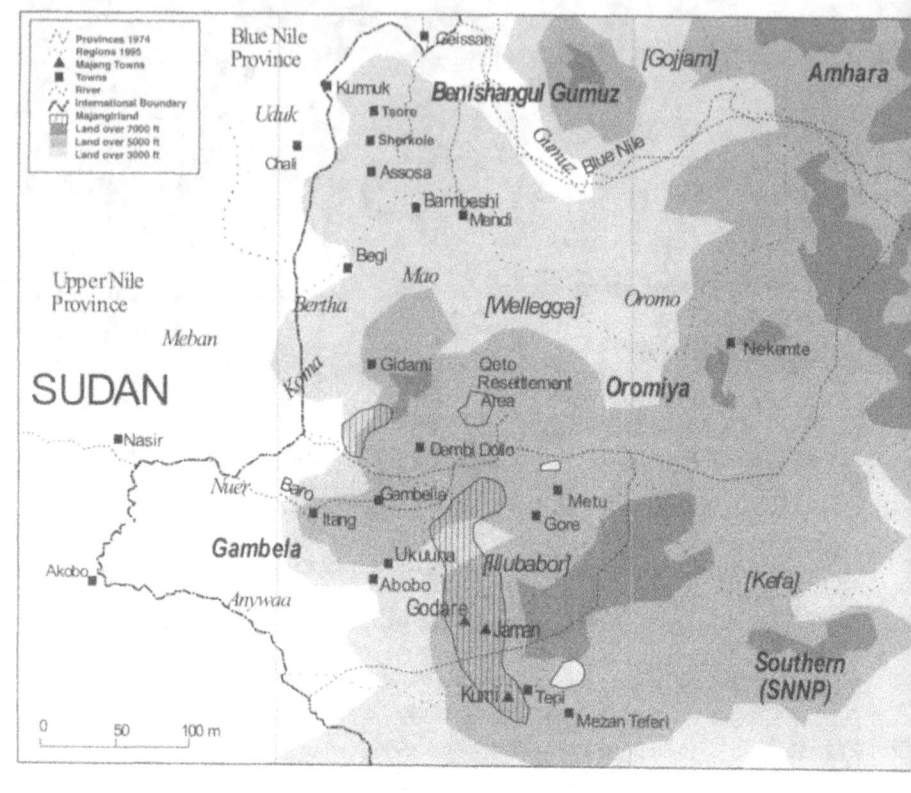

Map 1.4 *The western plateaux & valleys*
Note: *Former provincial names in square brackets*
(to illustrate Chapters 7, 12, 14)

II

Looking Back on the Projects of the Socialist State 1974–91

Introduction

Donald L. Donham

The most striking process that affected local communities from the revolution of 1974 until the end of the 1980s was what Clapham in Chapter 1 called *encadrement* – the attempt by the revolutionary state to penetrate and control local communities in a qualitatively new way. My study of this process among the Maale was published in 1999 as *Marxist Modern: An Ethnographic History of the Ethiopian Revolution*. Among other things, I emphasize the unevenness of *encadrement* in Maale. In areas closest to administrative centres, local Maale politics from 1975 until the late 1980s was effectively dominated by a series of new institutions like peasant associations with their schools, clinics, and prisons, while in areas farther away, in contrast, pre-revolutionary political structures survived, even if in camouflaged form.

This unevenness in Maale was not simply a function of how easily coercion was exercised (though exercised it was, often in highly theatrical ways). It was bound up as well with processes that had splintered cultural identities among the Maale towards the end of Haile Selassie's reign – most centrally evangelical Christian conversion. Conversion provided Maale with its local 'modernists', newly educated young men committed to development and progress. Because religious change itself was a process that radiated out from local towns, Maale areas closest to local centres were the most diverse in cultural and religious orientation by the time of the revolution. One of the most consequential effects of this cultural splitting was the undermining of traditional methods of conflict resolution. This meant that local peoples had to take their conflicts to the revolutionary state. In this setting, the new state did not penetrate local communities simply by exercising power from above. It also grew from the grassroots as traditionalists, evangelical Christians, and the handful of educated Maale ex-Christians who were inducted into the Workers' Party came into inevitable conflict.

The unevenness of *encadrement* is illustrated on a much wider canvas in the following case studies by Tadesse Wolde, Alexander Naty, Ahmed Hassan Omer and Jenny Hammond. Among the Hor described by Tadesse – an agropastoral group in a border lowland of little strategic value to the state at the time – the revolution was hardly a revolution at all. Local political structures simply went underground. As in many areas of the south, Hor ritual leaders became targets of the new state (as they were among the Maale), but in general local politicians were successful in holding the revolutionary state at arm's length. New institutions such as peasant co-operatives became essentially 'fronts' controlled by traditional leaders chosen in the old ways. And as the Derg's attempts to control commerce in the south grew ever more counter-productive, some traditional arrangements such as the *fund'o* – a remarkable interregional trading network – actually flowered. All of this meant that when the Workers' Party state fell in 1991, the Hor were unmoved one way or another. When EPRDF representatives arrived and mistakenly addressed the Hor in Oromo, one of the more wily of local leaders arose to tell the newcomers (in Oromo) what they wanted to hear: he complimented them on being the first representatives of the Ethiopian state to address the Hor in their own language.

While the rise and fall of the Workers' Party state occasionally had its lighter sides, these transformations set off far more serious consequences in former *gebbar* regions of the highland south such as the Aari described by Alexander Naty (neighbours to the Maale). There, the arrival of the *zemecha* students lit an explosion of hatred from below, as young Aari took revenge against local wealthy northern landlords. Alexander Naty places the violence in the context of the local Aari memories of humiliation – military defeat, subjection to serfdom, and the loss of land. Because this humiliation was culturally encoded among the Aari as a kind of feminization, the 1974 revolution was experienced, not only as empowerment per se, but specifically as a masculinization – hence the disturbing import of the rape of former landlords' wives. Revolutionary euphoria did not last, however, and by the end of the 1980s – after most of the typical effects of *encadrement* had been felt in Aari – local people once again felt helpless and therefore feminized.

If we turn now from the *gebbar* south to peasant areas within the old Amhara core, we see that class relationships alone never determined peasant attitudes towards the revolution. In the south, it was mainly class overlaid with cultural difference that was explosive. In northern Shewa as described by Ahmed Hassan Omer, by contrast, local lords were cultural kinsmen to peasants. Indeed, the Emperor himself traced kin links there. Ahmed shows the particular consequences of being 'close yet far' in this

Introduction

context. Northern Shewa was culturally a part of the old Orthodox Christian core, yet in terms of economic level and ease of communication, it was far from the centre of power. Land reform in this area was greeted more ambivalently than in Aari, and indeed, resistance movements led by former local lords in the Ethiopian Democratic Union (EDU) were able to command some peasant support. After the EDU was eliminated by the Derg, supporters of the Ethiopian People's Revolutionary Party (EPRP) fleeing the cities brought the conflicts of the 'red terror' to the countryside. In all of this, northern Shewan peasants were caught in the middle. In a revolution carried out in their name, they gained virtually nothing but increased exploitation. Ahmed captures the poignancy of this experience through quoting peasant sayings that circulated through northern Shewa at the time.

This leads us finally further north to the case of Tigray as described by Jenny Hammond. There, in another portion of the cultural core, an organized resistance movement was finally successful in opposing the Derg's rule. The revolution of 1974 had occurred in Addis Ababa and eventually radiated out into the countryside along the routes and capillaries defined by former imperial centres. Hammond describes the reversal of this pattern in Tigray during the 1980s as the TPLF established itself, first in the countryside and only toward the end of the decade in the major towns of Tigray. Hammond uses the memories of informants, particularly women's, to give insight into the privations and hopes of ordinary peasants in the context of what was a brutal struggle.

The remaining studies by Taddesse Berisso and Alula Pankhurst capture the Workers' Party state at its most expansive and climactic moment. Scott (1998) has argued that some of the greatest human tragedies of the twentieth century have resulted when the state remaps its populations with what he calls high modernist zeal – high modernism being a kind of irrational faith in rationality, an ideological use of science. Such a characterization captures much of the overvaulting ambition of the two projects analysed by Taddesse and Pankhurst, namely villagization and resettlement.

Villagization involved gathering Ethiopian peasant farmers into regimented, compact settlements (like *ujamaa* villages in Tanzania), while resettlement encompassed moving literally thousands upon thousands of impoverished northerners into the south – sometimes by force. By the end of the 1980s, the state had created new villages laid out in ninety-degree grids for close to half of all rural cultivators (at least in areas outside Eritrea and Tigray). Taddesse details how state agents accomplished this end among the Guji Oromo. And close to a half-million northern peasants had been resettled to the south, a process that Pankhurst examines in Wellegga. As the Workers' Party lost its grip on central power, however, both of these projects quickly collapsed, both leaving their residues. As Pankhurst shows, some of those resettled in Wellegga chose to stay on in their new homes to create functioning multiethnic communities – ironically, in what would

Looking Back on the Projects of the Socialist State

become the supposedly ethnically homogeneous new region of Oromiya.

In this part, we are pleased to include studies not only of the former southern periphery but also of peasant communities in the old cultural core. This illustrates in a way that *Southern Marches* did not (see Jon Abbink's comments in Chapter 8) that centre–periphery relations were never simply reflections of geographic or 'ethnic' distance. Rather, they were fundamentally about the structure and distribution of political power – a power that sometimes fell just as heavily on northern as on southern peasants.

Two

Evading the Revolutionary State
The Hor under the Derg

TADESSE WOLDE

This chapter explains how the Derg and later the Workers' Party of Ethiopia were perceived by the agro-pastoral Hor (known more generally in Ethiopia as 'Arbore') of south-western Ethiopia and how they viewed the coming to power of the EPRDF. The study is based on field research carried out in Hor country from August 1994 to August 1996.

I shall start by introducing the Hor in the 'ethnographic present': their territory extends over the desert plain from the Tsamako country in the north to the Kenyan frontier in the south and from the foothills of the Hamar country in the west to the foothills of the Boran country in the east, and includes the surface of the dry Lake Stephanie (see Map 1.3, p. 31). Their settlements are situated at the northern end of Lake Stephanie on the Limo (Woito) river delta. They call their flat country Wando and they numbered 3,438 according to a census I took in 1995.

The Hor live in four main villages and numerous other temporary settlements, some of which are cattle camps. Each of the four villages has its own *qawot* ritual leader and an age organization based on division into generations. The named generation of the age organization in power administers each village, including its sorghum cultivation and its pastoral activities. The Hor obtain their subsistence from both sorghum cultivation and animal husbandry and live on the delta of the Limo, which is the most fertile area of the whole basin. By regional standards, the Hor are rich.[1] The senior and junior *qawot*s of the Hor claim to have influence over the natural and social order of the region, and this is acknowledged not only by the Hor but also by their neighbours.[2] Whenever they can, ordinary Hor or members of these communities enter the cattle gates of the Hor

---
[1] See Smith (1897: 262–3) for information on Hor life and the wealth of the Hor at the time of his visit in 1895.
[2] See Lydall and Strecker's discussion with Baldambe and with the *qawot* of the Hor for Hamar opinion of the power of Hor *qawot*s (1979: 112).

*qawot*s with gifts of animals, honey, tobacco, coffee, herbs, etc., to obtain blessings to enable them to have children, wealth and good health. Sometimes people come not just on their own behalf but as representatives of whole communities seeking prayers and blessings, not only for fertility and well-being, but also for peace with, or victory over, their enemies.

Each Hor village has its own place of assembly and separate cultivation plots. Areas of pasture are shared with a second village. Leaders of the generation of the age organization in power named *jald'aba* are responsible for the administration of pastoral activities, while those named *mura* are responsible for the administration of the redistribution of land each year, the irrigation of plots, and the cultivation of sorghum in each village. Together *jald'aba* and *mura* run the affairs of each village and ensure that order is maintained. Jointly they make sure that essential items for the performance of morning and evening prayers and of blessings and curses against Hor enemies are available in the *qawot*'s house. They also make sure that appropriate sacrifices are performed by those responsible in order to guarantee human, animal and crop fertility, and to secure victory over enemies in warfare. The leadership makes sure that the transfer of power between generations is carried out peacefully and that the generation in power has been properly rewarded with gifts of coffee and a series of honey-wine drinking parties before it hands over power. This transfer of power and acquisition of control over the assembly place by the new generation is regarded as important for the proliferation and the well-being of the Hor in general and for acquiring the status of warrior and father, which entitles a man to arrange the marriages of his sons and daughters. Power is transferred once every forty years. An opening ritual called *nger* enables the new Hor generation taking power to become warriors, while a later ritual called *chernan*[3] entitles them to become socially recognized fathers arranging the marriages of their sons and daughters. These two phases of initiation are also the time for the senior generation to retire into grand-parenthood. They move out of the centre of the *nab* assembly place to the peripheral grandparents' *nab*. They also surrender their resting shade (men's house) on the eastern edge of the village and give up power over animals to the new initiates, their sons, who become fathers of the land.

Agricultural land is communally owned and redistributed annually. Outsiders who assume Hor identity and other outsiders who have the consent of the leaders are given land both as individuals and as groups. There are no landlords in Hor country, whether locals or outsiders. The Hor do not allow visitors other than their bondfriends from other groups and those traders who take traditional trade routes to come to their villages. In order to travel to Hor villages the prior knowledge and agreement of their leaders is essential.

---

[3] See Miyawaki (1996: 39–65) for an account of the second phase of the ritual. Miyawaki attended the *chernan* ritual of the most junior village of Gandaraba in 1991. The villages of Egude, Murale and Kulama had already gone through their *chernan* before it was held in Gandaraba.

Evading the Revolutionary State: Hor

The Hor under the Derg

A police station was established in Hor country in the late 1950s, and became known simply as Tabya ('station' in Amharic). By 1974, it was still the only state institution in the area. A German volunteer had lived and worked there for a while.[4] Currency had not been introduced into this area, and annual taxes were collected in livestock. Exchange in the market and elsewhere was carried out by bartering animals, honey, coffee, bullets, firearms, beads, small stock and other products. Thus Hor traditional life went on largely undisturbed by the setting up of the police station.

The only northerners in Hor country were some policemen, assigned to Tabya for a minimum of two years' service.[5] Because of the transitory nature of their stay, they were not interested in investing in friendships with the Hor and did not make any effort to grab land or acquire other forms of wealth, as is the practice in agricultural and coffee-growing areas. They did not mix with the Hor, as mixing with unbelievers, commensality and other social intercourse was considered unhealthy. The Hor, on the other hand, viewed the soldiers as a shield against dangerous neighbours, though at the same time they understood that they were being looked down upon. They feared the power of the police to disarm and imprison them. They also remembered previous harsh encounters with northerners.[6]

When the Derg came to power after overthrowing the government of Emperor Haile Selassie, there was neither jubilation nor surprise in Hor country. There was, however, a fear of possible brigandage by police and northerners living in the region. This fear was deeply rooted in previous experiences. Brigands had unleashed their terror in the region in the past whenever the opportunity arose. This was also the case after the death of Menilek, when Haile Selassie went into exile, and when the Italians left. Whenever regional governors appointed by Menilek and later on by Ras Tafari were transferred to other regions, it was common practice to loot the region and take as many slaves as possible before leaving.[7]

This time, the problem was primarily one of communication. In Hor country, life went on as normal.[8] There was no radio programme transmitted

---

[4] No one could tell me why the German volunteer had left Hor country. Bali Argido of Murale (the third head of the South Omo People's Democratic Organization appointed for Hor country while I was in the field) had been sent to school in Arba Minch by the German.
[5] At an earlier time, a northern Amhara man named Tesfaye, possibly a fugitive, had joined one of the clans, married a Hor woman and lived in the Arbore section. He left for Sidamo later, leaving his children behind (his clan had contributed to the cost of the marriage).
[6] See Darley (1969: 55–7), Garretson (1986: 202, 205) for the experience of Maji; Hodson (1927: 146) for a statement about a similar incident in Kefa in 1917.
[7] Later, in 1991, the fall of the Derg was followed by widespread looting in the major towns in Ethiopia. Road transport ceased, as vehicles were hijacked by both fleeing officials and the rebels. NGO warehouses, firearms depots and police stations were looted. Prisoners went straight home because their guards abandoned their duties. Members of the armed forces and police retired to their homes with firearms. Firearms were sold everywhere. Inter-group conflict rose considerably. The Boran and Hor went to war against one another in May 1992.
[8] Futterknecht (1997: 172, 182) mentions incidents of nearby conflict, raiding and famine.

in the Hor language. Only the police owned transistor radios anyway.[9] Police at Tabya did not know how to react to these events, but they feared a possible attack by the Hor. Decrees by the new government were announced on the radio on a regular basis, mainly in Amharic, Oromo or Tigrinya. Major decrees were translated and transmitted in some southern languages, such as Gamo and Sidama, but not in any of the languages of the pastoralists in southern Gamo Gofa. Because there were no landlords and not much information about what was going on at the centre, and obviously because the Hor had no clear idea about what the new regime planned to implement, there was no reason either for jubilation or for local resistance. Development Through Cooperation Campaigners (*zemecha*) were not assigned to this part of the province.[10] No 'ethnic' political organizations were active in the region, and no individual or group complaints were filed against any of the regional or national members of the recently overthrown government. No applications were sent to the new government for correction of ethnic and place names, wrongly used by northerners, as was the case elsewhere in the south. The preoccupations of the centre of Ethiopia with the national anthem, the name of the country and the colour symbolism of the flag were of no concern to the Hor. Similarly, Marxist ideas and the routes to be followed in the pursuit of progress were not on the Hor's agenda.

Attempts at elaborating the leading ideas behind the new administration were carried out in the Hor language by some very low-ranking members of the armed forces and the police, who made occasional visits to Hor country for this purpose. These were young recruits, given this responsibility because of their membership in the armed forces and the police.[11] Later, some junior cadres with very inadequate schooling joined in this work to explain about the revolution and its enemies. The Hor had

---

[9] Yukio Miyawaki, a Japanese anthropologist, gave his literate informants some transistor radios, which were eventually bought by salaried civil servants in Tabya.

[10] Campaigners were assigned to Geleb and Hamar Bako District only in the coffee-rich Aari area in Gazer and around the town of Jinka. Key Afer *wereda*, which included Hamar, Banna, Tsamako and the Hor, was, it seems, considered unsafe for campaigners and therefore avoided.

[11] From my own experience as a campaigner, these 'Disciples of Change' as they were titled in the early days of the Derg regime were ignorant of what was going on at the political centre and were not able to explain things as clearly as student campaigners were. They often had difficulty because of their lack of education. Their background as low-ranking members of the armed forces, and sometimes their non-standard Amharic accent, made them subject to teasing by student campaigners. They readily resorted to imprisoning those who raised questions, instead of giving them proper explanations. In many places, the rift between them and the campaigners grew into a chasm that was never bridged and developed into real enmity, which culminated in the white and red terrors. Many of these people were former prison guards and members of the police who had been antagonistic to students under the emperor. Such were the people assigned as leaders of the campaign at *wereda* level. Secondary school and university teachers were assigned with them, but the latter often sided with the students. Some of the officers whose views were not dissimilar to those of their subordinates were assigned to different parts of the country as *yelewt hawaria* or 'disciples of change'. Some lucky provinces and government ministries had well-educated officers with training in elite military academies assigned as 'disciples of change'.

enough enemies and could not accommodate more, so they suggested that the revolution had better take care of its own enemies and that they would take care of their own. Whenever the revolutionaries paid visits, the partly literate young Amharic-speaking Hor, of whom there were very few, were used as interpreters.[12] As the Hor are usually multilingual, the other groups of the area, such as the Assile Hamar, the Wungabaino, the Karmet and the Tsamako, were easily reached through these interpreters.

These young men helped to translate some of the songs of the revolution into the Hor language and sang them to a small group of young children around Tabya who were attracted by them. The occasional meetings which the officials and political cadres called at Tabya gave the interpreters an opportunity for employment in political work. Other non-literate Hor, such as Korranki Ghino in Tabya, who spoke Amharic through long contact with the police, were also used as interpreters. When Assile and Wungabaino Hamar were required to attend official meetings, they were ordered to come to Tabya. During such meetings, the mediums of communication were Hor, Hamar, Boran and Amharic. Usually two interpreters were used.

The Hor say that these meetings were long, with boring subjects, and that they often coincided with herding or agricultural work. Officials were easily offended when they saw Hor participants lying down on their backs and headrests during meetings. The Hor would say that it was their ears that were listening and that there was no point in sitting up straight. The subjects discussed varied from the elaboration of election procedures to outlining the duties and responsibilities of elected officials in the local *qebele*s (the smallest administrative units). The nature of the new decrees and their implications, the achievements of the revolution at its various stages and the machinations of the counter-revolutionaries from the point of view of the government were also explained. Other subjects over the years included fund-raising for the war effort, the recruitment of fighters for training, and elections of various sorts to fill vacant *qebele* positions. Often the advantages of farming and settled life were emphasized.

The agendas of meetings varied from time to time and depended on events at the political centre, which lacked relevance for local people in this area. The Hor case did not fit well with government policy for handling pastoralism and pastoralists. Pastoralism was seen as an abnormal life style. The government planned to settle pastoralists and to engage them in farming. The Hor were already settled in four villages and had sorghum plots for each village in close proximity. Neither villagization nor sedentarization, which were important parts of the Derg's policies towards agricultural and pastoral groups, could be implemented, as they already existed in Hor country.

Three things were interesting to the Hor: inoculation programmes for children, veterinary services for their animals, and military training and the access to arms which it provided. The first worked well, thanks to the

---

[12] The only literate Hor in 1974 were Horra Sura, Tadelle Sura and Bali Argido.

assistance of the Norwegian Save the Children organization. Veterinary services were available for a nominal payment. However, the Hor did not have cash but only small stock and bullets to use for payment, and this was usually not acceptable to veterinary workers. The military training went well, but was favoured only if it occurred close to Hor country and if the trainees could come back to Hor country afterwards.

In general, the Derg's meetings were too much for the Hor. They had their own *nab* assemblies every night at which daily business was discussed. They had to do something about the outside meetings and decided that only two or three senior elders from each village should attend Tabya meetings whenever such meetings were called.

Thus *jald'aba* elders of each village (two or three from each) were sent so that herding and other work could be carried out uninterrupted. The remainder of the participants at the meetings were residents of Tabya who were by no means Hor,[13] but many officials did not know, nor care very much, whether they were or not. Some of them did not even know that the Hor had other settlements.

In general, the Hor saw their society as being healthy and did not want the Derg to be involved too much in their lives. The *jald'aba* of the generation in power did not see any problem with the Hor way of life. Both *mura* and *jald'aba* leaders of the age organization took full control of herding and sorghum cultivation activities and made sure that law and order were maintained. Throughout its time in power, the Derg had no control whatsoever over the village life of the Hor. The *jald'aba* and *mura* had the responsibility of distributing sorghum plots and the smooth running of agricultural and herding work, as well as deciding on camping and grazing land and on watering. They also controlled the movement, encampments and well-being of the cattle. No outsider who secured the consent of Hor *qawot*s to live in the territory was refused such rights. Even groups such as the Assile and Wungabaino Hamar were allocated camping areas in Sura. There was no one in Hor who was prevented from gaining access to land. The state was never involved in any of this and was represented only in Tabya.

The Hor were, and still are, categorized into various hierarchically organized, bracelet-wearing clans with claims to varying degrees of mystical power.[14] Other, non-bracelet wearing clans with no such powers depended on the bracelet-wearing clans for their well-being. The heads of the senior bracelet-wearing clans acted as the senior and the junior *qawot*s of the Hor. The basis of Hor politics was clan and age organization, which worked in accordance with principles of seniority and juniority and of gender complementarity.

[13] According to a census of Hor I took in 1995, Tabya had only 103 houses, of which 16 belonged to Hor, 16 to non-Hor civil servants and police, 16 to Karmet and Boran, and 43 to Konso. The rest belonged to members of various other groups or illegitimate Hor who cannot live in any Hor village.

[14] See Schlee (1994: 22, 175–200) for the powers which clans among the Sakuye, Gabra, Arbore and Rendille claim to have.

Evading the Revolutionary State: Hor

Plate 2.1 Cattle gates of the Hor qawot *[T. Wolde]*

Not only were the Hor satisfied with their way of life, but they were also well aware that their neighbours envied them for it and admired them for the mystical power of their *qawots*. Hor *qawots* were regarded very highly, particularly for their blessing and cursing powers, for their rain-making abilities, and for enabling friendly neighbours to defeat their enemies. Hor *qawots* commanded, and still command, respect for enhancing human, animal and crop fertility and for facilitating relations between groups by keeping the main trade routes and bond friendships open and safe for users. To take advantage of this power, neighbours of the Hor use specific routes to come to Hor country in order to enter through the *qawot*'s cattle gates (Plate 2.1). They enter these gates with gifts to seek blessing, to acknowledge their allegiance to Hor *qawots*, and to express their gratitude for wishes fulfilled through the mediumship of Hor *qawots* in the past.[15] At other times they invite Hor *qawots* to their own territories in order to benefit from their powers.

Given this situation on the ground, the Hor elders did not see the necessity of allowing the Derg further into their lives. They believed that they were at peace with themselves and with their neighbours and that there was enough rain and harvest. Their children and animals were healthy. There was no wedge that outsiders could use to loosen what held the Hor together. And there were no Hor ready to go against their own community. The only thing that really lured the Hor to the revolution was the military training and the prospect of being issued with firearms.

---

[15] Various Hor neighbours have specific types of gifts prescribed by tradition that they bring to the Hor *qawots*.

Looking Back on the Projects of the Socialist State

Plate 2.2   Rufo Ali, qawit *[T. Wolde]*

The growing antagonism of the Derg towards traditional institutions and the news of the looting of the premises of Bammale Kayabo Kaytare, a *poqalla* of nearby Konso, by his own people sent a strong signal to the Hor. They had to be cautious in their relations with the Derg if they wanted to keep their traditional institutions. As their suspicions grew, they increased their distance from outsiders. No government, NGO or religious institutions were allowed to base themselves in Hor villages. *Jald'aba* and *mura* tightened control of their own affairs and restricted their own presence in Tabya to market days.

Hor suspicion increased when the equality of women and the education of both sexes was advocated. Education, the Hor argued, made their children hate animals and, even worse, made them 'Sidam', that is, lovers of meat for whose rearing they contributed no labour, or people who make others do things for them. This was how the Hor perceived and still perceive all civil servants, northern Ethiopians in general, and the personnel of NGOs and various Christian organizations. Girls in Hor are assets of the clans, of the age organization, and of the honoured senior men's association *luuba*. The whole community had a stake in a properly arranged marriage. The above categories had rights to shares of meat, coffee, tobacco and honey-wine. Weddings were central and were occasions that brought the Hor together. Educating girls meant surrendering these rights and sending Hor women into prostitution, which the Hor attribute to the Sidam. They did not want their sons and daughters to be Sidam and to hate cattle and small stock, which form the basis of Hor life. They wanted them to remain Hor and love animals, and this meant keeping children away from schools and from any outside influence.[16] When Duba Fora, of the Hor *qebele* or pastoral association, and Hora, the secretary, tried to register children at the new school, the Hor conspired and attempted to murder them. Duba Fora told me that they ostracized him until he finally gave up the idea.

Things that made the Derg appear worthy of suspicion increased as time went by. The senior Hor *qawot* Konso Ali's long staff, a symbol of his power, was taken away by force and broken in public. Konso himself was taken prisoner.[17] This was a major reason for the Hor to believe that the dangerous nature of the Ethiopian state had not changed. This, for the Hor, proved that the Derg was a continuation of the imperial state, a state that looted, enslaved and killed the people it had made into its subjects. Now Konso, the senior *qawot* of Hor, had become its current victim, not through any apparent fault of his own or his people, but only because state officials imagined there to be resistance or potential resistance.[18] After the imprisonment of Konso, the Hor were cautious in their relations with the Derg and were at the same time careful not to antagonize the people in power.[19]

A *qawot* must not live outside Hor country, must not shave or cut his fingernails, and must not drink from containers other than his own. He must not eat or drink *parso* (sorghum beer) from crops grown on lands other than his own plot. With his arrest, Konso was made to transgress taboos which the Hor believed would cause them as a community to suffer disaster. They feared it would lead to the kind of life experienced during Menilek's time and after the murder of Konso's father by the Hamar in 1952. This was a big blow and a threat to the Hor, and nothing could have been worse. Konso Ali's arrest clouded relations with the government and nurtured the growth of suspicion and distrust towards the Derg among the Hor.

After Konso's arrest his mother, Rufo Ali, was made *qawit* (Plate 2.2) in order to minimize any danger that might arise as a consequence of her son's arrest. She was made to care for the cattle gate of the senior *qawots* of the Garle clan and to be the mother of the Hor. She acted as a widowed woman would act, as caretaker of the extended family group (*wori*), to protect the Hor and their cattle with her cowrie shell belt.

---

[16] These Hor views were a consequence of the highly centralized curriculum in Ethiopian schools and an educational system which did not pay attention to the interests and specific values of the rural people of Ethiopia. The curriculum seemed to be designed with the objective of 'civilizing' rather than creating self-awareness.

[17] The Derg had a policy of arresting ritual leaders in the south, since it feared that they might ally themselves with northern landlords in resistance to the revolution. It did not check whether or not Hor *qawots* were landlords. But there may have been awareness of the power and influence they command in the region.

[18] The position of *qawot* is hereditary. A *qawot* accedes to power by sitting on the main seat of his predecessor and by the purchase of genitals severed from the war dead of specific enemy groups, in this case the Maale or Kore (Samburu).

[19] The fact that the Derg's men broke the long staff of the senior *qawot* was later used by the Hor to explain the fall of the Derg.

Looking Back on the Projects of the Socialist State
Resistance through subversion

From an early stage, the Hor seemed to understand that the only way to block the threat of the new regime was by adapting the new institutions to their own conditions and to take as much control of these institutions as possible. This strategy might not have been particularly innovative, but the Hor used it effectively to subvert rules, and even to share power with some institutions.

Government-sponsored public meetings or meetings of local *qebele* required the attendance of all the adult members of a *qebele*. However, the Hor redefined this to suit their own conditions and let some leaders of the age organization meet and make decisions on serious matters important to the Hor. Similarly, when they were required to set up *qebele* associations, they set up only one for all four villages instead of one in each village and based it in Tabya, away from the villages. They also received the officials who came to meetings in Tabya instead of their villages, thus leaving Hor villages free from any outside presence. In this way, they were able to domesticate and tame the institution that was meant to demonstrate the invincibility of the revolution. The *qebele* was made the equivalent of the traditional Hor assembly place, the *nab*, by the fact that the officials elected to it were members of the initiated generations or of the yet uninitiated who behaved acceptably.[20]

Membership of *qebele*s was another area in which they subverted the law. According to the law a married couple living in a house was supposed to be the administrative unit. The Hor redefined this to mean a *wori*, a number of married brothers and their parents who share the same cattle-gate. The *wori* became the unit instead of the house, and the cattle-gate the unit of taxation instead of married couples owning a house and a plot of land and animals.

Some of the organizations the Hor were expected to set up were simply not established in the villages. The youth association and the women's association were established in Tabya and were not the concern of the villages. A women's committee, required within the association, was not established in Hor villages but in Tabya, where the women who attended were the wives of policemen and of traders from other groups. Similarly schoolchildren were recruited from among non-Hor children living in Tabya.[21] Only some *jald'aba* attended the literacy programme instead of the whole population, who were actually required to attend.

The Hor way of adapting the functioning of the revolutionary government's structure to their own requirements did not stop there. In electing

---

[20] *Nab* assembly places are places in the centre of villages where the retired, those in power and the uninitiated have specified areas where each discusses matters of importance.

[21] At the start of the revolution, Horra taught about seven students to read and write with the assistance of Hor police and the Ministry of Education branch in Arba Minch. Some of these students gradually took control of *qebele* responsibility and were made members of the Workers' Party of Ethiopia towards the end of the Derg's period of power.

Plate 2.3 Lemma, *fund'o* leader and elders [T. Wolde]

representatives to the regional organization of the peasant association, the Hor elected initiated elders of the age organization, who would be able to enter the state structure, influence decisions and favour themselves. Duba Fora, a member of the Otgalcha generation, was elected in such a manner and sent to the sub-district, the district and finally the provincial organization in Arba Minch. This was a serious subversion of the Derg's directives.

Taking effective control of some of the Derg's new institutions and avoiding participation in others, such as the women's association, was not enough to contain the change the Derg was planning to introduce in Tabya. In addition to controlling Tabya's new administration, the Hor strengthened their grip on the traditional trading organization, the *fund'o*. This institution is responsible for the safe passage and security of traders and goods within Hor and between Hor and other groups, and also facilitates bond friendship and acts as guarantor of debt payments. Today the network of the *fund'o* and its appointed officials covers key areas in southern Ethiopia (see Map 1.3, p. 31). There are representatives in areas such as Konso, Hamar, Banna, and in Jinka town. The Hor maintained good relations with Lemma, the popular *fund'o* leader in Tabya (Plate 2.3), as part of their strategy of coping with the unpredictable nature of the state.

The Hor *fund'o*

The *fund'o* chief, Lemma, spoke for the Konso residents in Tabya and was partly incorporated into Hor society through his membership of a clan, although he was not initiated into the age organization. Because of his Amhara upbringing in Konso district, he was a Christian and refused to eat meat slaughtered by the Hor. He was multilingual, married to a woman from the Gawad'a group, and spoke Boran, Konso and Amharic. He was literate, an expert writer of letters and applications to courts, and a master of court intrigues and bureaucratic dealings.

The Hor *qawots*, as well as their ritual assistants (*kernat*) and the *jald'aba* and *mura* officials had good relations with him, and he had many Hor bondfriends, whose gifts he was not able to reciprocate fully. The gifts he was given were considered investments by their donors. When the donors required his skills, they could freely expect his assistance. Every Hor who had a court case in the *wereda* (district) town in Qey Afer, or a case with the police at Tabya or in the newly established *qebele* office, needed his expert advice and required him to write applications related to their cases. Lemma advised them on how to respond to confusing police interrogations and to related matters. Even literate Hor like Horra or other *qebele* leaders could not work without his advice. They let him draft letters for them or sought his advice on what they had drafted. He was the man behind Hor politics. I was told the people liked him because he knew 'how to say what officials needed to be told' and what others were interested in knowing. 'His words were sweet to everyone.' He was a man they could trust regarding matters involving the police and the *fund'o*, of which they were a part.

His house served as a tiny rendezvous where local *areqi* and *chaqa* were sold and where prominent Hor met to discuss regional politics and Tabya gossip. His house also served as a safe house for prominent Hor to keep their rifles until they returned to their villages. Both outsiders and Hor felt at ease at his house. In the mornings the backyard of his house was converted into a *fund'o* court in which case after case was discussed and decisions were taken. On Saturdays there was hardly any room in his house, as Hor men crowded into it to drink and Hor women piled up their shopping before setting off for their home villages. On Sunday mornings civil servants and traders in Tabya huddled together and queued up to drink *chaqa* especially made for the saving society or *iqub*, to pay their weekly contributions and to find out who was to receive the money collected. Gossip was exchanged and latecomers fined. In their free moments, elders from villages and the *fund'o* lazed around on the sand in the cool shade of the *garsante* tree, played *qora bolo*, a board game, and went for drinks with money taken as fines from those found guilty during the court hearings. Amidst all this coming and going, Lemma was everywhere.

The police liked him because he was considered Amhara, was Christian, and headed the weekly saving society, *iqub*, of which they too were members. They also counted on the help they might receive from him in writing letters of application to courts and even to their own officials. The police also looked to him as a mediator when they fell foul of their officers. His ability as an elderly man to settle disputes was extraordinary. He was respected because he was the link between the Hor, Konso district and the *fund'o* network. Furthermore, as a senior man knowing Tabya life and the region in general, his opinion was often sought by visiting officials, and he was also sometimes used as an interpreter.

As head of the *fund'o*, Lemma wrote letters to relevant *fund'o* representatives everywhere. Lemma and the *fund'o* office were indispensable to the Hor. Although he had so much influence, these matters were confined to Tabya, and Lemma's access to Hor villages was limited to attending feasts, at which he only enjoyed honey-wine, coffee and tobacco.[22] He did not have any influence over Hor life in the villages. When there was a *fund'o* case in which a Hor person was implicated, he would go to the village concerned with his *helitta* messengers and Konso members of the *fund'o* to join the *jald'aba* of the village in order to hear the case and make decisions. If the person implicated in the case was Konso and lived and traded in Konso district, Lemma's team would visit the victim and send their decisions regarding the matter to the relevant *fund'o* body in Konso. When the person implicated was a Hor person, the same procedure was followed, and the penalty was made known in the village for the guilty person to fulfil.

It could be said that the *fund'o* institution works more efficiently than the many police stations, courts and other state institutions that operate in the proximity of Hor, Konso and Hamar. Members of groups of the region who wish to become traders affiliate themselves to the *fund'o* as they begin to trade actively. The *fund'o* recruits its executive members, who represent it in the different groups, by carefully selecting influential personalities who can take care of the interests of the different groups in the region. The *fund'o* of Hor were appointed after sending a list of candidates from both Hor and non-Hor residents in Hor land (mainly Konso, and people of Boran origin) to the *fund'o* in Purqud'a in Konso, which approved the list and the *fund'o* team started functioning. The council employs the customary practices of sanctioning used by each disputant's group in executing its decisions and does not impose the practice of the office of the *fund'o* or that of the Konso which is exercised in the base area. In other words, it uses the range of legal structures and the customary laws of groups which are found in the region and which the network covers.

The weakness of the state institutions of law enforcement in protecting the lives and the property of citizens of the region, together with the continued marginalization of the pastoral peoples of the Ethiopian–Kenyan frontier, is evidence of neglect by regional as well as central political authorities. In this context, alternative traditional institutions such as the *fund'o* keep some regions going. This was evident during the fall of the military government in 1991 and the coming into power of the EPRDF. The rural south was in general peaceful, and local markets ran well. There was neither looting nor genocide. This was not due to fear or to the

---

[22] Unlike many northern Orthodox Christians, he did not abhor snuff, though he did refuse to eat meat slaughtered by the Hor.

strength of the government then on its way to power but to the fact that institutions such as the *fund'o* went underground when governments were tough on them and become potent when governments became weak and lost control in the region. If the rural south is peaceful to this day, it is by no means an expression of the strength and stability of the state but instead a product of the strength of such traditional institutions in the region.

Strengthening control over the *fund'o* & the *qebele*

To maintain their position in Tabya, the *fund'o* had to be protected from the corrosive power of the Derg, which was already active in other places campaigning against traditional institutions.

Once the Hor had made sure that the *fund'o* and their own clan and age organization were in good shape and had put their literate members in key positions, they had the whole of Tabya except the police under their control. Civil cases that did not exceed 500 *birr* (£50) were taken to the *qebele* court instead of the *wereda* court following the decree that established peasant associations. The *qebele* court, after seeing each case, referred it to the *fund'o* instead of to the police. The *fund'o* made sure that the property owed was paid and at the same time that the guilty were fined. The fine, usually small stock or *areqi* liquor, was consumed by the members of the *fund'o*. The fine for a Hor person sometimes consisted of just small stock. In this case the *fund'o* were simply honoured by *mend'etcha* goatskin strips, which were tied on to their wrists, and by feasting on roasted goat. When a Tabya resident policeman or a Konso or a Karmet person was involved, the fine was cash to buy 'a yellow jerrycan',[23] which was given to the *fund'o* and others. Such a penalty in *areqi* is regarded as equivalent to flogging.

The *fund'o* usually gave members of the *qebele* court, the *jald'aba* and even the police a share of the fines they collected from disputants. If the guilty party refused to abide by decisions of the *fund'o*, or when a guilty person became too powerful for his Konso group, the *fund'o* directed the case to Hor *jald'aba* for the person to be flogged by young Hor men or simply handed over to the police. However, many preferred the *fund'o* to the police.

With their influence over Tabya growing ever stronger, Hor *jald'aba* and *fund'o*, either on their own or together, demanded that the police pass criminal cases in which Hor and Konso were implicated to them in order to punish offenders according to Hor practices. The police usually agreed. Compliance with such demands was rewarded with bonuses of goats and cash collected from guilty persons. Whenever the police refused to co-operate, they were warned of the risks to their own lives and those of their families. The curse of the mystically powerful Hor *qawot*s was used as a

---

[23] This is called a *bicha taka* and is a small food-oil jerrycan previously used as container for relief oil brought for food for work projects, but here being used as a container for *areqi*. It is a popular unit of measurement which holds five litres of *areqi*.

threat. By the time of the fall of the Derg, the Hor and their control of Tabya had grown so much that the only real responsibilities left to the police were searching passengers and trucks that passed along the road. The *jald'aba* did not succeed in taking control of that aspect of police work. Tabya had grown into one of the most notorious passenger checking and search stations.[24]

Other government bodies, NGOs & Christian churches

After the Derg came to power, other institutions began coming to Tabya. Religious and non-religious NGOs based in Arba Minch started branches in Jinka, Dimeka and Tourmi. Catholics based in Arba Minch began working in Tabya from their new branch at Dimeka. At Tabya they constructed a warehouse for relief food, a co-operative shop for the *qebele*, which they provided with an initial capital of 5,000 birr (£500), and toilets. They also supported a few Tabya children, who were taken to a boarding primary school the Derg had established for the children of pastoralists in Dimeka and Qey Afer.

The Ministries of Health, Agriculture and Education assigned some staff members to work in Tabya. The Relief and Rehabilitation Commission cleared an airstrip for the transport of relief food during emergencies. The former house of a German volunteer was converted into a government clinic. The Ministry of Education constructed a shabby, wooden, warehouse-type school. The police constructed a similar four-room station, of which one room was a cell, another the chief's office, the third a store, and the fourth a communications room.

Two Norwegian NGOs, Radda Barna and Norwegian Church Aid, began operating in Hor country after the Derg came to power and established their offices in Hamar *wereda*. Radda Barna focused on health education, inoculation and agriculture (crop cultivation and veterinary service support).[25] Norwegian Church Aid particularly focused on well-drilling and succeeded in drilling two wells (one for one of the four Hor villages and the other for Tabya.) Neither the Catholics nor other NGOs were involved in any evangelizing missions. Only the Mekane Yesus and Qale Hiywot churches had representatives in Tabya, and neither of them succeeded at that time in gaining any Hor converts. Like the civil servants, they were restricted to Tabya and their work was focused on civil servants and a few Karmet and Konso residents.

[24] See Hodson (n.d: 138) on checkpoints at the beginning of this century in the south-west.
[25] Radda Barna set up a costly modern irrigation system without paying much attention to Hor irrigation practices and without assessing whether the Hor would use it when the project was phased out. They also seem to have failed to note the change in the course of the river over the years and do not seem to have properly mapped previous courses of the river over the last century.

The Hor did not allow any of the above into their villages.²⁶ To the Hor they did not appear as dangerous as the state, but they had to be kept at a safe distance. Of the NGOs, the Norwegian Save the Children organization was the closest to the Hor, as they visited children when epidemics broke out and when Hor were short of sorghum. They also supported a few schoolchildren who went to Jinka from Dimeka for secondary education. To the Hor these activities were not a threat, although they complained that the water from the wells, which they refer to as 'water of the Sidam', gave them a burning feeling when urinating. They used it only when they had no choice and preferred water from the Limo for brewing ritual drinks. The Hor did not compete with these groups, as they did not threaten their leaders or themselves as a group. Whenever the Hor needed favours (such as when sorghum supplies became short), they did not hide the fact that they sometimes gave stomach-fat necklaces and goat-skin bracelets to fieldworkers of Radda Barna.

Works of the Derg period

The NGOs' work was acknowledged by the Hor as part of the Derg's project in Hor country, and hence they gave the Derg credit for their achievements. A road connecting Hor country, Hamar and the Omo River in Dassanetchland was constructed and put to use. Another road was constructed south from the Tabya–Tourmi road towards Lake Turkana. The new road system linked the Hor with their trading partners and bondfriends in the north and south, as well as to the Mekane Yesus clinic in Gisma (25 kms north) and Arba Minch hospital. The *wereda* town of Qey Afer and the provincial town of Jinka²⁷ could now be reached easily. This put an end to the use of the old Italian road for vehicle transport, although this is still a major trade route linking Hor land with its eastern neighbours.²⁸ Using the new road, eastern neighbours could be reached quickly through Baqawle in Konso. These together make the Hor

---

²⁶ During my fieldwork, a representative of the Mekane Yesus, Ato Tayyé, lived in Tabya. He went to the villages carrying coffee husks for sale and carried with him illustrations depicting stories from the Bible. He managed to convert three Hor, two men and a woman. One of the men, Gosha, a son of Hando of Murale who had had one of his legs amputated, remained a convert, while the others abandoned the faith. The Qale Hiwot team lived in one corner of Tabya in a big compound. They sang praises on Sundays with a Konso schoolteacher, some Karmet and their own families; they did not have a Hor following. The Catholics did not live in Hor country. The Ethiopian Orthodox followers did not have a church but a committee and a feasting society for some saints' days. Civil servants and most Konso traders belonged to this group of believers, who also did not have any Hor members.

²⁷ For most of the Derg period, the Hor of the Gamo Gofa area (the Arbore section) were under the Arba Minch administration, while the Marle section was under the Awasa administration (see Map 1, p. xii). After the founding of the People's Democratic Republic, both Arbore and Marle were put under the administration of Jinka. When Gamo Gofa was divided and renamed North and South Omo, the Hor were made subject to Jinka, the capital of South Omo.

²⁸ The Hor say that it was originally constructed along the line of a footpath that was at that time a major trade route.

country in general, and Tabya in particular, a place from which routes radiate out. The presence already of regional institutions of trade and of trade routes gave prominence to Tabya and facilitated its role in the region. The construction of the roads helped to facilitate communication, but this service did not last long.

The desire for riches by provincial customs officials and various government ministry officials hindered the smooth functioning of the roads. Officials made it illegal to transport many items, such as wood, hides and skins, coffee, incense, sorghum and animals. As the Hor saw it, only 'Sidam' (civil servants and employees of NGOs) could do this. Other people who paid to transport such items had them confiscated and lost both their property and the cost of the transport. Products bought from the centre or from shops in the Omo town of Rate were often confiscated as contraband. Bribes were required to get these items through checkpoints. Through their representatives, provincial officials sought to control the roads, to search cars and to confiscate anything they needed. This was in addition to the search and confiscation that was a widespread practice at all police stations, such as Tourmi, Omo Rate, Tabya, etc. Not all these institutions established checkpoints at the same time and place. They sometimes delegated the police, a forestry official or an administrator from the coffee development office to do it on their behalf. Because of this, the Hor could not acquire the unhusked coffee they needed for bridewealth through official routes. Travelling became a real problem. It was a waste of time and money, and it was not worth investing in the construction of roads if they were only going to be barricaded by official institutions. Bribes could not be afforded. Passengers had to present passes and identify themselves or suffer detention. This was too much for ordinary travellers, and ultimately the state officials, NGO workers and long-distance travellers became the sole users of the roads. The Hor and their bond-friends retreated to their traditional routes, where the hyenas did not reign and there was respect for life, property and human dignity.

On the other hand, the Hor found the Derg different from all previous Ethiopian regimes. This was mainly due to the Derg's finding outside opportunities for a number of individuals. Some Hor were sent for training to other regions of the country. Some went to the Farmers' Training School in Agarfa, Bale Province, and studied veterinary medicine. Others were sent to the Taṭek Military Training Centre, west of Addis Ababa. One of these men remained as a military trainer in the centre. A few were sent to the Ogaden to take part in the Ethiopian–Somali war. A handful were sent to military training camps in Sile Wora Gammo near Arba Minch and to a smaller centre in Tertalle, after which they were armed and sent back to their country. Quite a few joined the army. A couple of Hor were sent to the central Yekatit Silsa Siddist Political School in Addis Ababa, and one of them was appointed deputy governor of the *awraja* of Qey Afer in the final years of the Derg.[29] By early 1991 there were about

[29] This was the first, and highest ever, government position held by a Hor person.

four Hor members of the Workers' Party of Ethiopia, one of whom was a member of the armed forces then serving in a base at a state farm in Dassanetch country.

Nevertheless, on the whole, the Derg was viewed as bearing a crude resemblance to previous imperial rule. Its unlimited power, its unpredictability and its ability to victimize its subjects at any time made this clear. The imprisonment of the *qawot* made this evident. According to the Hor, the Derg and its predecessors all belonged to the Sidam, regarded as a very large group that can always avenge themselves if attacked. The Hamar, by contrast, consider the Sidam to be female and cowardly and refuse to consider them as good enemies worth killing. To the Hor, the Sidam were violent, particularly initially (during incorporation). They killed many Hor, enslaved people and looted cattle. They are believed to have had good craftsmen who made firearms for them. They are comparable to Europeans. The Italians beat them and became rulers, but because of their great numbers and ferociousness, the Sidam beat the Italians in their turn and came to power. The Sidam do not raise cattle but like meat very much. They consume cattle that other people have raised. They do not like carrying their own belongings while travelling and do not cook their own food but make others carry and cook for them. They love taking bribes, other people's possessions and rifles. They like wearing clothes. They divorce their wives and pay money for prostitutes but also kill men who have sexual relationships with their own wives. They carry money, a thing that does not reproduce like cattle. They do not like to work but want others to work for them. They only like to eat, drink and indulge in prostitution. They do not love animals and do not like people who raise and live with animals. They do not eat meat that others have slaughtered. They are to be avoided. This was and is how the Hor view outsiders – civil servants, the police, NGO workers and missionaries or their representatives – that is, those who do not love animals and who do not work on the land. This was the Hor category of the Other. The Derg and others that have come to change Hor life are put into this category.

The Derg was also similar to its predecessors in its failure to recognize Hor leaders. This was made worse by the view Derg authorities took of leaders in general. They wrongly thought that all 'leaders of ethnic groups' owned land and property. While this may be the case in agricultural, especially coffee-growing areas, it was certainly not the case among the pastoral Hor or their neighbours. It was on this basis that the senior *qawot* of the Hor, Konso Ali, head of the Garle clan, was imprisoned. The Hor viewed this act as horrible and predicted that the Derg, like its predecessors, would fall because of this act. When the Derg's fall was announced, the Hor did not celebrate the occasion but said that it had fallen because it had broken the long staff, *ngawe qawot*, of their senior *qawot* and because it had taken him prisoner.

In all other respects the Derg can be said to be the Ethiopian government that succeeded in coming closer to the Hor than any previous government. Nevertheless its reach did not extend beyond the Tabya barrier set up by the Hor.

The coming to power of the EPRDF

The EPRDF took control of Addis Ababa in May 1991. It was not until mid-July that they reached Hor country in a long convoy, well armed. Top-level Derg party and government officials from Wellegga, Illubabor and Kefa had crossed the Omo River through Kullo Konta and reached Arba Minch. Preparing to flee to Kenya, and out of fear that they might be pursued by the EPRDF, they made Arba Minch a battleground. For two nights, the town was filled with the noise of gunfire. Fleeing officials in Arba Minch forced workers in petrol stations at gunpoint to bring petrol. Local officials in Arba Minch forcibly took four-wheel-drive vehicles in good condition from government offices and fled to Kenya across the desert through Konso and Boran and partly through Hor country. In Arba Minch town the firearms depot belonging to the provincial commissariat was intentionally left open by fleeing officials. Firearms and ammunition got into the hands of young boys, and these were easily purchased by rural communities. The provincial police archives, going back over forty years, were left to the wind. Beqawle town in Konso had its police station looted. At Kholme Gawad'a junction in Konso a cattle trader named Tsegaye was killed and his cattle taken.[30] All along the route, police at checkpoints went into hiding out of fear of revenge from those from whom they had taken bribes and whom they had mistreated in the past. Prisoners and criminals were set free, since, as the state structure began to crumble, prison guards retreated to their homes. Provincial administrative and party functionaries joined the run towards Kenya.

A voluntary committee set up in Arba Minch took control of the town and passed information about happenings in Arba Minch town to districts in North and South Omo, who in their turn passed the information down the hierarchy. Markets were guarded by volunteers.[31] Gamo country elders took control of the markets and assembly places. Families received returnees from the war front. The elders made sure members of other groups living on their territory were safe. National radio transmission was suspended, as was the telephone service. Only the VOA and Voice of Germany radio programmes kept the Amharic-speaking community informed about developments, and the programmes calmed the fear of chaos. What lawlessness there was was initiated from above, as fleeing Derg officials opened the doors to anarchy. In Jinka, however, only the administrators fled, while the police and their commanders remained on duty throughout, until the EPRDF officials arrived and suspended them.

[30] The *fund'o* discovered the murderers, who were later handed over to the government; they paid cattle to the family of the victim.
[31] Communication was aided by the Catholic Mission radio in Arba Minch.

In Hor country, however, there was neither panic nor looting. As during the Derg's ascent to power, they did not celebrate this new ascent, nor did they rejoice at the fall of the Derg. Instead they comforted civil servants working in Hor country, who were in a state of panic following the events of the fall and wanted to return to their respective places of origin. However, the Hor *jald'aba* ordered them to stay, lest they be killed in these uncertain times. The Hor leaders provided police and other civil servants with maize rations in the style of the Relief and Rehabilitation Commission that had provided the Hor themselves with relief handouts in lean years. There was lots of terrifying news of fleeing officials and university students being murdered by pastoral groups further south. Some Hor who were doing military service at the Omo Rate base and others in the civil service returned home as the civil service began disintegrating. These returnees made the Hor aware of the political turmoil at the centre and advised them how to cope with the situation.

When, in mid-July, the EPRDF contingent arrived, the Hor were thus ready to receive them. This time the intermediaries were carefully selected. Sura Gino, the man who attended the meeting of southern chiefs, and had formerly been given the title of Grazmach and made *balabbat* in Gardulla by Haile Selassie's government, received the guests. There were Hor students with him who had come back from Dimeka escaping possible chaos. Thus two generations were there, ready to receive the EPRDF officials: Sura, the *kernat* of the Otgalcha generation, who had witnessed the fall of many governments (the pre- and post-war Haile Selassie governments, the Italian colonial administration and the government of the Derg), and the young students who were in the process of becoming Sidam but were not yet fully Sidam. They were not fully Hor either. These students had no experience of government, nor any ideas of how the state functioned.

The scenario on the EPRDF side was no different. The contingent consisted of a group of armed Oromo without any marked hierarchy of status, which might have given the Hor a sense of their new authority as a government. The Hor wanted to feel the presence of someone who represented the new regime. Firearms alone were not enough as markers of authority: the Hor themselves had them. Sura looked for someone with authority, someone that seemed to be in charge like the fathers of the land, like the *jald'aba*.

The EPRDF assumed that all the people in the region were Oromo. With this in mind, the Oromo soldiers addressed the meeting in Tabya in their own language. This did not require interpreters, as the audience were Tabya residents who spoke Oromo. The soldiers briefed the meeting at length about the evils of the Derg period, the 'red terror', the repression of national movements and the sacrifices they had made to free the Hor from the grip of the Derg. As during the Derg period such talk of national politics did not interest the Hor. They were addressed as though they had experienced life under the Derg. The soldiers had no idea whether the

Evading the Revolutionary State: Hor

Hor had suffered torture or the infamous 'red terror'. But they asked the assembly in Tabya to inform them about Hor members of the Derg, the WPE and members of the intelligence services, so that they could arrest them. The Hor listened to what they were told. Sura rose up and told them what he believed the guests would like to hear, namely that all the Hor were pleased to receive them and that this was the first time ever that the Hor had been allowed to speak in their language, Oh Boran (the language of the Boran), and that it was the wish of the Hor that their guests should live in Tabya. However, Sura was sure that the guests were ignorant of the fact that the Hor were a different people and had their own language and that they were not Oromo.

The contingent did not stay in Tabya, as there was no dependable source of food and the *wereda* town was Tourmi. The desert heat and mosquitoes made Tabya life difficult for them. As they proceeded to the cooler Hamar lands in Tourmi, they took educated youths along with them, recruiting them as the new intermediaries between the state and the Hor. However, they were not needed immediately for service in Hor country but among the neighbouring groups. First, however, they were given Kalashnikovs and old uniforms of the Derg army, most of them being sent for training to Awasa, Arba Minch or Jinka. Out of the fifteen, three ran away, two with the new rifles, and about twelve others stayed for training as policemen, political cadres or soldiers.

In the meantime, life went on as usual after the guests had left Tabya and it was time for the Hor to initiate the new generation Melbasa/Melere into power. This was crucial for the Hor and the first phase, *nger*, was performed against this background. Soon afterwards, a year after the EPRDF came to power, and when the Hor of Marle had just finished the second phase of the initiation, the *chirnan* of Melabasa/Melere, the Boran attacked the Hor. According to Hor estimates, this cost the Boran 485 lives and the Hor sixteen lives and ten wounded. In these incidents, as in others before them, the state played no role. During its first few years, the present regime seems to have been pushed a bit further out of Tabya, towards Tourmi in Hamar country.

Later, the new intermediaries joined the police (there were eight in Tabya during my field research) and the administration in Jinka (two) and Hamar Banna *wereda* (two). The political organizational structure of the South Omo Peoples' Democratic Organization reached Tabya. Members attended the monthly political education lessons and contributed membership fees. In addition, each Hor village contributed goats for the South Omo Peoples' Democratic Organization and for the South Omo Peoples' Development Association. Whenever *wereda* officials came to Tabya, some of the intermediaries accompanied them. Officials visited Tabya for tax collection, military conscription and pre-election propaganda work (the main activities of the state at *wereda* level and below). Some of the goats contributed were slaughtered according to Hor tradition for stomach-fat necklaces or *mend'etcha* goatskin wristlets to honour the guests, and some of the goats were exchanged for beer and other drinks in the restaurants of Tabya to entertain the officials.

Under the new regime, the Hor are well represented. However, representation will be meaningful and effective only if it is tied to guaranteed land and other democratic rights, rights that have an enduring impact on the livelihood of the Hor and their neighbours. The Hor, like their northern neighbours the Tsamako, are under constant threat from the growing and unrestrained power of the privately owned Birale cotton plantation. They are constantly urging their sons in government to press hard for the resolution of the problem that exists with the Tsamako, who are the traditional owners of the land and the plantation. Established under the Derg's policy of a mixed economy, with arrangements made between the last governor of South Omo and the investment office in Awasa, this Tsamako land was taken over illegally without the consent of the Tsamako elders. Some underhand methods appear to have been employed in acquiring the land, in influencing the late ritual chief of the Tsamako, Dalle Armar, and in legalizing the conditions of use. Especially during the last few years, the size of the plantation has increased dramatically and river water has been appropriated without due consideration for peoples like the Karmet, Tsamako, Hor, Assile and Wungabaino Hamar, their sorghum plots and animals. The river has been polluted to the extent that the fish in it had already begun dying in 1995. The fall in the volume of river water has affected the size of sorghum plots, and chemicals used in spraying have been killing most of the bees, on whose honey the Tsamako and Birale depend for their livelihood. The effects on human and animal health can only be guessed at. Most of the flowering acacia trees were bulldozed in making the plantation, and the Tsamako's pastures and sorghum plots have been reduced considerably. The owners of the plantation and the Tsamako entered a serious confrontation in 1995, which resulted in military involvement by the government and the execution of nine Tsamako without proper legal procedures.[32] The problem had not yet been resolved at the time of writing. To avoid further disasters it is essential that the rights to representation of the Hor and their neighbours be tied clearly to land and other relevant rights, inorder that vulnerable groups be able to protect themselves from land theft and be able to survive.

The way the Hor perceive the EPRDF government will depend on the kind of relationship the state forms with them during its time in power. Depending on this relationship, they may allow the EPRDF to reach the village or may keep it at a distance. As far as I understand matters, the Hor have been hoping – and I have no doubt that, as subjects of the Ethiopian state, they still hope – that this government will guarantee them and their neighbours a good future, at least by assuring their rights to land, to unpolluted water and to administering their own affairs unthreatened by land-grabbers.

---

[32] For details of this incident, see Miyawaki (n.d.) and unpublished material distributed during the International Ethiopian Studies Conference in Kyoto, December 1997.

Three

Memory & the Humiliation of Men
The Revolution in Aari

ALEXANDER NATY

The relationship between the Aari people of the former province of Gamo-Gofa (see Map 1.3, p. 31)[1] and the socialist state in Ethiopia was conditioned by the interaction of two faces of political culture. In the first, the Aari viewed themselves as powerless vis-à-vis the state, but in the other they engaged in open rebellion, during which they reclaimed old notions of masculinity. This pattern of relationship also characterized relations between the Aari and the northern settlers. The context under which the Aari articulated the two faces of political culture differed. The Aari rebelled in situations where the state was weak and unable to exert any control. Acts of violence often followed their rebellion. They justified their violence by articulating a cult of masculinity. In contrast, they manifested powerlessness when the state grew strong. Underlying the culture of powerlessness was a sullen but silent opposition and a critique of state domination using a variety of political metaphors. This chapter will show how open rebellion by the Aari during the initial period of the revolution was eventually transformed into silent opposition against the socialist state.

Imperial background

Our understanding of state–society relations for the period between 1974

---

[1] The Aari language belongs to the Omotic family. Haberland (cited in Jensen 1959: 40) reported that they numbered less than 20,000 in the 1950s. According to Bender (1976: 15), there were 32,000 in the 1970s. A 1984 report from the Central Statistical Authority (CSA) reported the Aari population to be 107,764. This source also indicates that 75,878 live in the former Geleb and Hamar-Bakko districts in South Omo region, while a further 31,667 are found in the Gofa district.

The fieldwork on which this paper is based was supported by a Rockefeller Foundation African Dissertation Internship (Grant no. RF 87001, # 19). I carried out the field research from 1988 to 1990 at Sinigal and Zomba. I gratefully acknowledge the support of the Foundation.

and 1991 in Aariland would be incomplete unless we examined the condition of the Aari under the imperial regime. Examining the previous pattern of state–society relations also enables us to understand the reaction of the different segments of Aari society to the Ethiopian revolution. What follows is therefore an overview of the relationship of the Aari with the imperial state for the period between the 1890s until the advent of the revolution. Prior to their conquest and incorporation into the Abyssinian Empire during the late nineteenth century, the Aari existed as a politically independent entity. They depended on the cultivation of ensete, sorghum, maize and vegetables. In the higher lands they kept sheep, while cattle were confined to the lowlands. They were organized into marriage classes, clans, and caste-like divisions, within chiefdoms headed by ritual kings called baabi. These chiefdoms were Baaka, Kuree, Shangama, Beya, Siida, Woba, Bargeda, Gayl, Argen, and Gooza.

Territorial frontiers existed between the chiefdoms, though they changed from time to time depending on the military strength of the warring polities. In comparison with other Aari chiefdoms, the Baaka polity was militarily stronger, a fact which enabled the Baaka to wage wars and annex the territories of the neighbouring Shangama chiefdom. The Baaka also fought against the Maale people and annexed their territories. The warfare between the Baaka and the Maale is also mentioned in Maale oral history (Donham 1986b: 81). The Aari people in Gayl and Gooza fought against their neighbours the Gofa (Borelli 1890: 437). As a result of their frequent involvement in warfare, the Aari developed a political culture of bellicosity and a cult of masculinity. The presentation of the genitals of enemies taken in battle to the ritual king was an act glorifying the masculinity of the warriors. With the conquest, the era of political grandeur ended for the Aari and a period of subjugation began.

The encounter between the imperial army and the various Aari chiefdoms resulted in the defeat of the latter and the establishment of hitherto unknown patterns of socioeconomic and political relations in Aariland.[2] With the conquest, an institution of serfdom known as *gebbar sirat* was established, in which Aari families were assigned to imperial soldier-settlers (*neftennya*). Besides paying an annual tribute (in cash as well as in kind), Aari families had to perform labour services for the soldier-settlers. When families failed to pay tribute or to perform labour service, their members were taken as domestic slaves. Aari oral history indicates that a considerable number of Aari were taken as slaves when they could not pay tribute or undertake labour services. Similarly, many individuals were enslaved through a practice known as *leba sha*.[3] The impact of the serfdom

---

[2] For a discussion of the Aari interpretation of their conquest by and incorporation into the Abyssinian empire, see Naty 1994a.
[3] The *leba sha* was a mechanism for detecting thieves used in northern Ethiopia, and also in southern Ethiopia after the Abyssinian conquest of the late nineteenth century. The use of the *leba sha* in the *gebbar* areas of southern Ethiopia was exploitative and oppressive in that it was practised as a pretext to enslave the local populations. Although, with modernization during the reign of the Emperor Haile Selassie there was an attempt to abolish the institution,

system in Aariland was more or less the same, with some variations depending whether a region was a *qotari* or a *mad bét* area. The soldier-settlers who participated in the initial conquest of the region settled in the former regions; the personal troops (*bét lij*) of the provincial governors settled in the latter region. The *gebbar* institution had a greater impact on the *mad bét* than the *qotari* areas, due to factors that are beyond the scope of this paper.

Although shaped by earlier state structures and institutions, state–society relations during the later period of the imperial rule, especially during the reign of the Emperor Haile Selassie, were different. The development of commerce and the centralization of the state that took place during this period affected state–society relations (Donham 1986a: 25). With the introduction of commerce, the different Aari regions became economically vital. Previously, these regions and similar areas of southern Ethiopia had been viewed as wild areas inhabited by savage populations. The Ethiopian state used to send government officials who misbehaved there as a punishment. That peripheral areas such as Aariland were wild and thus appropriate places for the exile of political prisoners is reflected in the Amharic saying *amelu kekefa lakew Gemu-Gofa*, 'If someone misbehaves, send him to Gamo-Gofa.' The economic opportunities that arose during this later period generated a new conception of Aariland, as a region rich in resources that one could tap to make a fortune. Indeed, the saying *Gemu-Gofa birr yizaqal bakafa*, 'In Gamo-Gofa, one scoops money with a shovel', indicates the economic importance of the region where the Aari people are located. This view motivated individuals from other parts of Ethiopia to settle in Aariland, where they controlled the economy by setting up business enterprises such as shops, restaurants and bars in the various towns. The economic fortunes of Aariland not only attracted people from the distant north, but also from neighbouring districts such as Basketo, Gofa and Gamo, as well as Aari peasants from poorer areas such as Gelila. Many Aari from such places settled in coffee-growing areas such as Bio-Barka, Tolta, Gazer and Jinka. The movement of population into Aariland brought about the growth of small towns. This contributed to the eclipse and stagnation of the former garrison towns of Bakko, Shangama and Debre-Tsehai.

The centralization of the state resulted in the introduction of modern institutions in Aariland. Amongst other things, it introduced a system of public finance and accounting. The Geleb and Hamar-Bakko public finance office was established for the first time in 1941. The office, locally known as *gimja bét*, was located in Bakko, its function being to collect taxes from local peasants and to pay salaries to government workers. With the reorganization of the old imperial state administration after 1941, some fundamental socio-economic transformations took place. For instance, after 1941 many former tribute-paying peasants (*gebbar*) voluntarily became

---

3 (cont.) this did not happen until the Italian occupation. For further discussion of the *leba sha*, see Naty 1992, 1994b.

tenants (*chisennya*) by moving on to the lands of local landlords. The cause for this change in status was the burden of taxation. The amount of tax that each tenant paid to the landlord was less than what a *gebbar* paid. Moreover, the *gebbar* paid taxes that were commensurate with the amount of land they owned, regardless of whether or not they cultivated the land. Most *gebbar* did not cultivate the land they owned but simply ended up paying more money in taxes. For these reasons, they preferred to become *chisennya* so that they could pay less in tax. The consequence of the change of status from *gebbar* to *chisennya* was that government revenue from taxes decreased. By contrast, the landlords were able to attract more tenants. As a result, they received a greater amount of income from taxes. When tenants failed to pay the taxes, the landowners expelled them, and they also confiscated their crops. Due to such injustices, conflicts and accusations were widespread. Because Aari peasants did not know the legal provisions regarding property ownership in the Ethiopian Civil Code, *neftennya* lawyers known as *negere fejj* handled their cases, receiving money from the peasants in return for this service.

The condition of the Aari during the early and late imperial periods that I have discussed so far reflects a particular pattern of state–society relations, which can be generalized to other southern societies in similar situations. In *Southern Marches* Donham (1986a: 48) identified three types of political economy: the *gebbar* areas, the semi-independent enclaves, and the fringe peripheries. The relationship between the Aari and the imperial state was of the *gebbar* type. The pattern of relations in this region was exploitative and oppressive, as is indicated in the preceding short discussion. Such conditions had resulted in the development of a particular political culture – what I call the *gebbar* political culture – with reference to the pattern of relationships between the Aari and the imperial state. The *gebbar* political culture refers to the Aari's state of powerlessness in their relations with the state. Underlying this political culture is the deep-rooted social and psychological impact of conquest and of the *gebbar* institution on local attitudes, particularly on the spirit of rebellion and the notion of masculinity. The Aari expressed the *gebbar* political culture through metaphors which reflect their subordination to the state and use the imagery of domestic animals and gender.[4]

Despite the persistence of the *gebbar* political culture among the Aari, one should not take it as an immutable cultural trait. The recent history of the Aari shows that the *gebbar* political culture can be transcended, albeit temporarily. The Aari transcended it during the period of the Italian occupation in 1935 and with the advent of the revolution in 1974. Factors such as the politicization of external agents, amongst other things,

---

[4] The Aari used a variety of metaphors to describe their relationship not only with the imperial state but also with the socialist state. They often expressed these metaphors in Amharic, which most Aari spoke. These metaphors included 'becoming women' (*set honenal*), 'castration' (*tekolashetenal*), 'penis-shortening' (*qulachin atere*) and 'becoming sheep' (*beg honenal*).

contributed to this transcendence. In the rest of this chapter I shall examine the relationships between the Aari and the socialist state, focusing on the pattern of interactions for the period between 1974 and 1991. More specifically, I shall discuss the dynamics of state–society relations during the earlier and later phases of the revolution, which differed due to changes in the attitude and policies of the socialist state.

The Aari & the Ethiopian revolution: the early phase

State–society relations during the period between 1974 and 1991 must be examined in the context of the unequal power relations between the Aari and the Ethiopian state under the imperial regime. In addition, for their fuller understanding, we have to consider the reaction of the various groups in Aariland towards the revolution. These include the different segments of Aari society, the northern settlers, former landlords, student campaigners (*zemach*), district administrators and the police. Aari society was not monolithic but constituted of social groups such as elders, juniors and members of high and low caste groups. Because of the introduction of Protestant Christianity by the Sudan Interior Mission (SIM) in the 1950s, there were also Protestant Christians among the Aari. The northern settlers included descendants of pioneer soldier-settlers and those individuals who had settled in Aariland more recently. Included within this group were also people of non-Amhara background who had assumed Amhara identity through assimilation. During both the imperial and socialist regimes, district administrators and police in Aariland consisted not only of 'ethnic Amhara' but also of individuals from other backgrounds who were assimilated to Amhara identity either through intermarriage or education.

The revolution brought about the end of the unequal power relations that existed during the imperial regime. It introduced practices and ideas such as land reform, equality, and class struggle. The land reform altered previous socio-economic relations. Indeed, the mass support of the revolution, especially during the initial period that Dassalegn Rahmato refers to as the 'populist period' (1987: 157), was due to the land reform. Although many scholars have indicated the benefit of the land reform to Ethiopian peasants, particularly in southern Ethiopia, I caution against an overly economistic understanding of it. Such a conceptualization overlooks other contributing factors such as the concept of equality and class struggle for the popularity of the revolution. The Aari supported the revolution not only because of the land reform, but also for its appealing discourse. The discourse of equality was particularly appealing to the Protestant Aari, who were familiar with the Biblical notion of the equality of all human beings and who rejected Aari ideas of naturalized hierarchy. The ideology of class struggle was, moreover, compatible with the Aari concept of revenge.

Most Aari found the idea of equality attractive only in their relations with the northern settlers. Members of the high caste did not apply the notion of equality to relations between the two caste groups that existed in the society. Traditionally individuals from these groups were prohibited from eating together or having marital relations because of the fear of the polluting effect of the low caste groups.[5] The Aari considered the distinction between the castes as natural and immutable. The notion of equality between the two castes was alien to the traditional Aari concept of naturalized hierarchy. To abolish this social distinction, the student campaigners of 1975 slaughtered an ox and instructed members of the two groups to eat together to symbolize the equality between the two castes. Because high-caste Aari did not approve this action of the students, they performed purification rituals in an attempt to protect themselves from the misfortunes that this violation of the norm would bring about. It was believed that eating and living together, or having sexual intercourse with individuals from the low caste would cause a skin disease for members of the high caste.

Ideally the class struggle which the students had politicized was directed not only against the former landlords, but also against traditional authorities such as the ritual kings and religious leaders who were also the class enemies (from the perspective of the student campaigners) of the peasants. The Aari did not plunder these local figures for fear of their curse, although the peasant association confiscated their guns. In some villages they had to pay a certain amount of money to the peasant associations. The class struggle against the landlords included the plundering of property and raping of women. The looting involved the slaughtering of livestock not only of the landlords, but also of the northern settlers in general. A northern settler described his tragic experience to me in the following way:

> On that day, I was at my house. Many Aari peasants came to my compound blowing their horns. They requested me to give them one ox so that they could slaughter it and eat the meat. I told them that I was a poor person like them, but they refused to listen to me. Then I told them that they could go into the cattle compound and take whichever beast they wanted. The Aari entered the compound and took one whose name was Dammena. After they took the ox, the Aari asked me to take them to the house of a person named Gunjo. I took them to his house, and the peasants took also one ox from him. Later, they went to Muli's house and requested one from him too. Because Muli had many adult sons and relatives who could fight against the peasants, he refused to give it to them. One of Muli's sons was called Haile. Haile had a gun.

---

[5] Traditionally a child born of sexual relations between a low-caste and a high-caste individual was called *mingii*, meaning 'impure'. Keeping such a child in the society was considered to cause misfortunes, such as drought and famine. As a result, such children were thrown alive into a river to drown. After the incorporation of the Aari into the empire, such babies were adopted by the northern settlers. The Aari families who have been converted to Protestant Christianity have abandoned the practice of throwing babies away.

Memory & the Humiliation of Men: Aari

He loaded his gun and told the peasants that if they persisted with their request, he would fire at them. After some heated argument, Muli agreed to give an ox to the peasants. While some peasants went from house to house demanding cattle, others looted any livestock they came across on their way. When I told the peasant association leader about the plundering, he said that the situation was beyond his control. However, he wrote a letter, which I took, to the administrator in Gazer. Before I arrived, a *zemach* student and an Aari peasant came from Dordora. They had been sent by the peasant association chairman. They told me to return to my village so that the situation could be settled by the students. The *zemecha* were furious about what had happened to me. In a meeting they told the peasants to pay me 1600 *birr* in fines. I told them that I did not want to take the money. I refused to take the money because many other individuals had also been plundered. It was a situation which affected all of us. It did not make any sense for me to take the money.

The above quotation reflects the collective behaviour of subordinate groups during periods of revolution. Social scientists have been interested in understanding the behaviour of groups in revolutionary situations for sometime. Le Bon (1960) interpreted this phenomenon in terms of the irrationality of individuals in a crowd. Scott (1990), an advocate of the moral economy perspective, viewed such behaviour as a declaration of the 'hidden transcript', that is, the 'discourse that takes place offstage, beyond the direct observation of powerholders'. According to Scott:

When the first declaration of the hidden transcript succeeds, its mobilizing capacity as a symbolic act is potentially awesome. At the level of tactics and strategy, it is a powerful straw in the wind. It portends a possible turning of the tables. That first declaration speaks for countless others, it shouts what historically had to be whispered, controlled, choked back, stifled, and suppressed. If the results seem like moments of madness, if the politics they engender is tumultuous, frenetic, delirious, and occasionally violent, that is perhaps because the powerless are so rarely on the public stage and have so much to say and do when they finally arrive.

(Scott 1990: 227)

The intensity of the conflict between the Aari and the landlords varied; in some areas the tensions were mild, while in places such as Bio-Barka there were open confrontations. Bio-Barka is located in the former chiefdom of Beya. Being in one of the Aari regions which grew plenty of coffee, it had also a greater concentration of landlords engaged in coffee cultivation and trade. Due to the commercialization of coffee, conflicts over land were frequent in Bio-Barka, as a result of which Aari peasants were often driven off their lands. Such conditions prevailed throughout the later period of the imperial regime. The eruption of the revolution created an opportunity for the Aari to plunder their former landlords in revenge. This led to an incident in which many peasants died in Bio-Barka. Describing one incident, an Aari elder said:

> Many peasants from the different villages blew their horns (*shoora*) and gathered to plunder the former landlords in the town. The Aari wanted to loot the property of the landlords because they believed that they had become wealthy by exploiting the peasants. Therefore, the Aari wanted to reclaim the property of the landlords. This was the reason for the conflict in Bio-Barka. In this conflict many Aari died. Some individuals who died in the fighting were buried secretly without proper funeral ceremonies. The Aari did this because they were afraid that the landlords would kill the members of the dead families for revenge if such funerals were done. Most of the Aari who died in the fighting were from Gelila. Finally, the peasants and the landlords were reconciled through the efforts of the elders from both sides. I was one of the mediators from the side of the Aari. We requested each of the parties to lay down their guns, spears and knives on the ground. Then we slaughtered two sheep (one brought by the landlords and the other by the Aari). We instructed individuals from both sides to wash their hands in the blood of the sheep. In the end, both groups took an oath not to fight again.

The raping of wives and daughters of the former landlords by Aari was a reflection of their patriarchal gender ideology. The idea was that men ought to defend women, who are viewed as weak and powerless. Men's failure to live up to this expectation was considered to be an indication of their weakness and a sign of submission. While the Aari asserted their masculinity through such direct acts, the landlords were symbolically emasculated. Indeed, given the pronounced sense of masculinity that the landlords portrayed before the revolution, such acts proved their emasculation. It was clear that they were not living up to the often expressed Amharic proverbial saying *ya man rist man yewarsal, ya man mist man yewesdal*, 'he who is entitled to seize another man's land is entitled to take another man's wife'. In general, these acts of revenge signalled Aari insubordination – they were ways of humiliating the landlords publicly. Commenting on such acts of defiance by subordinate groups in general, Scott remarks:

> However satisfying the first act of refusal or defiance may be we must never overlook the fact that its satisfaction depends on it being public. The defiance, obsequiousness, and humiliations of subordination are extracted as part of a public transcript. To speak of a loss of dignity and status is necessarily to speak of a public loss. It follows, I think, that a public humiliation can be fully reciprocated only with a public revenge. To be publicly dishonored may lead to offstage discourse of dignity and secret rites of revenge, but these can hardly compare, in their capacity to restore one's status, to a public assertion of honor or a public turning of the tables, preferably before the same audience.
>
> (Scott 1990: 214–15)

Memory & the Humiliation of Men: Aari

The individuals who engaged in these acts of revenge were mostly junior men. Aari elders acted as mediators in restoring peace between the juniors and the landlords. In their effort to mediate, the elders used the idioms of 'consanguinity' and 'working together', the former implying affinity, the latter denoting co-operation between the northern settlers and the Aari in communal work groups. The *neftennya* often viewed themselves as culturally superior and refrained from participating in such work groups because working together also involved sharing food and beverages with the Aari, which most of the northern settlers disliked. The Aari considered the notion of working together as a mechanism for the 'naturalization' of the northern settlers: those individuals who fulfilled this criterion became naturalized into Aari cultural identity. Accordingly, the Aari did not plunder landlords who acquired such an identity. The tendency of the elders to discourage the juniors' acts of revenge was a result of the memory of counter-revenge by the northern settlers that took place after the defeat of the Italians in 1941.[6] Given this memory, it was the anticipation and fear of a possible future counter-revenge, and not so much the notions of working together or consanguinity, that motivated Aari elders to portray such a pacifist attitude. The juniors had no such memories. This lack of experience, and the need to vent their anger, pushed the juniors to be more violent and rebellious. For them, the initial period of the revolution was a moment to declare their defiance.

The northern settlers interpreted the revenge differently. Although to some extent they used the same idioms that the Aari elders used in their disapproval of these acts of revenge, the *neftennya* interpreted them as an act driven by Aari impoliteness and irrationality. This interpretation failed to look at revenge in the context of the legacy of oppression and exploitation that characterized the relationship between the Aari and the northern settlers. The Aari considered revenge to be a means of asserting equality by inflicting physical and psychological damage on their former oppressors. From the Aari perspective, it was a mechanism for regaining their cult of masculinity. Thus, for the Aari revenge symbolized the reversal of history and the inversion of experience.

The clash between the Aari peasants and the former landlords lasted for only a brief period. The individuals who engaged in the plundering in the various villages were arrested and imprisoned. Subsequently, state–society relations in Aariland began to take on a different pattern. The departure of the student campaigners and the alliance of the former landlords with the police left the Aari in a dilemma. Nevertheless, the Aari did not return to the previous socio-economic relations, although later unequal power relations began to evolve between the Aari and the socialist state.

[6] The Aari use the phrase *aushtra dannya* (or *jamaa dannya*) to refer to the initial period of the defeat of the Italians, during which the northern settlers took revenge against the Aari by plundering their property and reintroducing serfdom for a short period. The reason for the revenge was Aari collaboration with the Italians. The word *aushtra* refers to the Austrian Mannlicher gun, whereas *dannya* means 'judge' in Amharic. Literally the phrase means 'rule by the gun'.

Looking Back on the Projects of the Socialist State

Relations between the Aari & the socialist state, 1980s to 1991

The consolidation of the state beginning in the late 1970s and early 1980s led to the imposition of villagization and military conscription, and a system of taxation that affected the Aari people. Villagization was a programme in which Aari families were brought together in large compact villages that were patterned along grid lines. Villagization schemes had been established in Bakko, Alga, Arkisha, Gedir, Kaysa, Kuree, Sheepi and Zomba villages, all in Aariland. This was a strategy initiated by the socialist state to promote rural transformation. The state rationalized the programme by arguing that, if families lived in villages, it would be easier for the government to provide them with social services such as health, education and clean water (see Taddesse Berisso's study in this volume, Chapter 6). Despite this argument, the state did not introduce social services in all the established villages. Looking at villagization from a different angle, one can argue that the programme was a mechanism for state control and the surveillance of local populations. Most of the above-mentioned villages were accessible from Jinka because of the road system. This facilitated access to Aari households in terms of both extracting taxes and military conscription. Tax evasion and resistance to conscription were easier for Aari families when they lived in scattered homesteads. The previous pattern of relationships between the Aari and the socialist state began to change with the imposition of these policies. The relation of co-operation gradually changed into one of resistance.

The Aari resisted villagization for a number of reasons. First, it dislocated them from their original places to which they had deep-rooted cultural and sentimental attachments. The Aari bury their dead in the vicinity of their homesteads. Dislocation from their homesteads meant not only a physical and spiritual detachment from their land, but also from their relationship to the dead, particularly their important ancestors. Second, there were economic reasons. When families moved into the new villages, they were not able to take care of their coffee plants at their abandoned sites because of the distance. As a result, the coffee plants were damaged by baboons and other predators, and families could not generate income from selling their coffee; while the bushes planted at the new sites had not yet started yielding. Third, villagization violated certain existing societal norms. All Aari families, including members of the low caste, now had to live together in the settlement. This was a violation of the time-honoured norm of separation of the two castes as regards the pattern of habitation. In the Aari view, the breaking of this norm resulted in the spread of diseases in the villages due to the polluting effect of the low castes. Finally, some Aari disapproved of the scheme because, in their view, it encouraged conflicts and adultery due to the frequency of interactions between the inhabitants of the villages. One elder expressed his disapproval of villagization as follows:

Memory & the Humiliation of Men: Aari

The houses in the village are built too close together. Because of this, there are frequent interactions among the people in the village. A man entering the house of another person for a drink will often have a sexual affair with somebody else's wife. Another person enters somebody's house to ask for fire to light a cigarette and in the meantime he will have sexual intercourse with the woman in the house. A man entering another person's house under the pretext of drinking water or smoking a pipe often has sexual intercourse with the woman in the house. These conditions lead to fighting between men in the village. Eventually people would kill one another if we continue to live in this way. In the past, we used to live in dispersed homesteads. When we lived scattered, a man who was seen coming frequently to the house of another man in the absence of the owner was suspected of a sexual affair. But now it has become difficult to suspect because we live too close to one another. I am afraid that people will kill one another. There are individuals with different characters living together in the village. Some individuals want to have affairs with the wives of other men. Others are quarrelsome. Yet others are fond of stealing the property of other people. If you bring all these types of individuals together in one village, they will kill one another.

Military conscription also affected the pattern of relations between the Aari and the socialist state. The intensification of the armed struggle in Eritrea and Tigray required the recruitment of men from different villages in Aariland in order to send them to the war fronts. During the period of my fieldwork, I witnessed the conscription of young men in several villages. On 2 December 1989, I went to Jinka market, where I saw a truck full of men, new conscripts from around the Kuree region brought by force for national military service. They were singing to boost their morale. I imagined that they must have been told to sing because I knew that they had been conscripted involuntarily. On the same day, I returned from Jinka to Zomba and in Gazer saw another group of Aari conscripts being kept in a large compound and guarded with guns by peasant association militia. Some of the conscripts were from Zomba. Later, I was told that twenty men had been recruited from Zomba alone. Two trucks were waiting in Gazer to take the conscripts to Jinka, the district town. In front of the compound, waiting to send off the conscripts, were women, female relatives of the conscripts, some with babies on their backs. The women were screaming and crying with incredible desperation, as though they were paying their last respects. In the evening I heard the grief-stricken crying of the women in Gazer, and also the sound of the trucks starting up to take the conscripts to Jinka.

On 4 December, I came across a couple on my way from Gazer to Zomba. They were returning from Jinka after sending off their son who was among the newly conscripted Aari men. The couple lamented the conscription. While we were walking together and talking, we met a crowd of conscripts who were coming from Dordora, a village north of Zomba.

In front as well as behind them were armed peasant association militia. The recruits were being followed by women and elderly men. The women were weeping. It appeared as though a funeral procession were underway. We stopped along the road and stared at the passing conscripts, sharing their grief in a moment of silence until they disappeared from our sight. Then we continued our conversation. In an attempt to console the couple, I assured them that the conscripts would return safely from the war front. In reply to this futile consolation the man lamented that 'It is impossible to stop the sunrise and the sunset. We cannot stop the Ethiopian government from conscripting our sons for national military service.' This expression reflects the extent of helplessness that the Aari people experienced in their relations with the socialist state. The futility of my own statement was brought home to me when I recalled a saying in my own language (Kunama): 'When a swarm of bees goes to drink water, some do not return.' This saying is often used when a group of men goes off to war and some die in the battle.

These conscriptions were repeated throughout the Aari region in the 1980s. I shall describe one event in which I was personally involved. Two days after my arrival for a short visit to Gelila, a town in the then North Omo district, a militia round-up took place. The Bunke-Basketo officials had given Gelila a quota of seventeen new conscripts. Faced with this quota, the peasant association militia, red-eyed and drunk, had managed to find only ten. In their desperation, they had included even middle-aged and frail men among the conscripts. Two militia, frustrated in this tragic task of 'hunting down' their fellow villagers, suddenly approached me and demanded my identification. When I showed them my permit from the district office in Jinka, they refused to accept it. They insisted that I join the other conscripts who were being kept inside the peasant association compound and guarded by militia with guns. I resisted. In an attempt to extricate myself from the situation, I showed all sorts of identification, including my passport, but to no avail. Two of the elementary school teachers in Gelila saw my dilemma and came to help. After an hour of negotiation, I was finally released.

The Aari resisted these conscriptions for a number of reasons. One was that this recruiting were carried out by force. The usual practice was for district officials to request a certain number of individuals from the different peasant associations. When a peasant association could not fulfil its quota, militiamen often surrounded a village early in the morning and forcibly took whoever they came across. Sometimes the militia were not able to find young men. Consequently, they ended up taking even old men from the villages. Another reason for Aari resistance and disapproval of the conscription was that often individuals who were taken away died in the war, while others returned from it physically disabled.

The peasant associations faced a particularly difficult situation in their role as intermediaries between the Aari and the socialist state. It was they who had to recruit Aari men from the different villages. Failure to find conscripts resulted in penalties in the form of fines or imprisonment. Their persistent attempts at recruitment led to Aari discontent and resistance. At times, individuals threatened peasant association leaders. Trapped between the increasing demands of the state for conscripts and the resistance of the people, the peasant association leaders were helpless. The 1989 suicide of Worku Gebre-Mariam, chairman of the peasant association of Sinigal, was a direct result of this helplessness. In April 1989 state officials in Jinka requested conscripts from Sinigal, but when Worku tried to recruit, he encountered resistance. Caught between the resistance of the Aari and the demand of the state for conscripts, he killed himself.[7]

One of the features distinguishing the socialist state from the imperial state was its introduction of a more regularized and standardized taxation system. It levied taxes on all individuals who used the land, regardless of their sex, age or marital status. Aari used to evaluate various regimes as 'providers', who supplied goods and services, or 'consumers' who extracted taxes.[8] The Aari spoke favourably of the taxation system of the socialist state during the initial period of the revolution, when according to informants, peasants did not pay taxes. This, in my view, may have to do with the weakness rather than the benevolence of the state. Be that as it may, the Aari portrayed a positive attitude toward the state during the initial phase of the socialist regime, regarding it as their 'father'. The Aari used to sing songs in praise of the state invoking this fatherly image during the early phases of the revolution.

This initial positive attitude towards the socialist state was reversed during the later period because of increases in taxes. Informants reported that the annual level of taxes per household increased to 50 *birr* in 1988. During this later period, the Aari also contributed money to assist people who had been affected by drought and famine in northern Ethiopia. They also contributed money for the 'call of the Motherland' (*innat ager țirri*) to support the war being waged against opposition forces in Eritrea and Tigray. Unlike the imperial regime, which demanded taxes only from family heads (usually fathers), the socialist state requested taxes from all married households. Under the imperial state, married and unmarried sons who were under the control of their fathers did not pay taxes. The Aari regarded the socialist state as parasitic and often ridiculed a government that demanded taxes from women and unmarried girls and boys, referring here to the monthly membership fee that women and young people paid to their respective associations.

What were the consequences of not paying taxes? Those peasants who failed to pay were arrested and imprisoned. Such arrests were common

---

[7] Worku Gebre-Mariam was the descendant of a pioneer soldier-settler. During the initial period of the revolution, the Aari did not plunder his property because he was considered to have fulfilled the conditions of 'working together'. Indeed, he was appointed the chairman of the Sinigal peasant association.

[8] Based on this conception, the Aari favoured the Italian administration more than the Haile Selassie regime or the socialist state, although they also complained about the labour conscription that the Italians had imposed during the occupation period.

during the period of my fieldwork. When peasants were arrested for failing to pay taxes, their kinsmen had to borrow money in order to pay. Some Aari in Sinigal came to me to request money to pay their taxes. Sometimes peasant association officials also confiscated the livestock of those individuals who evaded taxes. On 25 January 1990, for example, I saw some men and women arrested in Metser (in Siida region) for tax evasion. I also saw some animals in the peasant association compound belonging to individuals who had run away, fearing arrest. Informants reported that if the owners did not come with the money, their animals would be confiscated. If the owners did bring the money, they could take back their animals, but had to pay a fine (2 *birr* per day) for the time the cattle had been kept in the compound.

In contrast to the Aari, some northern settlers had a different attitude towards taxation. One Muslim trader in Gazer (originally from Wello) remarked that the Aari inability to pay their taxes was due to their laziness and extravagance. According to this informant, although land was very fertile, the Aari did not work hard. When they did earn money, the man argued, they spent their earnings on drink. He pointed out that before he came to Aariland he was a poor person; but when he came to the region he became rich because he worked hard and saved his money. At the time of my interview, the person owned a shop and a grinding mill in Gazer. In his view, all peasants should pay taxes to the state. He justified this by invoking two Amharic sayings. One of these was *yetennya yakorfal, yegebbere yarfal*, which means 'He who sleeps snores, and he who pays taxes is secure', that is, as long as one pays taxes to the state, one will be safe. Conversely, if peasants do not pay taxes, then their arrest and imprisonment are justifiable. The second saying goes *yigebbir kalachuh jinjerom yigebbir, ye nigus aidellemoy emibelau midir*, which means 'If we have to pay taxes, even a monkey should pay taxes because it feeds on the king's land.' Implied here is the assumption that the land belongs to the state; therefore, the peasants who cultivate it must pay taxes to the state.

State–society interactions in Aariland during the later period of the socialist regime reflected unequal power relations between the Aari and the state. This resulted in the development of a feeling of helplessness among the Aari, a phenomenon that was similar to their condition during imperial rule. Indeed, the Aari drew certain parallels in the pattern of relations between them and the Ethiopian state under the imperial and socialist regimes. Informants often remarked that their situation under these states was similar to a marital relationship in which the husband assumed a dominant status in relation to his subordinate and subservient wife. Some informants used the metaphor of a serf–master relationship in characterizing Aari relations with the socialist state.

Although unequal power relations developed between the Aari and the state during the later phase of the regime, the Aari were more integrated into the Ethiopian state they had been before. Schools played a prominent role in the integration process. The socialist state built elementary schools in the different villages in Aariland. It also opened a secondary school in Jinka, although most of the students who were enrolled in the school were the children of northern settlers. The few Aari who had finished secondary education but had not passed the Ethiopian School Leaving Certificate (ESLC) were given training so that they could become elementary school teachers in the area. Some individuals were also recruited as civil servants in the various offices in Jinka. Compared to the imperial period, employees in government bureaucracies during the socialist regime were from diverse backgrounds. This is apparent when one looks at the names of individuals who were appointed as district administrators during the period between 1977 and 1984: their names suggest that they came from a range of peoples such as the Oromo, Wolayta, Konso, Gamo and Tigray.

Conclusion

An examination of the relationship between the Aari and the imperial Ethiopian state is important for understanding the pattern of their relations with the socialist state in the period from 1974 to 1991. The Aari had existed as a politically autonomous entity prior to their conquest and incorporation into the Abyssinian empire. The defeat of the different Aari chiefdoms by the Abyssinian army during the late nineteenth century resulted in the introduction of the *gebbar* institution, which in turn brought about the development of a political culture of powerlessness among the Aari. This political culture prevailed until the advent of the revolution, when the Aari had been able to transcend it with the help of the politicization of student campaigners. During the initial period of the revolution, the Aari rebelled against the old regime and the northern settlers. The Aari co-operated with the socialist state during the early phase of the revolution because of three important factors: the land reform, and the discourses of equality and class struggle. The imposition of the policies of villagization, military conscription and increased taxation during the later period of the revolution, however, produced resentment among the Aari, and led to their resistance. Thus, although the land reform and discourses of equality and class struggle had put an end to some previous inequalities, villagization, military conscription and taxation created new forms of power imbalance between the Aari and the socialist state.

Four

Close yet Far
Northern Shewa under the Derg[1]

AHMED HASSAN OMER

In spite of its proximity to Addis Ababa, Northern Shewa on the eve of the 1974 revolution remained one of the least developed and most traditional parts of the country.[2] It was sometimes referred to as *ye qirb ruq*, meaning that it was 'close yet far' from the political and administrative centre of Ethiopia (see Maps 1 and 1.1). It was *qirb* (close) because the distance from its border with Addis Ababa to its northern limit was only 295 kilometres. By the same token, it was *ruq* (far) for the simple reason that it did not have the basic infrastructural elements which would have facilitated trade and political relations with the surrounding areas and with Addis Ababa. Modern schools and health centres were not just limited, they were almost non-existent. For instance, Kombolcha, in southern Wello, which lay at a greater distance from Addis Ababa, had almost all the necessary modern amenities, such as an airport, factories, better schools and health centres. But the state had a special need to maintain a strong presence in Kombolcha since it provided the necessary link between Addis Ababa and the northern provinces.[3]

During imperial times, the state hardly penetrated to the grassroots level in Northern Shewa. For example, it was customary for landlords and petty government officials to expect and receive fattened sheep, butter and honey from the peasants as a gift, mainly on public holidays, at weddings and the like. If peasants did not submit these traditional gifts on time, they could be evicted from the land they had been farming for decades. All these factors contributed to making Northern Shewa a 'backward' region,

---

[1] My grateful thanks are due to Professors Alessandro Triulzi, Wendy James and Eisei Kurimoto, who have been sources of inspiration and encouragement in dealing with this research theme since the Kyoto Conference in December 1997.
[2] For more details, see Ahmed Hassan Omer 1994.
[3] Ibid.: 138–9; 'Ye Semén Shewa Hizb Ya Nuronna Bahlawi Hunetawoch', in Folder No 1001, File No. 3471 (Tegulet and Bulga Awrajja Archives).

politically isolated from the centre, despite its physical proximity to it.[4]

In addition, Northern Shewa has suffered continual droughts and famines. In the words of Sven Rubenson, the region is 'one of the most drought prone areas of Ethiopia' (1991: 74). We have, for example, a reference to drought and famine in this region as far back as the thirteenth century. It was also seriously stricken by the Great Famine of 1888 to 1892, which resulted in heavy loss of cattle and human life. In the twentieth century, the region has been affected by droughts that seem to recur at least once a decade (typical examples are those of 1974 and 1984). It is reported that in 1974 this region lost a total of 141,009 head of cattle and pack animals as well as 4,172 people. A decade later, in 1984, Northern Shewa is reported to have lost 192,311 head of cattle and pack animals and 6,004 people.

In the period covered by this study, Northern Shewa was a volatile region politically. The scarcity of natural resources such as pasture, farmland, and water pitted different groups against one another. Sometimes local governors aggravated these conflicts, as was particularly true in 1974, on the eve of the Ethiopian revolution.[5]

The initial phase: of joy, 1974–1975

In February 1974, the regional and district officials of Northern Shewa were facing serious student revolts as a result of the so called 'Educational Sector Review' proposed by the government. The central theme of this proposal was to focus the educational process at the level of the poor and backward national economy. It suggested that the country's educational efforts should be concentrated on the primary schools and be given a practical rather than academic orientation. The secondary school level was to retain its narrow status, while higher education was to be opened up only slightly. Post-primary education was to be funded only through loans which were eventually to be paid back by the students themselves.

This proposal was a stunning blow aimed at aspiring local elites, particularly in Northern Shewa where most people lived well below the poverty line (a condition aggravated further by the 1974 drought). During this period the student movement was particularly active in trying to persuade the government to cancel the proposed Educational Sector Review, which had become a serious national issue well beyond Northern

---

[4] Interview with the following five elders, December 1997: Ato Tesegay Welde Samayat (died 1998), a long-serving civil servant in various parts of the region; Ato Beqqele Sharew, long-serving civil servant and General Secretary of Yefat and Timmuga *awraja*, Tagulat and Bulga *awraja* and the former Shewa region from 1966 to 1982 (he is living in Addis Ababa); Ato Moges Mekonnen (Addis Ababa, January 1999); Ato Muhamad Siraj; and Ato Wube Wandimmu (Nazreth, February 1999).

[5] Informants: Ato Moges Mekonnen (Addis Ababa, January 1999); and Ato Beqqele Sharew (Addis Ababa, November 1998). Information also drawn from 'Northern Shawa Annual Report for the year 1987': Folder No. 6, File No. 314 dated 23/9/79 E.C. Cf. Hassan Omer, 1994: 108–39.

Looking Back on the Projects of the Socialist State

Shewa. Locally students living in the towns of Debre Sina, Molalé, Mehal Méda, Shewa Robīt, Aṭayé and Karraqoré all rose to demonstrate against government authorities, breaking windows and doors, demolishing furniture in schools and other government institutions, and disrupting the day-to-day running of buses and other vehicles by puncturing their tyres and breaking their windows. Government vehicles and particularly buses became the major targets. These student demonstrations were so intense that they eventually contributed to the revolutionary situation developing in Ethiopia. In fact, the leaders of the student protest were the sons and daughters of the local elite, who were mainly the high officials of the region of Northern Shewa.[6] This is well illustrated by a famous couplet of the time, representing the views of a local government official in the region:

አርመን፣ ኮትኮተን፣ ያሳደግነው ቀጋ፤
ዘመም ዘመም አለ፤ እኛኑ ሊወጋ፡፡

The wild rose we nurtured to grow
Weeding and turning its soil,
Began to sway back and forth
To prick us with its thorns.[7]

During this period, the student revolts in Northern Shewa began to influence the peasantry. To begin with, student–peasant relations can be seen in two ways. It cannot be denied that many of the students participating in the revolt came from a peasant background. Secondly, from 1965 onwards, students at Haile Selassie I University were spreading the ringing slogan of 'Land to the Tiller' and this began to reach the consciousness of peasants. Therefore, it was no surprise that the vast majority of peasants began to take an interest in the revolution. In contrast, officials, landlords, rich merchants and the traditional local elite as a whole were opposed to it, as they feared losing their status and privileges.[8]

The popular uprising was appropriated by the Derg on 28 June 1974. This gave the revolution a new dimension and scope. After removing the Emperor from power on 12 September 1974, the Derg came up with the new slogan, *ītiyopia tiqdem* (literally 'Ethiopia first') to consolidate its power (Bahru Zewde 1998: 221–4). The way the Derg deposed the Emperor was a matter of concern to a majority of the people in Northern Shewa, particularly the elderly. Although they did not condemn the change per se, they did not approve the way the Emperor had been deposed. In fact, it

---

[6] Informants: Ato Dejene Girma (Ataye, December 1994); Ato Kinfe Damṭé (Addis Ababa, December 1996). They both participated in the student revolts in the region in 1974. On the details of the Educational Sector Review, see, amongst others, Kiflu Tadesse 1985: 167; Teferra Haile Selassie 1997: 88–9.

[7] Informants: Ato Tawabe Ze Yohannes (died 1995); Ato Mammo Shebru (died 1993).

[8] Informant: Qés Shewayyé Tekle Yohannes (Addis Ababa, October 1998); cf. Fantahun Tiruneh 1990: 60–1, Bahru Zewde 1991: 195, 1998: 204.

was the massacre of Ethiopian cabinet ministers and high government officials on 23 November 1974 that produced the first real opposition to the Derg. Although support for this opposition was scattered throughout the country as a whole, it was particularly evident in Northern Shewa, which had produced a large section of the imperial elite and high government officials.

Although the 23 November massacre was a hasty measure, the Derg was soon able to move beyond its initial blunder. This was illustrated by two important decrees issued by the Derg: the 'Development Through Co-operation Campaign' (*zemecha*) of 21 December 1974; and 'Land to the Tiller' on 4 March 1975 (echoing the slogan popularized by students ten years earlier). These two important steps were able to break the initial barrier that had been created between the Derg regime and the general public.[9]

The active participants in the Development Through Co-operation Campaign were university and high-school students and teachers. Members of the campaign went to various parts of the country, including northern Shewa. Although their main objective was to make the peasants literate, they also tackled other projects with the help of the peasants, constructing small bridges, feeder roads, health centres and modern stations to demonstrate improved farming methods. In lowland areas, such as Rassa and Cheffa in Yifat, and Tachbét in Merhabété, where water was scarce, the *zemecha* students dug wells and small irrigation channels. The construction of a 'Friendship Bridge', on the River Weezar near the village of Mehal Méda in Menz in February 1975,[10] and the clearing of three feeder roads at Wuchalé near the village town of Muketuri in Selalé during the same period are cases in point.

It was widely known that the Weezar had always been a fierce killer during the rainy season around July and August, when thousands of cubic metres of water came roaring through its deep gorges. When the rains were heavy, the river became impassable, and travellers sometimes had to wait for days. Although it would have been impossible for rural communities to undertake such a project on their own, the bridge was built as a result of the co-ordinating activities of the campaigning students, who managed to persuade the government Local Community Development Office and the Baptist Mission of Ethiopia to help. Mr C. Staton, an engineer living in Addis Ababa affiliated to the Mission, visited the site and drew up a design. About fifty students were able to go around the nearby rural villages, calling for the assistance of peasant families. From the very outset the work became an example of co-operation and friendship. The local people gave wood, stone, lime and cement, as well as their labour.

---

[9] Informant: Ato Laqew Shewayyé (Atayyé/Epheson, March 1998); cf. Kiflu Tadesse 1985: 309–10.

[10] Informants: Ato Alemu Muluneh (Debre Berhan, March 1997); and Ato Tegegne Feqadu, both members of the campaigning students (Addis Ababa, October, 1998); details on the Friendship Bridge were published in the *Ethiopian Herald*, 7 September 1975.

The students, besides their role as the organizers and co-ordinators of the project, also contributed by giving valuable labour service. The campaigners also covered the expenses for two masons and for the crushed stone that formed the foundations of the bridge.[11]

The second thing the students achieved together with the co-operation of the local people was the clearing and gravelling of three feeder roads extending for a total of 53 kilometres. These three feeder roads branch out from the village town of Muketuri into the district of Wuchalé, serving as an important outlet for peasants to take their produce to the major markets in the surrounding areas. Prior to this, the peasants had a difficult time with communicating with the other village towns in the district.[12]

The role of the campaigning students was not limited to teaching the local peasants and initiating and constructing various rural projects, but also extended to playing an active role in implementing the Land to the Tiller decree. The news of the official announcement of the land decree spread quickly throughout the country, including the remote sections of rural society, by word of mouth. This was the joyful high point for the peasantry. Prior to this, peasants all over the country, who had heard only rumours about the forthcoming land-reform decree, would walk from place to place in the hope of gathering more precise information. An anonymous peasant farmer from Northern Shewa is reported to have gone to visit his landlord in Addis Ababa for a few days under the pretext of medical treatment. As luck would have it, on the very day the land-reform decree was promulgated, this peasant and his master were having a friendly chat over coffee – with the radio on in the background (Kiflu Tadesse 1985: 309; cf. Fantahun Tiruneh 1990: 60–4). When the radio suddenly announced that the land-reform decree was to become a reality, the peasant farmer's attention strayed from his landlord's conversation. The landlord, equally alert but worried about the peasant farmer overhearing, turned off the radio and pressed on with his friendly chat. The farmer, however, kept an ear on the news still very audible from a neighbour's radio. The agitated landlord went to his neighbour and asked him to turn off the radio or at least to turn it down. The neighbour, well aware of the landlord's embarrassment, increased the volume instead. Thus the poor peasant was able to learn the essence of the land decree and return to his native village in Northern Shewa with the happy news.[13]

To make the land decree better understood by the peasants and to implement it practically, the *zemecha* students had to work very hard indeed. Their efforts were so persistent that they quickly gained the hatred of the members of the formerly landowning upper classes in the region. In fact, some of the *zemecha* students were actually killed by outraged landowners. For instance, in Menz and Gishé the campaign station head

---

[11] Ibid.
[12] Informants: Ato Takkele Taddala; and Ato Lemma Olana (students) who participated in the campaign). Master Sergeant Gezaw Welde Mikael, Derg Member, inaugurated the completion of these three feeder roads. See *Ethiopian Herald*, 24 January 1978.
[13] Informant: Ato Moges Mekonnen (Addis Ababa, February 1999).

Tayyé and his friends were assassinated at a place called Kimmir Dengay in the district of Qeya Gebriél.[14]

Indeed, the students were hated not for their practical accomplishments but mostly for what they taught and inspired among peasant farmers. For instance, the formation of peasant associations and of peasant economic co-operatives were a practical challenge to the former landowners, depriving them of the power to exploit local people and instilling in the peasants relative self-reliance and independence.[15]

The number of skilled workers in the Ministry of Land Reform was so limited that the whole responsibility for implementing the land reform fell directly on to the shoulders of the students. To say the least, their knowledge of land-distribution, its associated problems and the laws governing it was limited. They were not given even preliminary guidelines on this difficult and complex issue (Kiflu Tadesse 1985: 311–12). To illustrate this point more clearly, the activities of students around Yifat in Northern Shewa can be cited. Abbabbu, Adane and Berihun, students who were native to the region, were assigned to go there. They had lived there all their lives along with their parents. Hence, they knew something about the difficulties and problems encountered by local farmers. Many other students who were with them were not so well informed about the rural situation. Most of these other students had come from the major towns and cities of the central parts of the country. By contrast, when Abbabbu and his friends were asked about the land issue, their views were clear, precise and to the point. They unanimously said that the Derg government was being too hasty in launching the 'Development Through Co-operation' decree. They also stated that the right students had not been properly selected in the first place, nor had they been given the necessary guidelines in order to carry out their duties.[16]

The second phase: from joy to betrayal, 1975–1977

The land reform decree of 4 March 1975 had a major impact on every section of Ethiopian society. To begin with, the proclamation seems to have been a pre-planned move by the Derg to erode the position of its opponents and consolidate its newly achieved power, merely paying lip service to the long awaited land-reform decree without actually making it reality. Nevertheless, the decree had both short- and long-term consequences for Derg rule. The short-term consequence was that the land-owning classes were highly displeased from start to finish by the Derg's harsh steps because the pillar of their pride and joy, the main source of

---

[14] Informants; Qés Shewayyé Tekle Yohannes; Ato Mohammad Sayed (Shawa Robit, January 1994).
[15] Ibid.
[16] Informants: Ato Ababbu Takkele, Ato Adane Garramaw and Ato Berihun Laqew (Atayyé, December 1994).

their wealth and political power, was taken away by a stroke of the pen. This was true everywhere in the country.[17] For instance, in Northern Shewa alone a total of 30,473 *gasha* of land were confiscated by the Derg from 1,615 prominent landholding families. Of these, about 30 families owned about 25 per cent of the total. The following table shows the breakdown of the *gasha* and their owners in the various districts of the region.

Table 4.1 Landowners in Northern Shewa, 1976

Name of localities	No. of land owners	No. of *gasha* of land
Tagulat	218	1,870
Merhabete	179	790
Yifat	128	1,200
Salale	175	400
Menz	503	15,000
Bulga	412	11,213
TOTAL	1,615	30,473

Source: Annual Report of the Shewa Branch, Ministry of Agriculture in 1976, dated 24/06/1976.

The final consequence of land reform was to force many of the able-bodied members of Northern Shewan land-owning families to flee to Menz in early April 1975 – from which they began an organized resistance against the government. Menz was selected by the members of the anti-Derg resistance movement for two reasons. First, it is a rugged area containing naturally fortified places which are useful for defence and also comparatively remote from the main Addis Ababa–Dessie highway. Secondly, local people in Menz were willing to give the resistance psychological, moral and practical support.[18] It is a well-known fact that the Emperor and other prominent members of the imperial elite traced their origins back to this particular area (Mehtema Selassie Welde Masqel 1965 E.C.).

The prominent sites of resistance in the Menz region were Afqera, Badogé and Gishé. They were situated around the Jema, Wenchit and Qechiné river valleys, which were full of natural caves for hiding men and material. Moreover, this area was exactly opposite Werre Îlu, another centre of resistance in southern Wello. As these two centres of resistance were within easy reach of each other, it was not difficult for members of the resistance movement to cross the Qechiné river and join forces for combined campaigns. Although the anti-Derg resistance movement was underway at each of these three sites, it was only with the coming to Afqera of the two famous Birru brothers, Merid and Mesfin, in late April 1975 and the rise of Bruké Demissé in Gishé at about the same time that the resistance movement developed its momentum and importance in the

---

[17] Informants: Ato Adane Gerremew; Qés Shewayyé Tekle Yohannes.
[18] Ibid.

region as a whole.[19] This fact is clearly demonstrated by a local elderly observer, who recounted this development in the following manner: 'When it was heard that the sons of Ras Birru Welde Gebriél had come to Afqera and Bruké Demissé had risen to prominence in Gishé, all the people who had lost their land, without exception, whether young or old, all flocked to Afqera.'[20]

The basic message of the quotation is that it was the Birru brothers and Bruké Demissé who were able to turn Afqera and Gishé into a strong central base against the Derg in the region. Time after time, they devastated the Derg forces that were sent against them. To cite one such incident, thirty heavy military truckloads of soldiers were annihilated at a place seventeen kilometres away from Molalé town on their way to Afqera in July 1975. A second force was sent to Afqera in early September consisting of 2,000 heavily armed soldiers with an additional 700 local militia to back them up. They were again totally defeated and the few soldiers who remained were in complete disarray. It was at this critical stage that the Derg was forced to use air power to reverse its losses. Even this drastic measure did not suppress the strong resistance in Afqera. The continued defeats of the Derg's soldiers were directly responsible for boosting the morale and psychological support of the resistance movement led by the two Birru brothers, so much so that the local peasantry became more or less their total supporters.[21] As a result, the Derg army which campaigned to Menz was ridiculed by the local people in the following manner:

> አንተ የደርግ መልዩ ለባሽ!
> አንተ ንጉሥክን ነካሽ!
> አንተ አህያ በላ!
> አፍቀራ አፋፉ ላይ ፈለጉለሽ መላ!!
>
> Fellows of the Derg in uniform!
> You who bit [the hands of] the king!
> You who feed on the flesh of donkeys!
> [The rebels] dealt with you effectively
> At the edge of the Afqera plateau.[22]

---

[19] Informant: Qés Shewayyé Tekle Yohannes. The strategic importance of Afqera can be traced back to the seventeenth century; for details, see Ahmed Hassan Omer 1987: 9; also Gabra Sellasé 1959: 55-68.
[20] Ibid.
[21] *Mestre Zemene Derg* ('Secrets of the Derg Regime'). (Personal collections of Ato Tsegayé Welde Semayet). His documents report on casualties on both sides, that is, the Derg and the resistance group. They also cover various themes concerning the Derg regime.
[22] Informant: Ato Tsegayé Welde Semayet. Eating the flesh of donkeys, pigs, horses or mules is not a tradition of the Ethiopians, whether Christian or Muslim. Soldiers in the area used to eat fruits, beef, and tinned sardines; the rural peasantry were very suspicious of any food packed in tins, which they thought might be pork, donkey or horse.

Looking Back on the Projects of the Socialist State

It should be noted that the resistance movement in Northern Shewa was not unique as such but was a common feature of many areas of the old cultural core and beyond. There were pockets of resistance in Gojjam, Kefa, Sidamo and Wello. There were three important sites of resistance in Wello, namely Werre Îlu, Kutaberr and Awsa. Around Awsa, a province inhabited by the Afar, student campaigners were attacked by the local population. As a result they were forced to go to Dubti, one of the main Afar towns in the region. On their arrival, they found that the traditional Afar chief, Sultan Ali Mirah, had already fled the country in June 1975 for his personal safety because of his fierce opposition to the Derg. The Afar Sultan was able to escape without the knowledge of members of the Derg such as the notorious Major (later Brigadier-General) Getachew Shibeshi, who had gone there to propagate the virtues of the Derg and to restore peace and security. Nevertheless, the opposition was sustained by local petty chiefs and the Afar people of the area in general. In fact, the opposition was so fierce that the Derg was forced to use heavy artillery, armoured cars and tanks. In spite of killing so many of the local Afar people, Derg forces were unable to suppress the resistance. As a result, the Derg officials in the region considered poisoning water wells in the region, but were stopped by the central government itself (Kiflu Tadesse 1985: 316).

Compared to all these other areas of resistance, Afqera was by far the most important. To begin with, it had become the major rallying point in the country for all landlords, former government officials, the former ruling elite, dissatisfied members of society and finally even highway robbers. Moreover, being situated close yet far, Afqera was able to create extreme tension and fear among the leadership of the Derg. As long as Afqera continued to remain the rallying point for all opposition elements in Ethiopia, the very survival of the Derg was in question. As a result, the Derg was forced to use massive air power in Northern Shewa against its own civilian population. In spite of these harsh measures, the two Birru brothers were able to move in and out of Afqera and cause severe damage to the Derg forces and local militia in the region, until they and many of their followers were eventually killed in air strikes in October 1975.[23] This heralded relief and joy in military circles that was perhaps equal or even greater than the happiness expressed at the rout of General Siad Barre's Somalis, who invaded eastern and southeastern Ethiopia two years later.[24]

The destruction of Afqera and the death of the two Birru brothers changed the power base of the resistance movement of the region. The other sites of Badogé and Gishé henceforth became the new bases of the resistance movement under the fresh leadership of a local dissident, Bruké Demissé.[25] Although they would not gather the fame of Afqera, Badogé and Gishé soon became new centres of resistance to the Derg and were

---

[23] *Mestre Zemene Derg* ('Secrets of the Derg Regime'); see note 21 above.
[24] Informant: Ato Tessemma Welde (Addis Ababa, April 1999), a journalist and civil servant who participated in the campaign against Somali and witnessed Siad Barre's defeat.
[25] *Mestre Zemene Derg* ('Secrets of the Derg Regime'); see note 21 above.

able to reunite the rebel forces of Shewa with those of Werra Îlu. Moreover, Bruké Demissé was able to rise to the supreme leadership of the resistance movement, thus successfully replacing the Birru brothers. The Werre Îlu and Menz (Gishé) camps were thus able to work smoothly together and to inflict heavy losses on the Derg. This was because the resistance forces in both regions were also able to move from one region to the other when attacked by the Derg and to regroup there in order to counter-attack the enemy. For instance, in April 1977, Derg forces were trapped on the banks of the Qechiné river by a joint manoeuvre by the Menz and Werre Îlu resistance. In this incident alone, 150 soldiers were killed and over 250 wounded.[26] Co-operation between the two resistance movements was very close, so much so that a local elder explained their relationship by saying, 'They were like the left and the right hands of a man working on an assigned task.'[27]

At this critical stage, the changing role of three important political forces became evident, namely the local peasantry, the resistance, and the Derg army. To begin with, the Derg was able to gain the support of the local peasantry by proclaiming that this was the time to join hands with its soldiers to destroy the 'remnants' of the defunct ruling elite. The resistance consisted of those who had gone to Afqera, Badogé or Gishé in order to regain their lost political and economic power. This group also wanted to get the local peasants on its side in order to fight the Derg. In gaining the support of some of the local peasantry, their propaganda was no less effective than that of the Derg. They told the peasants that the Derg was not trustworthy because it was responsible for the death of the Emperor, who had been anointed by Almighty God and that that therefore peasants could not expect to benefit in any way from such evil. They said further that the Derg regime lacked faith and was out to destroy the ancient Ethiopian Orthodox Church, the pillar of Ethiopian Orthodox Christianity, to which a large section of Northern Shewa peasants belong.[28]

The third and last political factor was the local peasantry, which constituted the largest part of society and which was deployed by both the Derg and the former landowners. The local peasantry was exploited, both willingly and forcibly, by both groups. During the daytime, local peasants were forced to carry out the orders of the government, the peasant associations, and Derg soldiers stationed in the area. But at night, they had to do the same for the leaders of the resistance movement. In other words, both the Derg and the resistance used the wealth and energy of the peasantry for their own selfish ends.[29] Local peasants encapsulated the poignancy of their situation, their anger and sorrow, in the following saying that circulated in Northern Shewa: 'Some of us laboured day and

---

[26] Ibid.
[27] Informants: Ato Degennetu Kessemé (Saramba Mafud, December 1998), Ato Tsegaye Walda Semayet.
[28] *Mestre Zemene Derg* ('Secrets of the Derg Regime'); see note 21 above.
[29] Ibid.

night with the shiftas of Afqera, Badogé and Gishé, while the rest of us did the same with the soldiers of the Derg. We expended both our energy and property uselessly, to be left finally empty-handed'.[30]

Growing discontent: from the 'red terror' to the fall of the Derg, 1977–1991

The growing discontent in Northern Shewa was mainly due to three important events: the dark years of the so-called 'red terror'; the 1984 famine; and the massive forced conscriptions undertaken by the Derg to overcome the ever-growing internal threats.

As already noted in the previous section, by 1977 the peasantry in Northern Shewa had reached a desperate situation. This was due to the endless pressures exerted upon them by both the Derg and its opponents in the resistance. The peasantry was weary and exhausted and therefore ready to listen to any newcomer with the promise of any new hope.[31] This was provided by the Ethiopian People's Revolutionary Party (EPRP), which had started founding its local political network in Northern Shewa by this time. It did not take long for the members of the EPRP to obtain strong peasant support, which had three main aspects. In the first place, most peasants were ready to support the EPRP for the sake of a brighter future, so much so that they were ready to supply fighting forces to this end. Second, locally educated young people were ready and willing to accept the cause of the EPRP for the simple reason that they knew that their relatives and parents had been misused by both the Derg and the resistance. Third, not only were they ready to serve the EPRP directly, they were also active in recruiting the local peasants and urban dwellers in the nearby towns.

Within a short time rural areas of Northern Shewa were beginning to attract members of the EPRP who had been attacked and ousted from towns such as Desé and Kombolcha in southern Wello, from Aṭayé, Shewa Robīt, Debre Sina, Mehal Méda and Debre Berhan in Northern Shewa, and finally from Addis Ababa.[32] The following couplet is what local observers in the region, who watched the constant urban–rural flow of assorted dissidents to the area, seem cynically to have composed in order to describe the situation:

ዝናብ ጥሎ ጥሎ፥ ዳር ዳሩ ጭቃ ነው፤፤
የአፍሮ (የተማሪ) ነገር አልሰየም ገና ነው፤፤

---

[30] Informants: Ato Gezaw Beyyene; Ato Degennetu Kessemé; Qés Shewayyé Tekle Yohannes.
[31] Ibid.
[32] Informants: Ato Adane Gerremew; Ato Tsegayé Welde Semayat; Ato Kebret Endashaw (Aṭayyé, January 1998).

Close yet Far: Northern Shewa

With the rain pouring ceaselessly
The roadsides are muddy.
As for the students with 'Afro' hair
Their role is still unclear.[33]

This prophetic folk couplet was followed by another realistic one, which clearly expressed the activities of the Afro-styled students who were later directly involved in the ensuing anti-Derg guerrilla struggle around the various peasant villages in the region. They expressed their heartfelt admiration by appreciating the way they managed the resistance and used guns against the Derg forces:

በብዕሩ ብቻ ይባል ነበር ድሮ፤
አዙሮ ተኮሰ፤ ተማሪ ዘንድሮ፡፡

'Students fight with pens'
Was the adage of days bygone.
This year they fired guns
Training them [on their foes].[34]

The EPRP leadership also considered the region an ideal place from which political and military actions could be planned and launched against the Derg and its supporters in and around Addis Ababa. In addition, the leadership felt that a working relationship could be forged with such strongly anti-Derg movements such as the Ethiopian Democratic Union (EDU) and the Afar Liberation Front (ALF). This happened after the EPRP had been completely routed in the towns and cities, and a faction of its leadership had decided to launch guerrilla warfare from rural areas, using Northern Shewa particularly as one of their leading bases. However, this attempt also ended in complete disarray in 1978 to 1979.[35]

At this stage it would be interesting to ask why this failure was brought about in such a short period of time. Three important factors can be mentioned. First and foremost was the creation of factions within the main EPRP movement. Second, within the same period and region, there were family members who were affiliated with different political movements and organizations, which caused leaks of vital secrets to outsiders. For example, there were cases in which the father was a close supporter and possibly member of the EDU, while his sons or daughters were members of the EPRP or the All Ethiopian Socialist Movement (MEISON), or else associated in some way with the armed forces of the Derg. A typical case is the family of *balambaras* Aliyi Adam, living in Duguguru (Efrata Jillé). He was one of the leading landlords in the region and an ardent supporter and

---

[33] Informant: Ato Tewabe Ze Yohannes.
[34] Ibid.
[35] Informants: Ato Dejene; Ato Belachew; and Ato Gizaw Kebret, who had followed the developments of the time.

apparently a member of the EDU, while his eldest son, Muhammad Aliyi, was a militia member, supporting MEISON and the Derg regime. His younger son, Hassan Aliyi, was an active member of the EPRP and was killed by the Derg in 1979.[36] The third and final cause was the role played by the Derg and MEISON in forming the Provisional Office for Mass Organizational Affairs (POMA) in April 1976, and MEISON's decisive role in organizing rural and urban communities to support the Derg.[37]

To oppose this, desperate local members of the EPRP became more aggressive and publicly rose against the Derg and the POMA cadres who were their immediate opponents, beginning to shoot them in broad daylight. For example, they assassinated POMA cadres Ato Radwan Abubakr and Ato Assallifew Mengisté in July 1977. Soon afterwards, a broad-based 'red terror' campaign was unleashed by the regime. Lt. Solomon Zebené was especially dispatched to Northern Shewa to coordinate this massive campaign. The campaign was justified particularly after members of the EPRP, in collaboration with a certain Ahmed Mekkenna, a local peasant who acted as a double agent, had risen up in arms against local government officials. Worse still, by switching sides and working in close contact with Derg cadres, the same Ahmed was responsible for the murders of several EPRP members before he too was assassinated (by the EPRP in November 1977). The local tradition calls this period 'a time of utter darkness'. This was because so many young people in the region were brutally massacred one way or another. This was the turning point for the Derg. It was at this particular stage that the public attitude toward the government changed totally in Northern Shewa, from joy to hatred.[38]

The period from 1977 to 1984 witnessed some of the Derg's most intense and successful activities. For example, the Ethiopian defence forces defeated the invading Somali army in late 1977 and turned towards internal insurgents like the EDU, the Tigrayan People's Liberation Front (TPLF) and the Eritrean People's Liberation Front (EPLF). The Derg was also active in undertaking a few policies that were designed to win it popular mass support. These included the 'Green Campaign' of 1978, intended to bring about rapid economic development (Bahru Zewde 1998: 224). In Northern Shewa, the 'Green Campaign' was dismissed as the 'Yellow Campaign' by the people in a cynical mood of humour. This was because almost all the budget allotted to it was misused by government

---

[36] Informants: Ato Mohammad Aliyyi and Ato Mohammad Omer (Dugugguru, January 1998).
[37] MEISON's Clandestine Leaflet, *Ye Seffiw Hizb Demts*, no. 33; EPRP's Clandestine Leaflet, *Democracya*, vol. 4, no. 9.
[38] Informants as in notes 22 and 35, above. Among the leading victims of the 'red terror' were Desta Mulugéta, Fessaha Mulugéta, Tekle Armedé, Kebbede Deresé, Tsegayé Sabsebé, Mellese Shiferraw, Tito Kebbede, Sayed Aminu and Zinabu Tilahun, all of whom were massacred in front of their families. For details on this, see *Mestre Zemene Derg* ('Secrets of the Derg Regime'); note 21 above.

officials for drinking mead (*tejj*) and whisky – both of which were yellow.[39] The most successful act of the Derg in this period was the literacy campaign of 1979, aimed at raising the consciousness of the Ethiopian masses through reading and writing.

In consolidating its power, the Derg's crowning achievement came about with the dissolution of the Union of Ethiopian Marxist-Leninist Organizations (IMALEDH) and with its replacement by the Commission for Organizing the Party of the Working Class of Ethiopia (COPWE, or in the Amharic acronym, ISEPAKO) in December 1979. COPWE was a centre for recruiting and assembling individuals loyal to the Derg regime in general and to Lieutenant Colonel Mengistu Haile Mariam in particular. The public response to the formation of this organization was both positive and negative. Pro-Derg opinion justified it as a step towards the formation of the Ethiopian Workers' Party (WPE), which would later be formed in September 1984. On the other hand, those who opposed the Derg felt that COPWE was a mere agglomeration of self-seekers, adventurers and vagabonds out for self-enrichment and political power.

In Northern Shewa, some local wit nicknamed ISEPAKO *i se fashko*, meaning that COPWE was no more than an empty wine-flask of the Ethiopian workers. In orther words, the Derg had destroyed the 'wine', the able and popular politicians. This period was a time when the Derg felt immensely strong and overly self-confident, so much so that one of its popular slogans went: 'We shall put not only reactionaries under our control but also nature!' Nevertheless, this slogan was short lived indeed as the 1984 famine developed.[40] Before the famine was over, the people in Northern Shewa improvised the following apt and timely couplet, which turned the regime's slogan into a parody.

በሥራ-በምስት ኮሚቴ ሲያናፉ፡ ሲያናፉ፤
በሰማይ ደመና፡ በምድር ዝናብ ጠፋ፡፡

> While dozens of committees
> Kept blaring slogans,
> The sky lost its cloud cover;
> And the earth, its water.[41]

The 1984 famine was caused by a lack of rainfall and an invasion of pests in the region, mainly in Menz, Tegulet, and Yifat and Ṭimmuga. At its peak, the famine was responsible for large-scale dislocations of the population, followed by general destitution, starvation and death (Adhana 1988). The overall situation in the region is illustrated by the fact that mothers were reported to have abandoned their own children to death for a crumb

---

[39] Informant: Ato Tsegayé Welde Samayat.
[40] Ibid.
[41] Informant: Ato Meseret Tesemma; cf. Fekade Azeze 1998: 46, 140.

of bread or a handful of *qolo* (roasted grains). The following couplet captures the misery and suffering:

በሰባ-ሰባቱ መወለዴን ጠላኹ፡፡
በገዛ እናቴ ቆሎዬን ተቀማኹ፡፡

Because of the year nineteen-eighty-four,
I wished I had never been born.
I was robbed then by my own mother
[A precious fistful of] roasted grains.[42]

While the famine was raging in Northern Shewa, the Derg, particularly its Relief and Rehabilitation Commission (RRC), failed completely to cope with the situation. It was only due to the intervention of international donor agencies such as the Save the Children Fund (USA), the Catholic Relief Agency and the Christian Relief Development Association (CRDA) that shelter, clothing, food and medicine were provided locally (Haile Kiros Desta 1992). In so far as the government tried to solve the tragic problem of famine at all, it was simply by uprooting and resettling the starving people in Metekkel and Gambela areas – without any previous planning (cf. Chapters 7 and 12 this volume). As a consequence, 2,994 families from Yifat, 224 from Tegulet, 2,321 from Menz, 345 from Selalé and 1,131 from Merhabété were sent to resettlement camps. As a result, the people's confidence in the Derg, which had already been strained by the terrible years of the 'red terror', was completely shattered.43

The Derg, who had failed to manage the problem of famine, was harrassed and forced to fight bitter wars against insurgents in Eritrea and Tigray. It was during this critical period in 1983 that it was forced to implement the National Military Service and Civil Defence Proclamation (Teferra Haile Sellassie 1997: 249–50). This proclamation was not accepted by the population. It was particularly resented by the people of Northern Shewa, especially at the height of the famine. Northern Shewans felt that conscription was the final burden that would destroy them totally. However, the Derg had no alternative than literally to drag young people into national service. It was the beginning of the period of continued forced conscription of young people to fight the civil war, which continued until 1991.[44] The bitter hatred of the people is reflected in the following popular poem of the time:

---

[42] Ibid.
[43] Informants: Ato Mulatu Werqu (Mehal Méda, January 1998); Ato Aklilu Leggesse (Dabra Berhan, December 1998); on Metekkel resettlement, see, among others, Pankhurst 1997. See also RRC 'Sefera' programme file no. 001517 (Shewa Region Archives).
[44] *Mestre Zemene Derg* ('Secrets of the Derg Regime'); see note 21 above.

Close yet Far: Northern Shewa

አሀያም ተነሽ! ልበሺ ሱሪ!
ደርግ አይምርሽም፤ ትእዛዝ አክብሪ፡፡
ምን ዐይነት ዘመን ነው፤ ዘመነ-እምባዬቹ!?
ወልዶ ለዘመቻ፤ ሥርቶ ለመዋጮ፡፡

Hey you donkeys! Get your trousers on!
You'd better obey the Derg orders.
You won't be spared otherwise.
Oh! What kind of time is this, this time of pettiness and apathy?
We are forced to yield our sons to the fighting
And our earnings to the war effort![45]

Thus the people of Northern Shewa were made fearful, hostile, and desperate. It was at this juncture that they became totally defiant of the authorities, breaking the law and resisting any form of co-operation with the regime. That is why the Ethiopian People's Revolutionary Democratic Front (EPRDF) was allowed to pass through Northern Shewa without opposition, an act that facilitated its eventual success in capturing and controlling the Ethiopian capital, Addis Ababa, in May 1991.[46]

---

[45] Informant: Ato Tewabe Ze Yohannes.
[46] *Mestre Zemene Derg* ('Secrets of the Derg Regime'); see note 21 above.

Five

Garrison Towns & the Control of Space in Revolutionary Tigray

JENNY HAMMOND

From the perspective of Addis Ababa and its elite, formed under the empire, all Ethiopia beyond the capital assumed a peripheral character. To the centrality of the capital was opposed the marginality of peasants and countryside. This was not necessarily the view from outside the capital, however. An overview of the whole region not only suggests a 'hierarchy of centres' (Donham 1986a: 24), but also a hierarchy of peripheries. The southern marches were the most marginal, both literally and metaphorically, of the lands incorporated by Menilek at the end of the nineteenth century (Donham and James 1986). The perspective from there tended to homogenize the old 'Abyssinian core' of Tigray-Amhara into a unitary and predatory centre.

The view from present-day Tigray is very different (see Map 1.2, p. 8). Tigray over the last hundred years has been the site of opposing tensions – the centripetal thrust, inherited from feudalism, of sovereignty dispersed to the regions, opposed to centrifugal pressures to establish a centralized unitary state. Since the accession of Menilek, Tigrayans, perceiving themselves not only distinct in language and culture, but also the inheritors of a rare and ancient history, have seen themselves as resisting a determined and deliberate discriminatory process which left them in the last years of Haile Selassie the poorest and least developed province in Ethiopia.

Small towns in Tigray

The study of small towns defies the too easy dualism of centre and periphery. Provincial towns fall anonymously between the two poles of capital and countryside which have dominated both Ethiopian and Western thinking about Ethiopia. All the towns in Tigray are small, but some are smaller than others. Tigrayans define as towns settlements that are so small they would be called villages in the West. What they call villages (*kushet*) are small clusters of farms, which for administrative purposes are grouped together into *tabia*, an area often defined topographically. The Derg introduced the *qebele* as the administrative subdivision in towns, but only the five largest towns in Tigray were large enough to be subdivided into *qebeles*.

The conventional relationship in which the town was the centre of economic, political and administrative activity for a local area, was the norm under the imperial regime and from that period Tigray inherited a hierarchical administrative structure, the point of which was to facilitate expropriation. At the apex was the provincial capital, Mekelle; then the *awraja* (county) towns like Adua, Axum, Endaselassie (Shire), and Abi Adi (Tembien); below that level were the *wereda* or local districts, each with its *wereda* town. The distinction between town and village was, in broad terms, an economic one. The economy of the countryside was based on agriculture; the economy of towns was based on trade. The markets were in the towns, but most of the goods they stocked came from the surrounding rural area. Muslims were not allowed to own land and so became traders and merchants, sometimes weavers, usually settling in the towns. Women without other means of support made their way to towns to scrape a living by selling local beer (*sewa*) or through prostitution. These towns were often very small indeed and the only education and health care was in the larger towns.

Although town-dwellers, especially in the main towns, tended to emphasize the social distinction between town and country, in many ways the urban/rural dichotomy was, and still is, misleading. Tigrayans believe themselves to have been severely discriminated against in that before 1991 there was not a single industry in Tigray, even in the provincial capital, Mekelle. The lack of educational and employment opportunities was a factor in keeping the economy of the towns intimately connected with the rural economy. The small towns were brokers or mediators between the countryside and the administrative centres of Mekelle and Addis Ababa, but also functioned as mini-centres themselves in relation to their localities. If the tracks and paths could have been lit up and become visible they would have shown the usual radial relationship we expect around towns. Many town dwellers had land and relatives in the surrounding countryside to which they went back and forth; peasants came to market to sell small surpluses or handicrafts or charcoal or wood fuel for cash to purchase cooking oil, ploughshares, anything they could not make themselves. Livestock of all kinds wandered the streets. So it is impossible to talk about the towns except in some kind of relation to the rural area, either of interdependence or opposition.

At the same time the larger towns were connected in an economic web to the centre. A glance at the road map before the 1990s makes clear the centralized and expropriatory relationship between Tigray and the capital.

Looking Back on the Projects of the Socialist State

The only all-weather roads until recently, built by the Italians at the end of the thirties on the site of older tracks, join the main towns of Tigray (still very small) in a chain northwards to Asmara in Eritrea, and southwards to the cities of Gondar and Dessie and to the capital, Addis Ababa. These roads both symbolized and facilitated the centralizing and autocratic process which, until the defeat of the Derg in 1991, is the context to any proper consideration of the transformative interactions between small towns and the rural area during the time when the revolutionary movement was consolidating its relationship with rural people. The useful distinction for us to make during this period is the one between towns on the one or two all-weather roads, which for that reason attracted a larger population, especially Mekelle, Adigrat, Adua, Axum, and Endaselassie, and towns connected by rough roads and tracks. The extension of TPLF control in the countryside was a process that by degrees confined the Derg to those towns which were connected to each other by all-weather roads and also to their supply lines in the south. At the same time the dissident forces were extending the network of dirt roads to facilitate connections between the small countryside towns and their own supply lines in the Sudan.

As dissent in one local area after another spread into civil war in the early 1980s, a new political relation in the towns began to supersede the economic relation, inherited from the past, between the areas of agricultural production and the larger cities to the south. Urban communities were still looking both ways, but found themselves caught between the repressive Derg in the towns and the revolutionary process in the countryside. The hierarchical administrative structure, based in different types of town, broke down. The towns remained as physical entities of course, but the political, social and economic relations that had sustained them disappeared under the impact of the civil war and were replaced by a new structure of relations.

During the period 1975 to 1991 the traditional relationship between town and countryside became inverted on the whole. The student movement in the towns which had contributed to Haile Selassie's downfall had been sensitive to the conditions of deprivation and near or actual starvation in which the peasants lived, and their 'Land to the Tiller' slogan had been the inspiration for widespread urban unrest. The students who started the TPLF had been active in this movement and central to their revolutionary strategy was the political work of converting the scattered and poverty-stricken farmers in the rural area of Tigray into a politically-conscious mass able to defend its interests. Thus in this strategic sense the countryside became the centre and the towns the periphery.

The students who went to the wild area of Dedebit in western Tigray in February 1975 were all from urban backgrounds in Tigray.[1] Most of them

---

[1] Although they are customarily referred to as 'students' in the legends already forming about this period, there were exceptions: one had been a member of parliament in Haile Selassie's Shengo, another the Director of the High School in Adua, a third a soldier in the imperial army.

had also been to university in Addis Ababa where they had been influenced by the radical student movement and where they also joined the Tigray National Organization (TNO), a precursor of the TPLF. So the strategic importance of the rural people to the making of a revolution began as an idea in the heads of urban students in Addis Ababa, was argued out in the Dedebit forest in the early months of 1975,[2] but became a political reality only through a long process of negotiation and working with the people themselves.

> The first year of the struggle almost all the fighters were from a purely urban background, let alone all the problems of jungle, shortage of water, wild animals. They had never even heard the roar. Also we had only five guns ... These were our entire arms! There were not only hardly any guns, but no ammunition either. But they had big sacks of books, Marxist books![3]

They were ill-prepared to survive in the countryside. Fighters from this time tell interesting anecdotes about the training given by the peasants to the urban fighters, whose education had not included rural survival skills. If the peasants had not intervened, they probably would not have survived. Local farmers told them where to find water, which wild roots and plants were edible and which were not, how to identify and react to different wild animals. They also rebuked them for inappropriate behaviour in the forest.

> Do you want to throw away your lives? You came here to fight for people, but you are throwing away your lives by laughing, by sitting beside the stream where everybody can see you.[4]

This image of peasants as teachers, although logical in the circumstances, is in itself an inversion of conventional expectations of the relationship between educated and illiterate, urban and rural. The liberation war demonstrated the conflict between different models of social organization. Neglect of the peasant majority had inflamed opposition to the Emperor. The Derg's approach to rural problems was even more authoritarian, despite its socialist claims, whereas the TPLF identified the rural population as its major resource. Before 1991, their basic strategy was to make military, political, social and economic transformation work together to win the confidence and support of the peasants. Although the fighters as they grew in numbers became the mobilizers of the peasant communities, they never became their teachers in any simple way, despite the number of

---

[2] Asgede Gebreselassie, farmer, ex-soldier and founding member of the TPLF, interviewed in Endaselassie in Tigrinya, 16 April 1991. 'They would spend twenty-four hours discussing reading and debating what were the political differences between us and the Junta, or between us and MEISON or EPRP. They would talk for hours about how they should approach the people, how they should organize the people. This was their daily work – politics!'
[3] Ibid.
[4] A founding member of the TPLF, interviewed in English in Hagere Selam, 29 March 1991.

Looking Back on the Projects of the Socialist State

Plate 5.1 New roles for women: ploughing near Awhie *[J. Hammond]*

schools of different kinds the TPLF set up in the liberated area.[5] Nor was it an easy task to win the confidence of peasants whose past relationships with officials and townsfolk had taught them to be suspicious of all outsiders. The fledgling TPLF set up a special department *kefli hezbi*, translated as the People's Department or the Mass Bureau or Department 08, to mobilize the people by addressing their special problems. The burning issue was land. Fighters in 08 had to convince sceptical peasant farmers that the land could belong to them, but that they, the farmers themselves, had to devise a just way of sharing it.[6] It was not enough to complain about injustice and exploitation; they themselves had to help devise alternative systems which could meet their needs and none of these benefits would be permanent unless they worked to replace the state itself.[7]

The TPLF had seen the Derg's 'revolution' as a betrayal of the intense popular movement for political change. They described it as a 'palace' revolution: essentially, little had changed in the field of power relations. One largely Amhara elite had replaced another, but the two big issues, rural poverty and the problems of a multinational state remained unresolved. By contrast, the TPLF began a process of education and training,

[5] The TPLF set up a wide range of schools and training establishments in the 'liberated areas'. These included elementary schools in the small towns, political schools for fighters and cadres, two women's schools, a music school and many agricultural and technical training centres.

[6] Sobeya in Agame was the first area where the TPLF implemented land reform. On 3 March 1997 I interviewed those involved in the first land reform, some of whom were current executives of Sobeya Tabia administration.

[7] Hassen Shiffa, taped interview in English in Mekelle on 8 January 1997, Mekelle.

Plate 5.2 Roman, Aregash & Mebrat, three of the first women fighters [J. Hammond]

both formal and informal, to convince formerly powerless sectors of society of their vision of an alternative society and mobilize them to action.

The organization of elected local committees preceded land redistribution and it was from these essentially practical situations that elected local government councils (*baitos*) developed. These councils worked with the TPLF to draw up the *serit* or 'laws of the *baito*' which recorded and safeguarded the terms and conditions under which land could be held and passed on. They also made new marriage provisions, included minimum ages of marriage and equal rights (in theory at least) to choice of partner, divorce and the division of property for men and women; and they prohibited all discrimination on the grounds of religion, gender or place of birth. The councils also collaborated with the Front on the provision of an expanding network of rural health clinics and elementary schools, all of which had hitherto been associated only with larger towns. So, during most of the 1980s, the countryside was not only the arena of agricultural activity for the 95 per cent of the Tigrayan population who lived there, but also the main arena of the political and social activity of the TPLF.

Awraja towns & *wereda* towns

Government activity, by contrast, was based firmly in the provincial capital, Mekelle, and the more important *awraja* towns, the largest of which were on the main roads. In these were situated the infrastructure of state control: administrative buildings, garrisons, police stations, courts and prisons. Government forces and officials became more or less confined to the *awraja* towns as they came under increasing pressure from guerrilla forces in the surrounding countryside.

The TPLF at this stage had little to gain from military assault on the towns; the towns were strategic for the Derg, but not for the guerrillas. Not only their military strategy, but also their political strategy ruled this out. The successful capture of a town would make them and the townsfolk an easy target for air attacks and thereby alienate the population. Exceptions were towns situated far from main roads and therefore from logistical support, but which were large enough to attract Derg occupation and from which garrison harassment became an obstacle to the political priorities of creating a reliable and supportive rural mass base. For this reason the TPLF fought a campaign from 1978 to 1979 to take Abi Adi, the main *awraja* town of Tembien in southern Tigray. The Derg was forced to abandon it and retreated to Hagere Salam, a small *wereda* town on the top of a mountain range about fifty kilometres from Mekelle. It was ill-adapted to support a large garrison of Derg soldiers, as it had only one small generator, no running water, and shortages of fuelwood. The people of Hagere Selam strongly resisted Derg occupation in 1979 and in the reprisals and pacification that followed many people were killed and houses and crops burned. Although the Derg evacuated the town in 1989, the scars from that time have taken a long time to heal, particularly because the effects of soldiers' assaults on women and the attraction into prostitution of the poorest women had disintegrating effects on family structure.[8]

By contrast, the tiny market towns at the *wereda* level, often at a distance from main roads, had by the mid-1980s been absorbed into the 'liberated' areas. However, government forces did not renounce these areas without a struggle. Sheraro in Shire was the site of a prolonged tussle with the Derg.

> In 1983 the Derg invaded Sheraro again. There had been a heavy war around Sheraro which lasted for five or six days. TPLF pushed the Derg back, so the Derg bombed the market place, killing thirty-one people and injuring many more. Then they brought in reinforcements from Eritrea. This included about forty tanks. At this point TPLF retreated and the Derg took the town. Everyone left the town except for some very old people who were unable to travel. They were treated badly. They were denied the right to go to their land because they were suspected of spying or running messages to the TPLF. The Derg stayed for thirteen months. The people simply left the town and made a temporary settlement in the hills. We had shops; we taught school lessons and went on as normal. But the Derg had many problems. They had no way in and out of the town, because the surrounding area was liberated and they had no food. The Derg was totally encircled, imprisoned.[9]

[8] Gebrehiwot Agaba, interviewed in English in Oxford in April 1995.
[9] Tsahytu Fekadu, interviewed in Sheraro, 12 January 1987.

Garrison Towns & the Control of Space: Tigray

Sheraro demonstrated that a town is defined by its inhabitants and their activities, rather than by its physical structure. This experience taught the Derg not to take isolated towns remote from logistical support, and the TPLF that siege tactics were more effective than armed confrontation or capture. The absence of roads was a contributory factor to the Derg withdrawal by the mid-1980s from most of the rural areas of Tigray. Yet air bombardments continued to be the typical experience of towns like Sheraro during the remainder of the 1980s.

Throughout the civil war only the Derg had airpower, supplied by the USSR in the form of MiG 21s, MiG 23s and helicopter gunships. They used this technology to harass and punish small towns in the TPLF-controlled area. One particularly severe atrocity was a massacre on market day in the market-town of Hausien on 15 August 1988, where bombardments by MiGs and helicopter gunships continued for seven hours with casualties of more than 2,000 people. Yet the absence of the Derg on the ground allowed people in these towns to maintain a cohesive culture of resistance. Every family had air-raid shelters dug out of the road in front of their houses, making walking at night rather hazardous as none of these small towns had any form of street lighting. MiG raids took place only during daylight hours, so that, as bombing raids were stepped up, more and more town activity took place after dark, including markets.[10] After the first few years the TPLF intelligence system became sufficiently sophisticated to enable it to warn small towns of an impending attack, so that the people could, as in Sheraro, decamp into the hills. In the last three years of the civil war, towns like Adi Hagerai and Adi Nebried in the western lowlands, one of the earliest liberated areas, spent increasing amounts of time camping outside the town, at least in daylight.[11] This necessary strategem was disruptive of the agricultural economy and destructive of the TPLF Agriculture Department's projects for food self-reliance.

The towns as frontiers

Within the five main towns, the urban populations became adjuncts of military garrisons and of the paraphernalia of Derg control. Instead of being the centres of commercial and administrative activity for a region, the towns became peripheral, in effect the frontiers. The main towns were for most of the Tigrayan population the site of the enemy in much the same way as towns occupied by colonizing powers. As frontiers, they also became contact zones or places of encounter between different and opposed forces of several kinds, primarily of course the forces of the Derg

---

[10] I visited many of these small towns in western Tigray between December 1986 and March 1987, such as Sheraro, Adi Hagerai, Edaga Hibret, Adi Nebried. All markets were held at night.

[11] Zafu Tsehaiye, woman Chairperson of Adi Hagerai *baito*, interviewed in Tigrinya in Adi Hagerai, 17 June 1989.

and the TPLF. Towns are thus shown to be not essential entities, but a relation, and in this period what had seemed their essential nature in relation to the countryside was radically transformed.

Their reduced function as places of encounter between Derg forces and a subjected population had serious consequences for the towns and for the sense of identity of their inhabitants. Identity is not an essential entity any more than towns are and the people in the Tigrayan towns at this period were the site of multiple and contradictory subjectivities, developing hybrid identities in a 'third space', as Homi Bhabha describes it, in response to the forces of compulsion and intimidation which were constructing their lives on a daily basis.[12]

In the garrison towns the experience of living with repression made them very different from the smaller towns. Perhaps the most extreme example was the 1978 'red terror' campaign by the Derg. This brutal attempt by the government to eliminate opposition, particularly of the two opposition parties, the Ethiopian People's Revolutionary party (EPRP) and the All Ethiopia Socialist Party (MEISON), took the form of arbitrary slaughter of mainly high school and university students in larger towns throughout Ethiopia in 1977.[13] The 'red terror' spread to the five main towns in Tigray in 1978, but not to the countryside or the small towns.

> There were teachers, like Fikre and Fasil who were shot and their students were killed. They were about twenty years old. The mothers were forbidden to weep or to take their children's bodies which were buried in a mass grave. We knew where they were buried and it's now farmland, but since the TPLF took that land for the first time last year, some parents have taken the remains and given them a church burial.[14]

It was certainly an irony that the Derg strategists presumed that the absence of schools, the prevalence of illiteracy and the dispersal of the population outside the larger towns also meant the absence of a radical opposition movement which they would need to eradicate. Yet it was in the villages that intensive TPLF education and agitation created the politically-conscious support which would eventually defeat the Derg.

The prohibitions on visible mourning and the inability of most families to buy back the bodies of their dead children in order to bury them[15] have

---

[12] In a published interview, Homi Bhabha explores some broadly similar ideas in the different context of post-colonialism and multiculturalism. 'The concept of a people is not "given", as an essential, class-determined, unitary, homogeneous part of a society prior to a politics; "the people" are there as a process of political articulation and political negotiation across a whole range of contradictory social sites' (Homi Bhabha 1990: 209).

[13] Zafu Abraha, woman fighter, interviewed in English in London on 15 August 1989; Kidane Asahegne, hotel manager, interviewed in Gondar on 15 March 1991; Garamo, teacher, interviewed in English in Gondar on 15 March 1991.

[14] Berhe Gebremichel, hotel washerwoman, interviewed in Tigrinya at hotel in Endeselassie on 16 June 1989.

[15] Mourning was forbidden on the grounds that it indicated sympathy for counter-revolution. The bodies were left on the streets for a time as a warning to the populace and relatives were allowed to bury the bodies of the victims only if they could pay the price of the bullet that killed them, but this was a rare privilege as the charge could be as much as a month's

Garrison Towns & the Control of Space: Tigray

meant that the experience was repressed in Tigray until after the Derg evacuation in 1989.[16] For those students who did not or could not 'go to the field' to join the revolution the repression worked. From then on many of them tried to avoid political involvement.

Most of the students by that time were terrorized by the 'red terror' so they tried to keep out of politics. Some of them started to support the Derg out of fear and were even recruited as cadres and military officers. Only a few youths continued to support the TPLF.[17]

After I was released I hated politics intensely. I didn't want to be involved in any political affairs. There was a study group in our school for that Party [the Derg party, i.e., the Workers' Party of Ethiopia], but I never went there. In the women's *qebele*, they were asking why I didn't go, but since those politics were the cause of my being in prison, I wouldn't go any more. They continued watching me.[18]

The authorities became increasingly repressive throughout the 1980s. The immediate hinterlands of towns were also under Derg control, but their status was more ambiguous than the towns themselves. Derg troops and police were sensitive about possible infiltration from areas outside their control and this meant that the five large towns, which remained in their control until 1989, were ringed with mines and guardposts. Access to towns resembled border crossings between distinct states. Peasants from the surrounding areas were subjected to body searches, beatings and abuse at checkpoints and were frequently arrested as 'political' prisoners suspected of being in touch with the TPLF. Freedom of movement became increasingly restricted. The towns thus became progressively cut off from the rural area. Natural economic relations were severed, town markets declined, civil transport links ceased to operate and the towns became islands of Derg control within a largely TPLF-held province.[19] Finally, when the roads became too insecure for convoys, nearly all supplies had to

---

[15] (cont.) salary for a teacher.

[16] In the second half of 1991 there were a series of deliberately managed confrontations in different Ethiopian towns between the parents of the 'red terror' victims and those locally responsible for their deaths, such as *qebele* officials, WPE members and local security police. These meetings took place in *qebele* halls where the bereaved, carrying photographs of their dead children, took turns to describe the circumstances in which their children lost their lives and demanded that the accused give an account of themselves. They were broadcast on Ethiopian television three times a week. I saw several broadcasts of meetings in Addis Ababa and Gondar.

[17] Berhanu Abadi, student, teacher, 'underground'member of TPLF, imprisoned for six years in Mekelle Prison until he was released by the TPLF in the Agazi Operation in February 1986, after which he became a TPLF fighter. Interviewed in English at the top of Min Minay valley, Western Tigray, 23 February 1987.

[18] Aster Fitiwy, teacher, Vice-chairperson of the Derg Women's Association, imprisoned in Mekelle prison for two years and seven months, interviewed in English at her home in Mekelle on 8 June 1989.

[19] Aregash Adane, member of the *baito* Department of TPLF, also on TPLF Central Committee, interviewed in English in a mobile camp near 'Beloved Sand' in Tembien on 9 June 1989.

be flown in by plane. A false economy, totally dependent on the garrisons, grew up. Only Derg personnel and employees like teachers and civil servants had salaries. Dependent services mushroomed – restaurants, bars, small shops and hotels, and of course prostitution.

> They gave us no help against poverty, just hand-outs of wheat and flour from foreign places. We became very poor with no ability to do anything for ourselves ... There was no improvement. All the people were gathered in different associations, which were always working in the interest of the Derg. The money of the associations was taken by the Derg, and the people were heavily taxed. Because of this we could not build houses or improve our homes or living standards, so there was much poverty. There was no money, no consciousness; everyone dependent on the wheat of the Derg and a few salaries.[20]

The experience of occupation was broadly similar in all the garrison towns. The presence of government troops and authorities, particularly the hated and powerful Workers' Party of Ethiopia (WPE) gave rise to pressures of an opposing and incommensurate kind. This was a war zone, but it was also a contact zone. Of course the population learned to 'live with the Derg', as they express it. The needs and demands of the garrisons and Derg representatives necessitated not merely copresence, but interdependence and interaction of all kinds, generating a multitude of interlocking understandings and practices on a daily basis but within radically asymmetrical relations of power. In general terms it was a life of coercion and fear of reprisal for real or imagined resistance activity.

> In 1984 with the establishment of the Workers' Party of Ethiopia (WPE) a lot of Tigrayans were being sent to prison. I was frightened about what would happen to me. Many who had been elected in Farmers, Womens and Youth Associations were being sent to prison. Here in Mekelle there was an underground organization. The officials suspected the Tigrayans of resisting the rest of the party so they arrested them as the best solution. I knew one person, the chairman of the mass association, who was tortured at that time. He was hung upside down and beaten with stick on the soles of his feet. He confessed to being a member of TPLF, although he wasn't.[21]

There was a marked escalation of rape, enforced prostitution, and violence to women.[22] The townspeople resorted to different survival strategies. Some collaborated fully with the authorities; government employees on the whole

---

[20] Lemlem Mahmas, wife of Derg militia and *sewa*-seller, interviewed in Tigrinya in Axum on 5 June 1989.

[21] Aster Fitiwy, 8 June 1989

[22] Abrahet Teklemuze, prostitute, aged twenty, interviewed in Tigrinya in Endaselassie on 15 June 1989; Rahma, prostitute, aged twenty, interviewed in Tigrinya in Endaselassie on 4 June 1989; Lemlem Mahmas, aged 28, wife of Derg militia, interviewed in Tigrinya in Axum On 5 June 1989; Aster Fitiwy, teacher and ex-vice-chairperson of Derg Workers' Association in Mekelle, interviewed in English in Mekelle on 8 June 1989. For fuller testimonies see Hammond 1989, 1999: 144.

equated survival with their salaries; women became prostitutes or capitulated within more regular relationships with officers and officials to whom they bore children.

There were women whose husbands were killed, and who lived all the time in terror. After their husbands died they used to marry the security officials. They never loved them, but said, 'Since they've killed our husbands they'll probably kill us.'[23]

Others kept their heads down and tried to escape notice:

> In every part of society we had bad memories. We tried to erase these by joining in government infrastructure, so many clubs. This made it easy for the officials to get their hands on the girls easily. They mentioned a club and pulled out whatever boy or girl they wanted. The officials organized it so that they could always track down a girl they wanted and know where she was. We were always occupied with organizational work and they were always calling us out of the safety of our families, so we were exposed to whatever they wanted to do.[24]

Every strategy involved concealment, evasion and the adoption of false identity to accommodate to the occupying culture. Yet, although the rural community was demonized as supporters of TPLF, the majority of townspeople had branches of their families in the villages. Many had relatives from both inside and outside the town who were TPLF fighters.[25] These relationships as well as the natural economic ones mentioned earlier had to be denied and concealed. The wives or parents of young people who had escaped 'to the field' to become fighters were particularly vulnerable to the authorities. For that reason many of the early fighters adopted field names to protect their families. However the arbitrary paranoia of the Central Intelligence Department (CID) meant that imprisonment, torture and death were a routine hazard.[26] Mekelle prison was bursting at the seams.[27]

---

[23] Aster Fitiwy, 8 June 1989.

[24] Askale Yohannes, mother and ex-bank clerk, interviewed in a mixture of English and Tigrinya in Mekelle on 8 June 1989.

[25] The TPLF made a distinction between fighter and soldier, both shaping and reflecting the significant distinction of definition, aim, and discipline. The Tigrinya word for 'fighter' used by the TPLF was *tegadalai*, *tegadalit* (f.).

[26] Mulugeta Gessesse, political administrator of Derg Farmers' Organization; Berhanu Abadi, supporter of TPLF, student, later teacher; Haile Abraha, teacher under the Derg, promoted to PR job in Ministry of Education; Assefa Bekele, teacher under the Derg, Chairperson of *qebele* Association; Sahle Gebregziber, supporter of TPLF, student at Addis Abeba University; Tesfai Lema, employed in Ministry of Culture in Addis Abeba. All were Derg prisoners in Mekelle prison and released by the TPLF in the Agazi Operation in February 1986. I interviewed them in English in caves in western Tigray on 4 January 1987. For selected transcriptions of their testimonies see Hammond 1999: 33.

[27] Mulugeta Gessesse, Berhanu Abadi, Haile Abraha, Assefa Bekele, Sahle Gebre Egziher, Tesfai Lema, interviewed on 4 January 1987 in the caves of the Propaganda Bureau in western Tigray. These men, all Tigrayan but of differing political affiliations, had all been prisoners in Mekelle Prison until released in the TPLF Agazi Operation in February 1986.

Looking Back on the Projects of the Socialist State

It was very common for women to prostitute themselves in the hope of release for their imprisoned relatives.[28]

Many torture survivors I have interviewed were loyal employees and supporters of the Derg. Tigrayan 'intellectuals' were particularly under suspicion, although 'intellectual' was usually used to denote educated people. They were to discover that identifying themselves with the politics and institutions of the Derg brought no guarantee of protection.

> The Youth Association chairperson was arrested and tortured; they told him to tell them the names of the TPLF members. 'I know Aster. She is a member of the TPLF. She is contributing five *birr* a month to the TPLF.' But it wasn't true. I was the vice-chair of the [Derg] Women's Association. They called me and accused me: 'Are you a member of the TPLF?' 'No, I am a member of COPWE, and I am working in the Association.' 'You are a member of the TPLF underground.' But I was not. Then when I said no, they never believed me. They beat me and kicked me cruelly. They tied my hands and my legs and put sticks under my knees and hung me upside down. So I confessed to being a member of the TPLF because I could not bear their cruelty. 'Yes, you are controlling other TPLF members. Tell us their names.' But I didn't, because I didn't know them. They tortured me continually for four days, and my right hand was painful for two to three months afterwards. I felt pain around my womb, so I was admitted to hospital and then they sent me to the main prison.[29]

> They started to torture me and I couldn't stand the torture. I decided if they refused to accept I was not a traitor, then I would become a member of that forbidden organization, starting from now. 'Who else?' they said. 'No one. I was a member all alone.' 'Give us the names of all the others, of your comrades.' I told them so many times, but they tortured me again and I gave them the names of all my friends.[30]

The TPLF 'underground'

On the other hand, it could be argued that government paranoia was justified. The TPLF did operate clandestinely in the Derg-controlled towns. Their political objective was to mobilize support and encourage a political consciousness in their supporters appropriate to their aims. They also had the more mundane objective of the acquisition of essential supplies. Their military equipment was acquired through engagements with government forces or from raids on army installations, but they also

---

[28] Yeobmar, prostitute, aged sixteen, interviewed in Tigrinya in Endaselassie on 4 June 1989.
[29] Aster Fitiwy, 8 June 1989.
[30] Assefa Bekele, interviewed in caves in western Tigray on 4 January 1987. He was a teacher and chairperson of the *qebele* under the Derg until he was arrested and imprisoned in Mekelle prison. In February 1986 he was among those released by the TPLF in the Agazi Operation.

needed non-military equipment and materials which could not be provided by the small towns in the liberated area. Convoys managed by the Relief Society of Tigray (REST) brought in goods from the Sudan, many of them humanitarian supplies like food and pharmaceuticals, but also including diesel and spare parts. For other routine essentials, however, like torches, batteries, soap and sanitary towels for women fighters, they had to rely on increasingly dangerous missions to the Derg towns.

The TPLF had networks of agents working in cells, often unknown to each other. A wing of TPLF, later to become a fully-fledged department, was formed in the first year to infiltrate the towns for information, for supplies, for 'agitation' against the government and to spread propaganda on their own behalf. They became skilled in disguise as peasants, traders or even beggars – an ironic counterpart to the different kinds of 'disguise' adopted by townspeople. The high school students in Mekelle and Adua were a strong source of support. Zafu, in the following extract, describes the double life of working in the underground in 1977. A few months later she fled to the TPLF with many others, narrowly escaping the 'red terror'.

> We had to collect information. We had many meetings. The sort of questions we discussed were: Why are we oppressed? What is TPLF? What are its aims? How can we help our organization?[31] TPLF's only help at that time came from the underground supporters, so we raised money for TPLF through knitting. Most of us girls used to knit – making jumpers, bedcovers, pillows to sell. We also had to be in the Derg's Women's Associations. All the women were told they had to spin cotton and make garments to be collected and sold to make money for the Derg. But, underground, we were getting the materials from the Derg and selling half for the TPLF and only half for the Derg. We also made tea and coffee and different kinds of bread and sold them. Our mothers too gave us money to make things to sell and gave food for TPLF, like flour or lentil stew. If it had been discovered we would all have been arrested, and our mothers too.[32]

In fact many young students working for the TPLF's clandestine operation paid dearly for supporting a vision of a different and more democratic society.

> Before I was married when I was a 6th Grade student in 1975, I was an underground TPLF contact for three years until 1978. Then I was arrested and sent to prison in Tembien. I lived in Abi Adi then. Seven of my friends from school were killed. The Derg used to come to the market place and kill people.[33]

---

[31] The common appellation for the TPLF by its supporters throughout Tigray, both during the civil war and still today, is *wedebna* ('our organization'). In the extracts quoted in this paper most if not all references to 'TPLF' would be *wedeb* or *wedebna* in the original Tigrinya.
[32] Zafu Abraha, woman fighter, interviewed in English in London on 15 August 1989.
[33] Asgaria, interviewed in English and Tigrinya in Adua on 12 June 1989.

In these various ways people suffered displacement from their normal identity and relationships.³⁴ The effect was a kind of schizophrenia, not helped by the cultural repression of the government's enforced Amharization policy, inherited from the imperial regime. Tigrinya was forbidden in schools or for any written or formal transaction. This was unenforceable in the rural area, where in any case there was widespread illiteracy, but in the towns Amharic largely displaced Tigrinya in any public use. It therefore made a deep impression on students that the TPLF wrote their programme and leaflets in Tigrinya, unfamiliar to educated Tigrayans. A Tigrayan university graduate of that time, about to join the underground, found written Tigrinya an unfamiliar challenge.

> I found it hard to understand it! This was the first clandestine programme I had ever had in my hand and it was in Tigrinya. It was duplicated. There were new words and new expressions, so it took me a number of days to read and understand that programme. I was happy. I felt honoured in fact. I had that feeling that I was about to do something very important, that I was about to be engaged in a very difficult task.³⁵

For those who were able to escape in time, the 'red terror' was the cause of hundreds of young people fleeing to the field and brought to a close the first period of TPLF's clandestine activity.

> Before the 'red terror' I was working with the TPLF as a cell member, but afterwards the cell started to deteriorate. Even though I was a TPLF sympathizer there were so many spies. I believed in TPLF's politics, but I had to keep it very secret. Although my behaviour was very restricted, I tried my best. The Derg was propagandizing about the TPLF, 'If you fall into their hands they will torture you, they will cut off your hands, they will brand you with a hot iron.'³⁶

Through the 1980s the TPLF's subversive activities in their second phase became much more sophisticated as they became much more hazardous.

TPLF 'spectaculars'

Town and country were opposed territories and the Derg and the TPLF made forays into each others' territory for strategic campaigns. Derg campaigns were conventional in military terms, using mechanized and infantry brigades and airpower. It was not until the Battle of Dejenna in 1988 that the TPLF began consistently to get the upper hand in military engagements. However, before that time they occasionally penetrated the

---

³⁴ Lisane Yohannes, member of TPLF underground in Gonder, interviewed in English in Khartoum on 9 May, 1991.
³⁵ Lisane Yohannes, 9 May 1991.
³⁶ Askale Yohannes, 8 June 1989.

Garrison Towns & the Control of Space: Tigray

towns in a different way from their routine clandestine missions. They were able to turn the towns into an arena for spectacular engagements which they turned to propaganda advantage. These provided vivid instances of small but successful confrontations with a hostile and brutal administration, which raised the morale of supporters and began to create those myths of invincibility among townspeople which much later in the war were to become a factor in demoralizing the army.

I will give three brief examples. All had practical and immediate aims and are well-documented factually within Tigray, but operated in the public consciousness in a number of versions, like myths.

The first happened in 1975 and was the first operation of the fledgling revolutionary movement only a few months after its founders went to Dedebit to start the revolution. Musie, one of the most important members, was captured in Adi Daro and imprisoned in Endaselassie. In a carefully planned and executed operation, his companions seized him, shackles and all, and carried him off into the wild. The fact that it was the height of the rainy season, that the countryside was deep in mud, the rivers in flood and 'Sehul' Gesesse Ayele, the head of the band, barely escaped drowning added details to the legend, but the fact alone was enough to serve as a propaganda coup for the TPLF.[37]

The second, a month later, was the spectacular robbery of the bank at Axum.[38] Again, this was an example of precision planning and timing. The rescue of Musie had been carried out by fighters moving in from the outside; the Axum Bank operation was carried out in collusion with underground agents in Axum itself. It was successful on several counts – it was carried out without mishap, although there were some unpredictable turns of event which make it very funny to hear in detail; they carried off a huge amount of money which financed the organization for a considerable time and enabled it to expand its recruitment; it provided a second propaganda coup of even greater proportions and made the TPLF well-known and respected throughout Tigray. Because Axum had been the cornerstone of Ethiopian religious and political traditions, it was a particularly significant choice for such an operation. The incursions of TPLF cut across Axum's traditional and mythic status.[39] In fact the story of the Axum Bank opera-

---

[37] Asgede Gebreselassie, interviewed in Tigrinya in Endaselassie on April 16, 1991, was the main source for this information. He was a founder fighter of the TPLF in Dedebit and justly reputed for his exceptional memory. Other sources were: Yemane Kidane, interviewed in English in Mekelle on March 25, 1991; Aklilu Tekemte, interviewed in English in Mekelle on February 6, 1991. Fuller eye-witness and participant accounts may be read in Hammond 1999.

[38] Asgede Gebreselassie, 16 April 1991.

[39] Axum's historical role as the centre of the Axumite Empire has left an astonishing heritage of ancient monuments, including the famous stelae or obelisks, which have been central to the insignia of TPLF. Axum is also the sacred centre of the Orthodox Christian Church, where the Church of St Mariam is sited, the Treasury of sacred objects, and the alleged Ark of the Covenant. The Ethiopian monarchy claims descent from King Solomon and Queen Makeda (Saba or Sheba) through their son Menilek I and Axum was the traditional seat of her power.

tion has gained mythic status of its own; it was taught in fighters' training camps as the heroic story of Deseligne the Mule, after the organization's famous first mule who carried away the loot on his back.[40] The stelae, the Church of St Mariam and the Axum Bank are all monuments of different kinds which reverberate in important and overlapping ways in the culture of Tigrayan people.

The third example of an urban spectacular was in February 1986, when a small force of fighters penetrated Mekelle and, through a series of imaginative diversionary tactics, were able to release 1,800 political prisoners from Mekelle Prison. It was called the Agazi Operation after one of the founder fighters who had been killed a few years earlier.[41]

These operations and others were extremely important in building up a reputation for the revolutionary movement which contrasted in every way with that of the Derg. As time went on stories began to filter into the towns of defeats of landowners, disbanding of bandit gangs, land reform, democratic reforms, and so on which, whether they could be substantiated in fact or not, provided a vivid contrast with the daily experience of living with the Derg garrisons. Of course there was scarcely a family that did not have members who were fighters and in Tigray the TPLF were not only believed to be efficient, but also humane. Their reputation even managed to penetrate Addis Ababa, on the whole well-instructed by Derg propaganda. In Addis during the time of the Derg, an efficient blue and white private minibus/taxi service was introduced. They were popularly known as *weyala* (a corrupt name given to TPLF fighters) because they were so swift and so quiet; the taller taxis, on the other hand, which had two benches facing one another on which people sat with their knees almost touching were called *weyeyit,* after Mengistu's forced political discussions.

The Derg evacuation

The opposed nature of the Derg and the TPLF was also reflected in their military tactics. The Derg had a huge conventional army, was armed and supported by the USSR and advised at different times by the Cubans, the East Germans, the Israelis and the North Koreans. It had mechanized brigades and airpower. The TPLF on the other hand had only the weapons it had captured from government forces, including anti-aircraft weapons during the 1980s, but had no heavy artillery or airpower. This resulted in a kind of stalemate. In 1985, for example, the military government sought to exploit the devastating 1984 to 1985 famine by launching the Eighth Offensive, a three-pronged attack using mechanized and

[40] The story of Deseligne the Mule was told me by Mebratu in Mai Hanse, western Tigray, on January 21, 1987. Asgede Gebreselassie gave me the factual version on 16 April 1991. See Hammond 1999: 109, 299

[41] Drar Gebreyesus ('Ambassa'), Commander in Agazi Operation, and Berhane Aberra ('Meley'), Commander of diversionary force. Both were interviewed in Tigrinya in Mai Humer in western Tigray on 30 January 1987.

infantry brigades, backed by airpower, against the rural areas in the south east, centre and west of Tigray.[42] This campaign did enormous damage and further alienated a hostile population, but it could not in fact win against the guerrilla tactics of the TPLF which, supported and protected by the Tigrayan people, continuously ambushed and harried the Derg troops. The roads became increasingly insecure for the army who were forced to retreat to the garrison towns. Yet it was hard to see how the TPLF could 'win' or how such tactics could gain its ultimate objective of final defeat and replacement of the military government by a democratic alternative.

After a period of political and military stagnation in the mid-1980s, the TPLF conducted a prolonged reappraisal and reorganization of its military strategies and tactics. These were first put to the test in March 1988 when it began an offensive against the major Derg-controlled towns and drove government forces out of Endaselassie, Axum, Adua and Wukro. The Derg evacuated Adigrat and retained its garrison only in the provincial capital, Mekelle. In the capital the goverment started a massive recruitment drive in preparation for a retaliatory offensive, but the TPLF response was to evacuate the newly-won towns after three months to protect their inhabitants from attack on the ground and from the air.[43] However the Derg army suffered a string of defeats. 'They had a problem of commanders and other logistical problems. They had an untried army, freshly recruited soldiers, and the logistical support was quite poor and therefore slowed them down.'[44]

The Derg forces, after briefly reoccupying and evacuating in turn the towns of Axum and Adua, retreated to Endaselassie and consolidated their position. Axum and Adua were immediately reoccupied by the TPLF. On 10 February the garrison left Endaselassie under orders to recapture Axum, but, strung out on the narrow mountain road, was attacked, totally defeated and dispersed in a nine-day battle. On the 19th the Derg retreated from Adigrat to Asmara and on 21 February from Humera in western Tigray to Gondar. On 27 February the Derg evacuated Mekelle, leaving Tigray to the administration of the TPLF.

In the eight days between the Derg defeat at Endaselassie and their evacuation of Mekelle, witnesses reported 'extreme tension' among the remaining Derg forces in Mekelle. Askale Yohannes, a book-keeper in the Bank of Ethiopia, was told in confidence to expect an attack at short notice and that it was a matter of duty 'to save the money in the bank'. Both deposits and employees would be transferred to Addis Ababa.

> This was a secret. No one must know it or it would create chaos. They told us this on the Saturday and gave us an appointment for Monday at eight in the morning. They cheated us. They all left on Sunday, so this

---

[42] Disaster Monitor No 3, June 1985, published by the TPLF.
[43] Gebru Asrat, TPLF General Command, interviewed in English at Asrega Training Camp, western Tigray, on 2 June 1989.
[44] Gebru Asrat, Asrega Training Camp, 2 June 1989.

Looking Back on the Projects of the Socialist State

> meant that many workers who were told that they would leave, didn't leave ... Then the Amharas went on foot to Dessie, Asmara and Addis, but all the Tigrayan workers refused to go. Actually it wasn't state money but that of the SOS, Ethiopian Orthodox Church, Ethiopian Evangelical Church, merchants, and the Ethiopian Cartological Secretariat, as well as the private money of workers. I myself had 300 *birr* in that bank.[45]

Townspeople describe a scene of deception, disorder and panic. Many people suspected an imminent attack by the TPLF, but reactions depended on individual political affiliations. Officials, Party members and civil servants feared for their lives if the TPLF took over the town.

> Everyone who was not in the know was being tortured mentally. They saw them streaming to the airport by car and on foot. The household equipment of the key persons, the colonels and the generals, was all packed up and taken. The higher ranks and officials took away all their household furniture, but the lower ranks were not allowed to take anything to Addis. Some other members of the Party left their children and their homes and went alone to the airport, crying and very frightened. They were never told the secret that they were pulling out of Mekelle and going to Addis. They thought that the TPLF would kill them if they stayed behind. They went to the airport, some holding suitcases, some holding their children's hands, and some children were left behind. When they reached the airport the policemen and guards were kicking them and calling them names. The army chiefs listed the prostitutes' names instead of the members of their Party. The leaders called the prostitutes' names for the airplane and the party members were kicked by the guards. Many wives are now left behind because they are not Amharas.[46]

Meanwhile, in parallel with the military struggle, the TPLF had been making political preparations for the expansion of the war south of Tigray. The leadership had been confident of liberating the whole of Tigray and now they expressed confidence in the project of spreading the revolution to the rest of Ethiopia.[47] They had been co-operating with the Ethiopian People's Democratic Movement (EPDM), a mostly Amhara liberation movement operating in Wello and Gojjam, since it splintered from the EPRP in 1981.[48] At the beginning of 1989 the two organizations formed a united front called The Ethiopian People's Revolutionary Democratic Front (EPRDF). Long-term Oromo members of the EPDM also decided to form their own organization and were joined by Oromo prisoners of war who were hostile to the Derg and were willing to fight for a democratic political system. After months of discussion and argument, they formed a

[45] Askale Yohannes, 8 June 1989. SOS is a children's charity set up in 1984–85.
[46] Aster Fitiwy, 8 June 1989.
[47] Meles Zenawi, interviewed in English in a mobile camp on 9 June 1989.
[48] The EPDM has been renamed since the end of the civil war in 1991 as the Amhara National Democratic Movement (ANDM).

third member of the coalition, the Oromo People's Democratic Organization (OPDO).⁴⁹

Life after the Derg in the liberated towns

The Derg had evacuated Tigray, but the problems of occupation remained. Some of these were in the nature of psychological scars which, as in the rest of Ethiopia, would take many years to heal. However many more immediate problems faced the incoming TPLF. The evacuation had split the population roughly along class lines. The great majority, although not all, of those who had salaried jobs for the government – teachers, health workers, skilled workers, civil servants and officials – left with the troops and Derg Party members, uncertain otherwise how they would survive. In many cases they left their families behind. Government troops in Mekelle had destroyed with tanks the electricity generating station, so there were problems of power and water supply.⁵⁰ This in turn affected the functioning of the hospital, where the most serious implications were for the refrigeration of drugs.⁵¹ In Adigrat they had burned the clinic to the ground. Rings of anti-personnel mines at random intervals up to five kilometres from the town were killing and injuring herders (mostly children) and their animals. The towns, under virtual siege, had survived through services to Derg garrisons and officialdom. With the sudden departure of the military and civil administrations this dependent economy collapsed. Economic links between towns as well as with the countryside took some time to restore. Some sectors of the urban people were deeply suspicious of the TPLF. Aster, a teacher and Party worker for the Derg, had been nevertheless arrested by government authorities and tortured in Mekelle prison:

> I was there during the Agazi operation, but refused to go with the people who released the prisoners, because I didn't know anything about them and I thought they would kill me ... We turned off the light and hid ... After I was released I hated politics intensely ... I decided not to go with the Derg but to stay with the TPLF, but even now hatred for politics never leaves my head... ⁵²

---

⁴⁹ A fourth organization in the EPRDF coalition was the Ethiopian Democratic Officers' Revolutionary Movement (EDORM), composed of Derg officers who were prisoners of war. The TPLF became very skilled in using the PoW camps to re-educate politically the inmates and win them to the revolutionary cause.

⁵⁰ The EPRDF learned from this experience of the destructiveness of the Derg. Two years later when EPRDF forces were preparing to capture Bahr-Dar and Gondar they appealed by radio to all skilled workers in the factories and power installations, to staff and students in hospitals and to teachers and students in schools and colleges to protect their workplaces from Derg sabotage. At the same time they promised to continue paying their salaries. I was present at some of these salary pay-outs in Bahr-Dar.

⁵¹ Dr Yemane, interviewed in English at Mekelle Hospital on 10 June 1989.

⁵² Aster Fitiwy, 8 June 1989.

Looking Back on the Projects of the Socialist State

The bank clerk, Askale Yohannes, a TPLF supporter, had a very different attitude to the incoming TPLF:

> After six hours the TPLF came. First of all I couldn't find the words – they came through the castle at about six in the evening in a single line. The light was getting dim and there was no electricity. Many people expressed their joy and they were ululating. Everybody came out of their houses and showed their joy. From that instant we believed that they could destroy the thieves and agents, and as time went on the cheers and ululation increased. They came down, men and women fighters together in a single line, and dispersed to different corners. 'Everyone be calm. Be settled. We are here now. Be glad.'

Despite the inevitable reductiveness of brief extracts from long and complex life histories, these two snapshots by two educated women, one a teacher, the other a bank employee, one skulking in fear of the guerrillas, the other describing their welcome, are some indication of contrasting responses to the experience of living in a Derg town.[53]

Yet the towns had to be fed and administered. In the countryside the TPLF had mobilized the people to share in the decision-making and in the solutions to problems. The TPLF strategy in the towns, drawing on their practical experience in the countryside, was to tap the resourcefulness of the townspeople in much the same way. On the whole the townspeople look much more sophisticated than the traditionally-dressed farmers from the countryside. They tend to wear Western dress; the women have straightened hair; the more educated speak English. Yet, compared to the self-reliance and political consciousness of the long-liberated rural areas, the townspeople in 1989 had the mental habits of dependency and fear. My research in the towns in May and June of that year provided many examples of the inversion of the conventional relationship between town and country. In one town meeting I attended in Mekelle the townspeople were being addressed by peasants from the surrounding countryside, where they had been running their own affairs for some years. They described their own experience of self-reliance and self-determination and told their audience in no uncertain terms to pull themselves together and start thinking how to solve the practical problems they faced. Problems were gradually solved in a typical TPLF manner through widespread consultation, not just with leaders but in meetings involving the whole of the population, category by category, merchants, workers, youth, women and so on.[54]

Out of these meetings a series of solutions began to emerge. By the May following the February entry of the TPLF into Mekelle, the links between the towns and the surrounding countryside were beginning to be restored. Restrictions on freedom of movement had been lifted and farmers were

---

[53] For much longer extracts from these testimonies see Hammond 1989.
[54] Tadelle, TPLF Administrator of Mekelle until the formation of the *baito*, interviewed in English in Mekelle on 10 June 1989.

coming in large numbers to the markets. In June I observed for myself the expansion of Mekelle market week by week, not only in terms of numbers but also in the range of goods available. Residents of the other recently-liberated towns told the same story. A woman community leader in Adigrat observed that 'under TPLF the most important thing I have observed is that the rural area and the towns have become interconnected again.' She continued:

> Before, the rural people had a problem. If they came here they were always questioned by the soldiers if they tried to enter the town. If they didn't have pass papers they were often arrested and imprisoned. So the relation between town and rural areas became very distant. Now we are closer. Relatives visit each other, and country people come to bring their produce to sell in the market. We are creating a hand to hand movement between the towns and the villages.[55]

However, until national reconstruction began after the final defeat of the Derg, there was a shortage of jobs, a shortage of resources to pay workers and a shortage of essential goods. In the rural area, although the TPLF was the *de facto* state, there had never been enough resources to pay salaries. The large body of fighters who staffed the TPLF Departments and 'worked with the people' as health workers, elementary school teachers, agricultural advisers or local administration officials, received their 'keep' at a minimal level in kind, either from the TPLF itself or, if they were based in the commmunities as teachers or health workers, through the *baito*, the local elected council.[56] The same system was extended to the newly liberated towns. There were only four conventionally-trained doctors/surgeons for the four million or so population of Tigray. They were now brought from the field hospitals into the town hospitals. The three hospitals (in Mekelle, Adigrat and Axum) which before had been open only for the military were now opened to civilians, staffed by TPLF trainees.[57] The same system was used to staff elementary schools, but there were never enough trained teachers to reopen the secondary schools and shortage of resources kept all facilities at a minimal level for the remaining two years of the war.

In June 1989 a credit system set up by the TPLF encouraged merchants to buy a fleet of trucks to begin operating across the Sudan border and

---

[55] Radiet Gebretensai, interviewed in Adigrat on 11 June 1989
[56] The concept of 'fighter' (*tegadalai,-it*) under the TPLF was very different from 'soldier' (*watader*). Although all fighters were armed and given basic military training, they were not all combat fighters. The essential criterion for a fighter was dedication to the liberation struggle in whatever role s/he could contribute. After the first few years those with education were usually employed in one of the departments.
[57] From the late 1970s the TPLF set up first battlefield clinics, then field hospitals and clinics, including a training hospital. From the early 1980s these were extended to a network of clinics for the civilian population based mostly in small towns, which were staffed by fighters with at least 6th grade education, who were given an initial six months training in the diagnosis and treatment of the main epidemic diseases and in first aid. Every year there would be additional training of two or three months.

between the towns. However, some goods, like metal products, were in such short supply they needed to be manufactured cheaply within Tigray itself. The scarcity of metal goods may have been connected, at least in part, by the arranged exodus of Falashas to Israel. Falasha communities, mostly in Gondar and some in western Tigray, were among the chief producers of metal goods, particularly ploughshares. Metalworkers, like potters and other handicraft workers, had long been victims of superstitious discrimination and therefore tended to belong to certain alienated communities. Outside these groups, there were strong prejudices against taking up metal-working as a trade. A TPLF project in the now-empty Mekelle prison attempted to solve several of these problems at one go by establishing the prison as a metal-working training centre. The raw material was furnished by defunct military vehicles and old oil drums. A payment was made to the trainee blacksmiths as an incentive for those not from traditional metal-working families. By the time I visited it early in 1991 the project was in full swing, turning out not only ploughshares in quantity, but cooking stoves, buckets and other metal objects.[58]

If progress on material and economic problems was necessarily slow, the re-education of the townspeople was moving ahead with some speed.

> During this period the people didn't know about TPLF, so they had to be educated about them – their aims, what is democracy? What is a *baito* – a people's *baito*? Therefore it was necessary to establish a temporary *baito*. It would have been very difficult to establish the real *baito* straightaway, because of the *qebeles* and the people who worked in the *qebeles*. So if you want to organize a *baito* you have to clean up all this, you have to know who is for and who is against the Derg. After all this the people at least knew one another and it was on this ground that the real *baito* was established.[59]

I spoke earlier of the Derg and the TPLF having different models of social organization and therefore different priorities for social, economic and political action. The practical experience of the townspeople after 1989 brought these differences home to them. Consultation and participation were unfamiliar phenomena to them and habits of passivity took time to shift, but the year's preparation for the first *baito* was a time of significant change. Gebru Asrat described this time as 'laying the basis for development':

> At present we cannot solve the majority of the problems which are facing the population; we cannot solve the problems of drought and backwardness, the low level of education and skills. Therefore our present work is laying the basis for development.[60]

---

[58] Metal workshops in ex-Mekelle Prison, visited on 16 February 1991
[59] Abrahet Woldu, pharmacist and member of the Controlling Commission of Mekelle *baito*, interviewed in English in Mekelle on 19 February 1991.
[60] Gebru Asrat, Secretary for Tigray and TPLF Central Committee, interviewed in English in Mekelle, on 10 April 1991

Garrison Towns & the Control of Space: Tigray

The main way of laying the basis for development was the formation of the *baito*. These local elected administrative councils and the executives they elected had long been the foundation of political and administrative life in the countryside, including the small *wereda* towns. The *baito*, the *baito* laws (*serit*) and the evaluation of tasks and progress (*gemgem*) were the triad on which rural participatory democracy was founded. The liberation of the larger towns made it possible to start introducing the same system into the main towns. The TPLF interim administrators saw the *baito* as 'a means of rebellion and struggle' against the old order and they were convinced that a somewhat higher standard of education, advance information about rural *baito*s, and the experience of undemocratic administrations under the Derg would make combine to make progress towards local democracy in the towns much swifter.[61]

In Mekelle an elected assembly elected a temporary *baito* to be responsible for administration until a proper *baito* could be established. Preparation began in October 1989 and the first real *baito* was in place the following year in November 1990. The *baito* assembly of four hundred elected representatives first had to debate the new laws, drawn up beforehand by an elected committee, and then elect an executive to carry out local administration. Laws prohibiting discrimination on the grounds of religion, gender, or place of origin were finally accepted. The fiercest arguments were on the marriage age and on property shares in the event of divorce. As in the countryside many years before, it was the women who were the force for change in the marriage laws. The new Women's Association at first encountered some resistance from women because of their negative experience of the Derg Associations, but by the time the assembly convened it was strong enough to lead the women's struggle. About 40% of the assembly were women. The men wanted the minimum legal age of marriage to be no older than fifteen; women wanted it to be eighteen to safeguard the education of girls. The women eventually won the vote. The other confrontation between women and men was on women's share of the wealth of the family, especially in the event of divorce. Muslim women were particularly active in this debate.

> The most important point was her participation in gaining wealth. Muslim women participate more than Christian women in shops, in the market, in the terrible business of selling small things all day long in the market. If he is a weaver, she has her own part. He does the weaving; she does all the rest of the process.[62]

Muslim and Christian women 'fought together', in the words of the Chairperson of the Women's Association, and the assembly finally accepted that all property should be divided equally between husband and wife on

---

[61] Tadelle, interim TPLF Administrator of Mekelle from February 1989 until the formation of the *baito*, interviewed in Tigrinya in Mekelle on 10 June 1989.
[62] Hiwet Ayele, teacher and chairperson of Mekelle Women's Association after 1989, interviewed in Mekelle on 10 April 1991

divorce. This had been included in *serits* (codes of law) in the rural villages for some years.

There was a marked contrast between my first visit to Mekelle just after the departure of the Derg and a second visit two years later early in 1991. The war fronts had passed south first to Wello, then to Gondar and Gojjam. Tigray had had two years of peace, except for air attacks, and these had decreased since the disintegration of the USSR had forced a tighter armaments budget on the Derg. On the first visit the twin symptoms of dependency and fear in the people were still marked. My subjective impression on the second visit was that even their posture had changed; instead of scuttling down the streets, as if afraid to be noticed, people walked upright and looked you in the eye. In some areas of town people were stopping me to make sure I knew that where I was walking freely had previously been restricted to Derg officials, soldiers and Party members. They spoke of the *baito* with pride.

The end of the civil war opened up the towns to many changes. The constitutional change to a federated Ethiopia had obvious implications for towns in Tigray. Regionalization itself marked a redefinition of relations between the old centre of Addis Ababa and the regional periphery which is likely to modify the importance of the capital and, along with elections, the power of the traditional governing elite. Mekelle was now the seat of state government and of the regional council. The war had been turned into an opportunity to bring benefits to the Tigrayan people in the rural areas, despite their sacrifices. However, real developmental improvements in agricultural production, education and health care had had to wait for the increased resources made possible by political and economic stability in peacetime.

So what effect did freedom from war, and relative peace and stability have on towns in Tigray? The towns' roles as economic and administrative centres began to be restored after the departure of the Derg forces, and federal autonomy of the region has given them an increasingly important role. But this is not to say that the old regional centre–periphery model of town and country has also been restored. It seems that a partnership of a new kind is being mapped out. On the one hand, the EPRDF economic programme has tipped the balance in favour of the small farmer, with a five-year plan of investment in small-scale sustainable agriculture.[63] On the other, there is not enough land, in Tigray anyway, to absorb the working population into agriculture, so a programme of gradual industrialization dispersed to provincial towns and encouraged by tax incentives has already started.[64]

In the war years, it seemed reasonable to speculate about whether the TPLF emphasis on working with rural communities in ways that were

---

[63] Resolutions of the TPLF congress, Mekelle, December 1994; Resolutions of the EPRDF Congress, Awassa, January 1995.
[64] Taped conversation with Gebru Asrat, President of Tigray Region, Mekelle, April 1994. Also in the first five-year development plan of the EPRDF government elected in 1995.

Garrison Towns & the Control of Space: Tigray

appreciated by those communities was not so much principled as opportunistic (because they needed to mobilize fighters in large numbers) or expedient (because that was the field of operations imposed upon them by Derg power in the towns). How much could the statements of principle by the TPLF be trusted? In 1989, when they won control over the whole of Tigray, the question to be asked was: Would the TPLF revert to an earlier secessionist phase, despite their assertions to the contrary? In 1991, with the future of Ethiopia in the hands of the EPRDF, the question was: Would the TPLF reverse their avowed priorities by leaving the countryside to its own devices and identifying with the interests of an urban elite, albeit a new one? In the event the TPLF did neither of these things, and their insistence on bottom-up politics in the villages and small towns of the countryside as the basis of all economic and social development continued into the post-1991 era.

Six

Modernist Dreams & Human Suffering
Villagization among the Guji Oromo

TADDESSE BERISSO

Villagization is one variant of the involuntary rural resettlement schemes often organized by governments. Like most involuntary resettlement programmes, it is characterized by an element of planning and control. Yet, unlike most resettlement programmes, it may not involve the moving of people over a significant distance. In most cases, it takes the form of a spatial regrouping of populations in areas where they are already living. Alemayehu defined villagization in Ethiopia as 'a process by which rural households formerly living in dispersed settlements are concentrated into nucleated settlements as a result of a government policy to reorganize rural settlements' (1989: 1). Pankhurst sees villagization as 'a process by which one creates a new spatial structure, by moving people together into a grid-patterned village' (1989: 2). Figure 1.2 (p. 18) shows the Derg's ideal plan.

The policy of villagization and thus the establishment of co-operative villages in rural areas, while new in Ethiopia, was not new globally. Various governments have embarked on villagization programmes at different times for socio-economic and political reasons. Despite its close association with socialist states and agricultural collectivization, villagization has also been attempted by some non-socialist governments. In Italy, for example, co-operative village settlements started before 1900 (Omari 1984). The *kibbutz* and *moshav* of Israel are other well-known examples of co-operative villages established by non-socialist governments.

In Africa, some colonial governments attempted to organize rural communities into villages to control the people and/or their produce. One of the earliest efforts at villagization occurred in north-east Rhodesia at the end of the nineteenth century, when the British South Africa Company ruthlessly moved people from isolated homesteads into villages in order to promote the economic interests of the Company (Key 1967, Cohen and Isaksson 1987). The British government also created strategic villages in the highlands surrounding Nairobi during the 1950s to deny recruits and supplies to the Mau Mau (Sorrenson 1976). Similar schemes, which collectivized the rural population into hamlets in order to prevent their active participation in their own liberation struggles, were attempted by the French in Algeria, the Italians in Libya, the Americans in Vietnam and the racist regime in South Africa (Gadaa Melbaa 1988a: 108).

The best-known cases of villagization are those attempted by socialist countries like China, North Vietnam, Mozambique, Algeria and Tanzania. Villagization in Tanzania in particular has attracted the attention of many scholars (see, for instance, Ergas 1980, Fortmann 1980, Hyden 1980, McHenry 1976). In countries that followed the socialist route to development, villagization was often considered a means of enhancing rural development in general and agricultural development in particular, being designed as a first step towards the complete collectivization of agriculture. But this ultimate objective has never been fully achieved. Even in the former Soviet Union, the People's Republic of China and North Vietnam, where great efforts were made to achieve collectivization, the practice is currently being reversed. In Tanzania, though collectivization was the aim, *ujamaa* villages never moved beyond preliminary attempts at collectivization. Despite massive efforts in the 1970s, even villagization appears to have failed. Using Goran Hyden's term (1980), Tanzania failed to 'capture' its own peasants.

A number of factors influence the success or failure of involuntary rural resettlement programmes like villagization. Cross-cultural studies of state-sponsored resettlement schemes indicate a lack of detailed and comprehensive planning, the selection of sites that are unsuitable with respect to location, accessibility and soil conditions, and the inadequate size of holdings as major factors underlying the poor performance, and even total failure, of resettlement projects (Oberai 1992: 94–5). Other contributing factors include the use of force in moving resettlers to new sites, a lack of legal frameworks – which in any case do not fully take into account the rights of resettlers – and the introduction of bureaucratic structures and organizations into the activities and ways of life of traditional societies (Mathur 1995: 17, Taddesse Berisso 1995: 126).

Villagization in Ethiopia

Beginning in late 1985, as part of its plan for 'rapid rural transformation' and greater control over peasant farmers, Mengistu's military government aimed to implement a villagization programme throughout the country, except in the war-torn regions of Eritrea and Tigray. Peasant farmers were instructed to dismantle their age-old scattered dwellings and move into grid-patterned villages, despite their unwillingness to do so. By the end of 1989, nearly 40 per cent of the country's rural population, numbering about 14 million peasant farmers, had forcibly been villagized. In March 1990, however, suddenly and unexpectedly, the government called a halt to most of its socialist programmes. Although the reform policy failed to mention villagization by name, the peasants understood it to include the option to de-villagize. Thus, as of March 1990, many peasants in different regions of the country were just as busy as they were during the villagization period in dismantling their huts from the new villages and reconstructing homes on the earlier sites (Tesfaye 1994: 1-2).

Villagization, according to Ethiopia's former military government, was a multi-purpose scheme whose central objective was to introduce a systematic land-use and/or recovery programme through collective and co-ordinated efforts (National Villagization Co-ordinating Committee 1987). Its aims were to move peasant farmers into villages where it would be easier to provide them with basic social and infrastructural services such as schools, clinics, water supplies, rural roads and milling. It was also said that the villagization programme would enhance extension services to peasant farmers and that this would enable them to raise agricultural production and productivity. Villagization was also given an important role in strengthening peasant security and self-defence, reducing rural– urban disparities and raising the consciousness of peasant farmers. In general, the government considered villagization a panacea for solving Ethiopia's rural socio-economic, political, and environmental ills.

Was villagization really a viable scheme to solve Ethiopia's rural problems, as the government claimed? In this chapter, an attempt is made to assess the socio-cultural and environmental effects of the villagization programme on the Guji-Oromo of southern Ethiopia. The fieldwork upon which this study is based was conducted in 1990 and 1991[1] and covers one intensive case study and thirty-two village surveys selected from three *weredas* (i.e., Bore, Bule-Uraga and Adola-Wadara) of the then Jam Jam *awraja*, now Southern Oromiya (see Map 1.3, p. 32). A total of 224 peasant associations had been villagized in these three *weredas*, many associations supporting more than one new village.

An overview of the Guji-Oromo & their land

The Guji-Oromo (also known as Jam Jam or Jam Jamtu by some of their neighbours) are one of many branches of the Oromo that live predominantly in the Borana zone of today's Oromiya Region. They belong to the Eastern Cushitic language family and speak Afan Oromo, one of the most widely spoken languages in Ethiopia. According to the 1994 Population and Housing Census of Ethiopia, the Guji have a population of more than 900,000 people, with an average density of 35 persons per square kilometre (Central Statistical Authority 1996: 25).[2] They were conquered and incorporated into the Ethiopian empire by the forces of Menilek II (1889–1913) in 1896.

---

[1] Fieldwork was made possible by a Rockefeller Foundation African Dissertation Internship, for which I would like to express my sincere gratitude.
[2] This figure has been calculated from the *weredas* inhabited by the Guji.

Modernist Dreams & Human Suffering: Guji Oromo

Unlike some other Oromo groups that constitute a single section, the Guji form a confederation of three independent but closely related groups, the Uraga, Mati and Hoku. Traditionally, each section had its own territorial boundary and its own political leader in the form of an *abba gada* (elected political leader). However, the three groups were mutually interdependent. They regarded each other as blood relations, acted together in wars against neighbouring groups, helped each other during economic crises and conducted *gada* rituals together. Intermarriage was fairly common, and there were few cultural differences among them. Indeed, individuals or families from one group could move and settle in the territory of another group.

The Guji live on a vast tract of land covering a wide variety of altitudes. In the lowlands the population is spread thinly over vast savanna land, subsisting mainly on livestock herding and minimal maize cultivation. The middle altitude is always green, with lush vegetation and forests. Population density is greater at this altitude, and people practise a mixed economy of livestock-herding and crop-cultivation, including coffee. Villagization was carried out in this zone but was not as extensive as it was in the highland areas. The latter are located in the northern part of Guji territory, covering much of the Mati and Uraga lands. Here, rainfall is more frequent and there is widespread cultivation of barley. People are permanently settled in the highlands and population density is relatively high.

In general, the Guji have a mixed economy of animal husbandry and crop cultivation as well as bee-keeping for honey. They subsist mainly by cultivating grain, pulses, and *ensete* (*Musa ensete*). But their real wealth consists in their cattle, sheep, goats and horses. Emotions and pride are centred in stock. People who do not own cattle are not considered to be proper Guji (Baxter 1991: 9). Cattle are important, not only economically but also for social and ritual life. The social status of a person among the Guji finds expression in the number of cattle owned. The owner of many head of cattle is a respected person. Ritually cattle are used for sacrificial purposes.

Except in areas where they are bordered by hostile groups, in which case houses were clustered together for mutual protection against raids, the Guji used to live in *olla* (neighbourhoods) of dispersed homesteads before the villagization programme. A typical *olla* contained a single house or a cluster of two or three round straw houses separated from each other by a cattle kraal and by crop and/or grazing land (Hinnant 1977). It was quite common for family and lineage members to live next to each other. However, this has not always been the case. In pre-villagization days the heads of polygynous families used to spread their family members over different ecological zones to disperse their herds and/or to prevent fights among co-wives.

Traditional Guji socio-political organization was based on moiety, clan, lineage and family structures and on the *gada* system,[3] with a *qallu* (supreme

[3] For an intensive study of the Guji *gada* system, see Hinnant 1977.

Plate 6.1 Villagization (i): Guji Oromo [T. Berisso]

hereditary religious leader) at the apex. There are two non-exogamous moieties known as Kontoma and Darimu, divided into seven non-totemic and exogamous clans each in Uraga and Hoku, and three in Mati. Each clan is divided into a variable number of segments called *mana* (house), which in turn are divided into a great number of patrilineages. Before villagization, the Guji family was an extended patriarchal family, ideally containing a husband, several wives and as many children as nature allowed. If a man is an *angafa* (eldest son), he is expected to live with or near his parents along with his spouses and children. Most ordinary families, however, consist of a husband, wife and their children. With the villagization programme the extended family was completely transformed into a set of nuclear families, as every married person was provided with a house in the village.

The *gada* is an age-grade system that divides the life stages of individual men, from childhood to old age, into a series of formal steps. There are thirteen such steps in contemporary Guji society. Transition ceremonies mark the passage from one stage to the next. Within each stage, activities and social roles are formally defined in terms of both what is permitted and what is forbidden. The ideal length of time in one rank is eight years. In the past, the *gada* system assumed military, economic, legal and judicial responsibilities. Currently, however, its function in Guji has been reduced to ritual activities.

With regard to religion, the Guji have developed a very complex set of beliefs and practices. They believe in abstract concepts such as *waqa* (heaven) and in the existence of the *durissa* (devil). They have *woyyu* (sacred shrines), under which prayers and sacrifices are made to *waqa*. They also believe in the power vested in certain individuals and families. The *qallu*

Plate 6.2 Villagization (ii): Guji Oromo [anon]

(supreme religious leader) and *abba gada*, who are respectively considered as *woyyu* ('holy') and *worra qallacha* ('virile family') are among these individuals. Attached to these are rituals and religious practices, such as oracles, spirit possession, complex and varied forms of divination, and other traditional orders. However, with the introduction of the great world religions and modern civilization, Guji traditional beliefs are changing fast. There were mass conversions of Guji farmers to Christianity (especially Protestantism) and Islam, particularly during and after the villagization programme.

Villagization among the Guji Oromo

The implementation of villagization in Jam Jam *awraja* started in January 1986. The implementing authorities were successful in organizing Villagization Co-ordinating Committees (VCCs) at the *awraja*, *wereda* and peasant association levels. As of June 1988, it was reported that about 45 per cent of the *awraja*'s population was organized into 240 villages consisting of 20,000 household heads, i.e., about 105,000 people.[4] Despite this successful achievement (from the authorities' point of view), the implementation of the programme was not properly planned or organized. No detailed socio-economic or environmental studies were made before implementation. The peasant associations villagized were not selected according to the criteria laid down by national guidelines, and no proper education was given to peasant farmers to make them understand the objectives and importance of the programme.

[4] Jam Jam *awraja* Villagization Co-ordinating Council, 1988: 153.

Looking Back on the Projects of the Socialist State

Villagization in Jam Jam *awraja* suffered from a top-down approach. Government and party officials were the sole architects of the whole process of implementation: they applied a paternalistic approach in which they assumed the whole responsibility of deciding when, where and how to implement the programme. Peasant farmers were not given a chance to participate in site selection or in designing village layouts, houses, or garden plots. They were called in only to provide labour and materials for building villages.

Guji farmers were totally against villagization because of the nature of the programme in general and the way it was planned and implemented in particular. They started opposing it from the very day they were told about it. When they were later forced to implement it, they resorted to various kinds of violent and non-violent methods of resistance. Some individuals and families fled their peasant associations in order to evade the programme. Others made the implementation difficult by refusing (or delaying) the clearing of bushes from the sites where the villages were supposed to be built, by cultivating and sowing crops on the sites selected for villages and by displacing the marks made by designers on the sites. In some peasant associations, farmers contributed money and bribed officials not to select their area for villagization. There were also individuals who went to the extent of threatening designers and implementation authorities with armed force (see Taddesse Berisso 1995: 153–7 for a detailed study of Guji peasants' resistance against villagization). Because of such resistance, authorities in Jam Jam *awraja* themselves resorted to force to villagize peasant farmers and to make them stay in the villages.

In general, peasant farmers and their institutions in Jam Jam were not involved in planning or designing villagization, though they were directly affected by the programme. In post-revolutionary Ethiopia, mass organizations such as peasant associations were expected to be the main vehicles enabling people to participate in rural affairs that directly affected their lives. But local people were totally alienated in respect of villagization planning. The programme hardly encouraged grassroots participation. More than 96 percent of farmers interviewed responded that they did not participate in planning or designing the programme at the local level.

A number of problems confront any attempt to make a sober assessment of the impacts of villagization on the socio-economic and environmental aspects of rural Ethiopia. First, it is difficult to isolate the impacts of villagization from the effects of other government policies and measures. Second, some of the impacts of the programme need a longer time to manifest themselves. The environmental impacts of villagization, for example, may not be seen quickly, since this is a long-term process in which effects often appear slowly. Third, it is difficult, if not impossible, to measure some of the sociopolitical and psychological impacts of villagization.[5] For instance, Mengistu's government claimed that villagization had

---

[5] The psychological impacts of villagization were very grave and require expertise in the field to reveal them.

helped raise the political consciousness of peasant farmers. It was said that the programme had changed farmers' lives, views and thinking, thus opening up a new chapter in the establishment of a modern society in rural Ethiopia (Mengistu 1986). Such claims are difficult to measure. Finally, information about the provision of social services, the number of people going to literacy schools, production and marketing information over time are virtually lacking. Given these problems, it is difficult to provide a complete assessment of the impacts of the programme, though some evaluation can be attempted.

Impacts on the economy

As mentioned previously, the Guji have a mixed economy of animal husbandry and crop cultivation on fertile land which covers a wide variety of altitudes. While cultivation is more important to Guji than they admit, animal husbandry was and still is their favourite economic activity. The implementation of the villagization programme had clear negative effects on their economy. The use of farm labour for house construction during peak agricultural seasons, increased distance between the farm and the homestead for most farmers after villagization, space limitations for garden plots and other sideline activities in villages, and the inconvenience created for animal husbandry were some of the ways by which villagization has adversely affected the Guji economy.

According to the *Villagization Guidelines*, villagization was to be implemented in a manner that would not hamper or disrupt the agricultural and marketing activities of the peasant farmers (Ministry of Agriculture 1986: 19). But the implementing authorities in Jam Jam *awraja* did not take the directive seriously, and the programme was implemented at various points in the planting, weeding, harvesting and threshing months. My field data revealed that of the 33 villages covered by this study, twenty (60.6 per cent) were constructed during ploughing, seven (21.2 per cent) during harvesting and four (12.1 per cent) during weeding. Only two villages (6.0 per cent) were built during the agricultural slack period. This affected not only farmers in peasant associations where villagization was implemented, but also all farmers throughout the *awraja*. Men and women, old and young, were all mobilized to participate in the villagization campaign. Farmers were not allowed to work their fields until they had finished their quotas for house construction. Consequently, there was a significant drop in crop production in Jam Jam during the years of implementation. Statistics are not available, but all peasants and peasant association leaders, Ministry of Agriculture officials and even government and party officials interviewed admitted a drop in crop production.

While this was a short-term effect in the *awraja*, the more lasting residue was the increased distance between farms and villages. This had tremendously negative effects not only on crop production but also on labour productivity. The distance between farms and homesteads in villages increased for the majority of peasant farmers (65 per cent). The longest distance recorded during fieldwork was eight kilometres. For a few households, villagization had, of course, shortened the distance between the farms and villages. But the average distance was still about three kilometres for the villages studied. As a consequence, time that could have been used for production was spent on the road. The long distances from farm to village also increased the incidence of crop damage by vermin, pests, and thieves. Increased farm-to-village distances also made it difficult for farmers to discover and treat crop and plant diseases on time or to use inputs like manure properly.

The limitation of space in villages for home garden and sideline activities was another reason for the drop in productivity. It is evident that many Ethiopian peasant farmers (particularly women and the poor) grow a considerable amount of produce in their garden plots for consumption as well as for cash. Villagization attached little importance to home gardens by allocating very little land (1000 square metres each) for such purposes.

Old garden plots far away from the new villages were therefore often abandoned due to the distances and difficulties involved in protecting them from pests. Most farmers had not even started growing crops and vegetables in the new villages. Some said that the space allotted to them was insufficient to grow anything. Others complained that their villages were built on dry pasture or forest land which would take years to become productive. Yet others complained that a lack of fences and fear of theft have prevented them from growing garden crops in the new villages. Farmers also found it difficult to engage themselves in other sideline activities like bee-keeping or raising poultry, due to the inconveniences created by villagization. In general, home garden production and sideline activities suffered badly from villagization. It is evident that efficiency of production by each farmer and an increase in acreage under cultivation were necessary ingredients for further expansion and improvement of agricultural production. But both conditions were constrained in Jam Jam by villagization, and consequently there was a drop in crop production.

No other government policy has ever threatened the Guji's favourite economic activity (animal husbandry) as much as the villagization programme. First, many farmers sold their livestock due to the inconveniences created by the programme and the fear that livestock might be nationalized. Second, animal mortality increased after villagization due to communicable diseases, shortage of fodder in the villages, cold, and attacks by wild animals. Third, the limitation of space and compactness of stables in the new villages did not permit farmers to keep many animals. Finally, traditional transhumance and *dabare* practices were disrupted because of the programme (*dabare* is the lending of cattle to poor kin and affines). As Hinnant correctly observed (1985: 804), *dabare* is not entirely altruistic, but rather part of a large-scale adaptive strategy. A man with sufficient milk cows parcels out his herd in a way that takes advantage of the varied topography of Guji land. To keep the entire herd in one area is seen as an invitation to predation, drought and disease.

Social & political impacts of villagization

Little has been said or written about the social and political impacts of villagization on Ethiopia's rural population. Almost all researchers have concentrated on issues related to the short- and long-term effects of the programme on the economy (for instance, Alemayehu 1989, Berihun 1988, Cohen and Isaksson 1987). Very few have considered the impacts of the programme on the provision of social services, individual and social relations, family structure, lives of women and children, marriage and death, religion, or traditional patterns of co-operative work organization like *dabo*. This section will briefly deal with some of these issues.

One of the major objectives of the villagization programme was to ensure that basic development infrastructural facilities and services were provided for the enhancement of the livelihood of the rural masses and their socio-economic uplift (Ministry of Agriculture 1986). Local administrators promised peasant farmers that the government would provide them with basic social services when they moved to new villages. The number of basic social services provided to peasant farmers by the government under the villagization programme in the 33 villages covered by this study is given in Table 6.1 on p. 126.

As can be seen from the table, there have not been any significant changes in the provision of social services to farmers through the villagization programme. Since its beginning in January 1986 until the end of 1991, only two elementary schools, one clinic and seven piped water facilities were provided to the 33 study villages by the government. Even these few social services deteriorated due to a lack of maintenance. Nobody appeared to be responsible for them. The overwhelming majority of peasant farmers interviewed continued to obtain services from the same sources as before villagization. In general, the government's promise to provide farmers with social services was unfulfilled.

The cultural impacts of villagization were more visible in areas of individual and social relations, traditional customs, family structure, and areas related to the lives of women and children. There were frequent conflicts among village dwellers arising from crop damage (in villages) by domestic animals, from the easy availability of alcohol, and from a situation that forced co-wives to live next to each other in villages. The Guji are a polygynous society. A man can marry as many women as he can support. Before villagization, tensions between co-wives were generally minimized by separating and settling them in different places (usually in different ecological zones). This strategy not only helped to defuse conflicts between co-wives, it also enabled husbands to take advantage of the differences in ecological zones for cattle-grazing and crop-cultivation. It also worked as a

Looking Back on the Projects of the Socialist State

Table 6.1 Planned villages in Guji Oromo

Wereda	Villages	School	Clinic	Piped water	Assembly hall**	MOA office**
	Furfusa Maro	-	-	-	-	-
BORE						
1	Gossa Wotiye	-	-	x*	x*	x
2	Alayu Dhibba	-	-	x*	-	x
3	Sutta Dhibba	-	-	-	-	-
4	Anno Qeransa	-	-	-	-	-
5	Qale Kuku	-	-	-	-	-
6	Qale Salato	-	-	-	x	x
7	Boroto Chichu	-	-	-	-	-
8	Sololo Qobo	-	-	-	-	-
9	Yirba Buliyo	-	-	-	x	x
10	Ajersa Kalacha	-	-	-	-	-
11	Kara Qulubi	-	-	-	-	x
12	Hiyo Komole	-	-	-	-	-
BULE URAGA						
1	Layo Taraga	-	-	-	-	-
2	Sonqole Kalato	-	-	-	-	-
3	Sonqole Hora	-	-	-	-	-
4	Dida Hora Burqa	-	-	-	-	-
5	Tebe Solamo	-	-	-	x	x
6	Tebe Haro Wato	x	-	x	x	-
7	Gadiyo Guratu	-	-	x	x	-
8	Bursa Dhokata	-	-	-	x	-
9	Afale Kola	-	x	x	-	x
10	Ballo Hanqu	-	-	-	-	-
11	Elalicha Dansuma	-	-	x	-	-
12	Guticha	-	-	-	-	-
13	Lubo Dama	-	-	-	-	-
14	Kochore	-	-	-	-	-
15	Andegna Okolu	x	-	-	-	-
ADOLA-WADARA						
1	Dole	-	-	x	-	-
2	Darartu	-	-	-	-	-
3	Gobicha	-	-	x	-	-
4	Koba Sorssa	-	-	x	-	-
5	Anfarara	-	-	-	-	-

Note: ** are services built at the expense of farmers
* are services provided by the Lutheran Mission

risk-avoidance mechanism in times of drought and animal disease. Villagization, on the other hand, congregated all members of a family in the same village, usually next to each other. This created major tensions, which sometimes erupted into fights between co-wives. Many such cases were reported during my fieldwork.

Very few alcoholic beverages were available in rural Jam Jam before the villagization programme. Guji farmers used to go to nearby towns, usually on the bi-weekly market days, to obtain alcohol (Hinnant 1977: v). Following villagization, however, many women started to distil *areqi* (local alcohol) and other alcoholic drinks to take advantage of the concentration of people in the new villages.

The availability of alcohol in villages in large quantities had many effects on farmers' lives. First, it became a source of income for women and for those who were engaged in the business. For others it became a drain on their pockets. Second, it became a major source of conflict in the villages. Many people fought and injured each other after getting drunk. Aware of the dangers of alcohol, a significant number of young people converted to new religions (Pentecostalist or Protestant Christianity, the Catholic Church or Islam). According to informants, had it not been for the conversion of many young people to new religions, the consequences of alcohol consumption in the villages would have been more grave and might have cost the lives of many individuals. However, the conversion of the Guji into different (and sometimes conflicting) religions may in the long run destroy their traditional solidarity of working together under the *gada* system.

Third, drinking dens became meeting and recreational places for many people in the villages. Instead of working their fields or tending animals, farmers started to spend most of their time in drinking dens, gossiping and backbiting. This obviously had adverse effects on time and energy which farmers would otherwise have used in productive activities. Fourth, alcohol and behaviour like crime and prostitution usually go hand-in-hand. There was a clear indication of the spread of such behaviour in some of the new villages.

Other sources of conflict in the villages related to livestock and children. Animals would break through fences into villages and destroy the few gardens that existed, as well as other properties. Children also became sources of conflict in villages. Sometimes they simply hurt each other while playing. They might also steal or damage property in neighbouring houses that had been left unattended.

In the area of customs, villagization had clear effects on farmers' privacy, mortuary practices, and indigenous work organizations such as *dabo*. Privacy is one of the most important dimensions of Guji culture. Guji say:

Dubbi namma gowati	A foolish man's affairs
worra geyya	would be known (heard) by his family
Dubbi worra gowati	A foolish family's affairs

Looking Back on the Projects of the Socialist State

olla geyya	would be heard by its neighbours, and
Dubbi olla gowati	A foolish neighbour's
subba (ardda) geyya	affairs would be heard by the community

As this saying clearly indicates, Guji keep most of their affairs private. The concepts of *buda* (evil eye) and *falfala* (sorcery) might have been invented or adopted to protect individuals' and families' privacy and property. Villagization, however, brought hundreds of people together in the same villages, where people could not keep their affairs and property private. What used to be private and family affairs became public.

The Guji do not believe in life after death and therefore do not worship ancestors, but they have great respect and affection for their dead relatives. They include the names of their dead ancestors when praying to *waqa* (heaven) and when blessing or cursing others. Traditionally, burial practices were dependent on the *gada* system. Men of different *gada* grades were buried in different places (some simply dumped on the ground, others buried in cattle kraals, and some others in *boro* or house shrines, depending on their *gada* status (see Hinnant 1977). For most common people, before villagization burial places were next to houses (on the right side for men and left side for women), where family members could easily take care of the graveyards. After villagization, however, burial places were established outside the village. Most peasant farmers interviewed complained that villagization had separated them from their deceased loved ones.

Villagization improved the opportunities to organize collective work parties like *dabo* by making people more easily accessible to one another. But people in most villages were divided into two groups after villagization: converts and loyalists. If one wanted to have a large number of people for work, one had to prepare two kinds of food and drinks. In pre-villagization days only *farso* (a thick and slightly alcoholic beer) had to be prepared to induce people to work. After villagization some young people did not consume alcoholic drinks and therefore other refreshments (e.g., milk or tea) had to be provided for them separately, causing extra expense.

Be that as it may, villagization increased interaction levels between people. For those who formerly lived in relatively isolated settlements, the positive aspects of the programme were significant. The social and economic benefits of a denser population were particularly important for women, enabling them to carry on small businesses like selling *areqi* in villages and to go to literacy schools. Villagization also made communication easier and faster within villages. People are easily accessible to assist each other in relation to accidents, death, illness and other problems. With regard to such assistance, Helen Pankhurst's observations on a peasant association in Shewa region could equally be applicable to the Guji area and most probably to all regions of the country.

> For the community as a whole, villagization increased the speed with which distress could be dealt with. ... Any call for help was more likely

to be heard and people were easily available when someone was ill, or a woman was having difficulties during childbirth and had to be carried to the clinic. This was an immense benefit for female-headed households whose members were otherwise particularly vulnerable.

(Pankhurst, H. 1992: 66)

Impact of villagization on environment & health

Of all the policy measures implemented under the Mengistu regime, villagization was the one that resulted in the most drastic impact on Jam Jam's forests and the environment. In Jam Jam, construction materials from the old houses were very rarely used to build houses or facilities in the new villages. This was most probably due to the relative availability of forests in the *awraja* and/or the problem of re-using old bamboo trees to build new houses. This resulted in excessive tree-felling, which caused deforestation in some areas. On average a minimum of 600 bamboo canes and 12 wooden poles are required to build a *sheka* (bamboo-roofed) house. Consequently, the number of trees needed to build over 20,000 houses (excluding village facilities like the kitchen, pit-latrine, stables and fences, where these were constructed) was 12 million bamboo canes and 240,000 other trees. Other vegetation was cleared in the process. It is also important to bear in mind that trees are the main source of fuel in the area. Since the disintegration of the villagization programme in 1991, devillagization has consumed as much forest, labour, time and money as villagization had done.

The concentration of people and animals in the villages has also adversely affected human health and the environment, due to lack of sanitation, overgrazing and soil erosion. At the time of fieldwork, an average of 2,800 animals were being grazed on about 400 hectares of peasant association grazing and forest land (totalling 800 ha.). The ratio of grazing land to animals was 1 to 7, suggesting that the stocking rate was greater than the theoretical carrying capacity of grazing land, i.e., 2–3 cattle per hectare.

Over-stocking in some villages caused roads to turn muddy in the rains and dusty in the dry season, exposing the land around the villages to soil erosion by wind and running water. Lack of garbage disposal, uncovered stables and pit-latrines, and mud and dust contributed to pollution which in turn became a breeding ground for various disease-bearing organisms. According to informants, hundreds of village-dwellers suffered from communicable diseases and cold-related illnesses. In one village alone (Layo Taraga), 35 people were reported to have died from meningitis, and sixteen people in Yirba Buliyo village because of typhoid. Animal diseases were rampant, killing thousands of beasts. There was no single study village that did not report the loss of animals in the new villages because of disease, predators, or cold.

Looking Back on the Projects of the Socialist State

Villagization as *encadrement*

The implementation of villagization in Ethiopia at national level started in late 1985, when the country had still not recovered from one of the worst famines in the country's recent history. The time and haste with which the government implemented the programme created widespread suspicions about its claim that villagization was primarily aimed at benefiting peasant farmers through the provision of services. Few studies have dealt with the political motives behind villagization.

Security problems gave birth to villagization in Bale in 1978 and Hararghe in 1984. Most of the villages in these regions were established as part of the government's programme to restrict insurgency and reduce secessionist tendencies among the Somali and Oromo inhabitants of Bale and Hararghe (Cohen and Isaksson 1987: 110, Harbeson 1988: 175). Thereafter, and particularly in the period between 1985 and 1991, Mengistu's regime was threatened by ethnic and nationalist movements. There were more than twenty liberation movements fighting the government in Ethiopia in the 1980s (*New York Times*, 6 January 1985). By recruiting peasant farmers from the areas they controlled, these movements had reduced the government's control over vast areas of the country. Mengistu's government thought that these movements could be controlled (if not completely destroyed) through the introduction of programmes like resettlement and villagization. Accordingly, peasant farmers were moved into resettlement areas and villages in an attempt to control them and deny recruits and supplies to opposition movements.

Regular meetings were held in villages, where party cadres agitated peasant farmers against liberation movements. In Jam Jam, peasant farmers were forced to have political meetings every month. A few loyal individuals from every peasant association were also recruited and armed by government officials to fight against the liberation movements. According to informants, villagization had made political meetings, tax and grain quota collection, and militia recruitment much easier for government agents than any previous programme. The villages served as strategic hamlets in the government's counter-insurgency effort. The programme was thus used for political control and for effective mobilization of the country's resources to fight against opposition groups.

There were other aspects. Villagization was introduced to increase the extent to which the peasantry was homogenized and integrated into the economic and political system. Yet another reason lies in the problems that governments face in ruling a peasant state. As Gerd Spitter has observed (1983: 131), 'There is nothing more difficult than to build up an administration covering millions of partly self-sufficient peasant households.' One way to solve this problem is to encourage peasants to become involved in the market economy. But for governments like Mengistu's, which ignored this option for ideological reasons, the problem of controlling peasant production was not easily solved.

An ideological commitment to the fair distribution of wealth on the one hand, and the practical need to supply the underpaid middle and urban poor with affordable food on the other, make the free-market option less attractive to such governments. Left alone, the peasant has many ways of fending off government control. In spite of its failure elsewhere in the world, villagization remained attractive to Mengistu's government because it offered a perfect means to control peasant farmers and their produce, enabling the government to stretch its politically repressive and economically exploitative tentacles deep into the countryside – the process highlighted as *encadrement* in Christopher Clapham's opening chapter to this book.

It is thus clear that behind the rhetoric of helping peasant farmers, the villagization programme was introduced in Ethiopia mainly to fight against national liberation fronts and to provide greater control over peasants and peasant produce in order to feed the army and the urban population, while at the same time helping Mengistu to prolong his hold on political power.

Summary & conclusions

According to Mengistu's government, villagization was introduced to enhance the socio-economic development of the rural population through the provision of basic infrastructural and social services. With a mixture of economic, social and political objectives, the programme was considered a viable strategy for enhancing rural development in Ethiopia (National Villagization Co-ordinating Committee 1987). To evaluate the extent to which these attempts were successful, one must turn to evidence from actual cases.

With regards to the economy, instead of increasing production or productivity, villagization turned out to be the major cause of economic decline in Jam Jam *awraja*. The use of farm labour for house construction during agricultural peak seasons, the increased distance between farm and homesteads, and the limitation of space for garden crops and other sideline activities greatly reduced production and productivity. Besides, since the programme was not accompanied by the introduction of modern agricultural technology and extension services (as promised by the government), no improvement was recorded in traditional ways of farming. Moreover, in most cases extension workers were used to implement government policies such as promoting producers' co-operatives, and raising funds for the war effort and the villagization campaign, instead of helping farmers raise production.

Even worse was the impact of villagization on animal husbandry. Due to the inconveniences created by the programme, the number of livestock in Jam Jam greatly declined. Some people sold their animals, due to the lack of space in the villages and the fear of confiscation. Thousands of animals died from communicable diseases, shortage of fodder or cold weather. Others were stolen or eaten by wild animals.

The social impacts of villagization in Jam Jam were not impressive. The government's promise to provide farmers with basic infrastructural and social services was largely unfulfilled. In addition, by bringing hundreds of peasant farmers together into compact villages, the programme created tensions and conflicts between individuals and families, eroding peasant farmers' privacy and indigenous customs, and contributing to increased alcohol consumption, adultery, and other social crimes like theft. Social conditions were not improved by villagization. Rather, the programme proved to be a major infringement of the basic human rights of the Guji by moving them into villages against their will.

Villagization scored more success in the political than the economic or social arenas. The grouping of peasants into compact villages gave the government easy access to them so that they could agitate farmers against opposition groups through regular meetings. It also made it possible for the government to control peasant production through grain quota deliveries administered by the Agricultural Marketing Corporation to feed the army and urban population, to collect taxes and 'voluntary contributions', and to recruit militias for war. However, these were all benefits for the government, not the peasant farmers. For the farmers, the success of villagization was negative, whether measured in terms of the expansion of agricultural output, improved social and infrastructural services, or environmental development. Thus the justification for the programme rests on political grounds rather than socio-economic or environmental ones. The objectives of socio-economic and environmental transformation were primarily covers for the government's political motives. Villagization in Ethiopia was a good example of how the centre manipulates and uses the periphery for its own political interests.

Thus, with all its economic, social, environmental and political ills, villagization caused devastating poverty to Guji farmers in a brief period of time. In Tefera's words, there was poverty without natural cause, crop failure without drought, starvation not witnessed by television cameras, and thus disaster without fanfare (Tefera 1988). For Guji farmers, villagization was the gloomiest of all government policies implemented in the country. Five years after implementation, the programme which was claimed to be of lasting benefit to peasant farmers could only be assessed in terms of its costs in human suffering.

Seven

Surviving Resettlement in Wellegga
The Qeto Experience

ALULA PANKHURST

More than half a million settlers were moved from northern to western Ethiopia in 1985. Those who were located in Village Three of the Qeto resettlement area in the former province of Wellegga (see Maps 1 and 1.4, pp. xii and 32), like all the others, have been subjected to tremendous changes in their lives in the decade and a half since they were resettled. The Derg's venture with resettlement was one of the most extreme attempts at social engineering in the name of Marxist ideals. Not only were the settlers uprooted from home and community, they were also villagized together with strangers and placed within the most stringent form of collective, with virtually no private holdings (for the typical pattern of settlement, see Fig. 1.1, p. 18). In effect they had been transformed overnight, from independent smallholders reliant on ox-plough cultivation to wage-labourers in a mechanized agricultural co-operative, earning work points translated into a share of the collective harvest. The experience was alienating and degrading, best summed up by metaphors about being treated like livestock.[1]

However, settler responses were not uniform or unchanging. In order to understand the dynamics of their decision-making we need to consider, first, the various categories under which they fall, and second, the profound changes that have taken place in the fifteen years since they were resettled by the Derg. Indeed the settlers have now lived longer under the current government, which has different priorities. At least in the twenty-village resettlement complex of Qeto, those who remained are now self-reliant, most are probably better off than they would have been in Wello, and some are even prosperous. However, the viability of ox-plough cultivation remains a major concern, as I learned during a visit to the

---

[1] The main findings of my original study can be consulted in A. Pankhurst 1992. The settlers used expressions such as: 'We have now become the oxen that have been yoked', and made references to being 'counted like sheep in the morning and evening' (ibid.: 149, 160).

133

resettlement area in 1999, a decade after I had been living and carrying out my original research there.[2]

In this chapter, I explore the different ways in which settlers have responded to changes in production and exchange since the time of their arrival. I also consider how settlers' identities have changed, notably with respect to the second generation and to current language policies. Finally, I question some prevailing assumptions about the impact of the settlements on the environment and the effects of regionalization.

Variations in settler responses

While the Derg's resettlement policy was officially portrayed as entirely voluntary, opposition groups suggested that it was carried out at gunpoint. Both views are partial and over-simplified (cf. A. Pankhurst 1992: 51–80). Between the two extremes of famine victims who, lacking food and having sold their livestock, welcomed the opportunity to start a new life, and those who were rounded up in market places and forced to leave, there were other, less overt and visible pressures and motivations. These included the propaganda campaign to convince people that famine would last and that the settlements would be idyllic. Cadres of the Workers' Party of Ethiopia sought to fill local 'quotas' for settlers, and some village level officials used resettlement to rid themselves of opponents. Resettlement was also used to move people off forest hillsides and areas designated for development projects. My findings also highlighted the role of family, peer, and community pressure (A. Pankhurst 1992: 98–107). Dependants were often not involved in decision-making, and spouses, parents, children, siblings, other relatives and friends found themselves leaving 'because of others', leading to a sort of chain reaction. Another significant category of settler not often recognized were victims of neither famine nor coercion, but a younger generation of mainly single men who were attracted by a sense of adventure.

Settler responses to life in the resettlement villages were related to the reasons for their move, the circumstances in which they left their homes, and whether they had relatives who remained behind. Thus those who left the new settlements in the early years, despite travel restrictions and the costs involved, included many who had been forced to join the scheme. There were also those who had been separated from family members during the journey or who had left close relatives behind, especially if the latter had managed to retain the former's land and were willing, and had been allowed, to return it to them (A. Pankhurst 1991).

However, in the early period settler responses were also related to the constraints and opportunities of the settlement environment. In theory the

---

[2] I visited the settlements most recently from 25–29 January 1999 (see A. Pankhurst and Hockley 1999). I wish to thank Tom Hockley and the Department for International Development in Addis Ababa for facilitating transport to the settlement.

socialist collective was meant to be egalitarian: planners assumed that earlier privileges and differences in holdings would be eliminated, since everyone received equal minimal household plots, and work was geared to collectivized production. This was the most radical form of the co-operative ideal ever attempted by the Derg, no doubt in the belief that an uprooted population would be more easily malleable.

In practice, however, the system was inherently unjust since it privileged the militia and literate elites. The former did little agricultural work and had their household plots ploughed for them when on duty away from the village. The latter held powerful leadership positions in the peasant association. For some of the militia, especially the younger generation, the resettlement experience was fulfilling: they saw themselves as leaving a situation with little future, where smallholdings and dependence on the parental generation for land and livestock provided little opportunity for self-improvement. Resettlement, by contrast, provided a chance for adventure and the prestige of being a bearer of arms, and was viewed with, as it were, a 'far-west' frontier mentality. For some of the literate young men, resettlement also provided a sense of purpose: they were given positions involving less arduous manual labour, such as health or agricultural workers, supervisors, etc. Those who were both literate and had military experience fared best, being able to exploit their positions of power for personal advantage.

Gender differences in resettlement were also visible. Women's lives were often disrupted by the resettlement more than those of men. Given virilocal residence in their homeland, many women had to choose whether to go with their husbands or with their parents if they were resettled in different areas, or whether to stay behind if either remained in Wello. In the resettlement villages women were made to work in the collective fields for points like the men, albeit for fewer hours and mainly during the weeding and harvesting seasons, which earned them fewer work points than men. They complained that they had not worked in the fields in their homeland, and men protested even more vehemently at what was seen as an invasion of the domestic domain. However, women did mention some benefits of the new settlements. They commented that they did not have so far to walk to obtain water and wood, and they appreciated the proximity of grinding mills. The availability of relatively good health services, at least once these were established and before they were scaled down, was also viewed positively by settlers, especially women. However, like the men, they suffered from a monotonous diet, the lowland climate and numerous diseases, as well as from the disruption of social life and living among strangers. Divorce and marriage were frequent, no doubt in part a symptom of the stresses of resettlement (A. Pankhurst 1994). Gradually, the practice of grooms offering brides clothing gifts was re-institutionalized and served as a sign of commitment to marriage: women who had initially married without these gifts, which represented a significant proportion of a young man's annual income, were able to claim that they were entitled to them. However, women who sought to leave their husbands found that the committee of elders, which comprised only men, generally tried to make them return to their husbands, in exchange for a dress in compensation if their grievances seemed legitimate. Single, divorced or deserted women were in some instances able to retain a household plot and gain a degree of independence, leading to complaints from young men that women were unduly privileged. Being household heads, single women could also earn full points, though they had to work long hours like the men. A few women who were literate obtained positions of responsibility, although the peasant association was run and controlled by men, and the women's association was powerless and its activities resented as impositions by ordinary women.

Differences in settler responses in terms of age and generation were also clear. Generally, as one might expect, the old found the experience difficult to come to terms with, feeling the loss of home and community, changes in diet, the climate and the environment, and lowland diseases more severely. Moreover, they were also excluded from decision-making, which was monopolized by the cadres together with the younger generation of the literate and the militia. Old men found themselves having to work as hard as young men and resented being ordered around by people of their children's generation. Sons were no longer dependent on their fathers and many showed little sense of obligation towards them.

A few elderly men found some sense of fulfilment in roles as mediators in informal conflict resolution, generally concerning marital affairs; half a dozen were even appointed to a semi-official Marriage and Divorce Committee set up by state agents to address the perceived problem of high rates of marital disputes and divorce (A. Pankhurst 1995). However, even this Committee had a young literate official as its secretary and deferred to the peasant association leadership. Some of the elderly men and women, especially those with children in Wello, simply wished to return to die in their homeland, and many did not survive the first decade of resettlement, with all its trials and tribulations.

Young people suffered least: those born in the settlements are better adapted to local environmental conditions, and those who left Wello as young children no longer retain vivid memories of, or nostalgia for, a different homeland. Households with young children, elderly parents or disabled members were initially not in a position to leave the settlements even if they wished to do so, since they could not risk, or afford, travelling. Whereas some with young children were glad to see them grow up with enough food to eat, others felt they were trapped in unpleasant circumstances.

Another category that suffered particularly from spiritual deprivation were devout religious men, both Christians and Muslims. The lack of freedom to worship and the condemnation of religion as unnecessary and noxious were sorely resented. One former priest, Memre Lissan, became a storekeeper through his skill in writing, but other priests and deacons left the settlement. Several Muslim sheikhs were also among those who left in the early days.

Surviving Resettlement in Wellegga: Qeto

Coping strategies in the face of state control of settler livelihoods involved engaging in petty trade, putting energy into small neighbourhood networks and socializing, notably around the drinking of coffee. Some pooling of grain took place through rotating credit groups, and friends formed small informal religious gatherings.[3] Dissent was difficult and liable to punishment under the rule of the cadres. Though many of the cadres were well-meaning and motivated, most were from urban professional backgrounds, and few seemed to have much understanding or appreciation of agricultural life. They often resented being expected to serve 'in the bush', and some found the experience alarming. Many were dedicated to their work, some believed in their mission, and a few were remembered with appreciation by settlers (A. Pankhurst 1992: 149–55). However, there were also cases of settlers being abused, and incidents of imprisonment, beating, rape and even murder.

The settlers had little choice but to accept the situation or leave. Group protests were rare. One day in March 1987 the women protested at work in the fields, which led to negotiations to persuade them to continue. On another occasion I witnessed an initial refusal by the men's work teams to eat boiled sorghum that had been brought to them in the fields, on the grounds that this was an infringement of the custom of eating at home. However, the most tenacious form of resistance was the pervasive weaving of social life and daily reciprocities among neighbours, notably at night. In the end those who could no longer take it had no alternative but to leave. Although the settlements were guarded, escaping was not difficult, though returning to Wello was more hazardous, since vehicles were controlled. Nonetheless many succeeded, partly through bribes or forged documents. Others sought wage labour in western Ethiopia, especially during the coffee-picking season. Some single women left and worked in bars in local towns, and a handful married local men, often as second wives.

So far I have argued that the settlers cannot be understood as homogenous in their responses to resettlement even in the early period. This was related both to reasons why they were resettled and to conditions in the settlements. Despite widespread discontent, notably with collectivization, differences according to age, generation and gender were significant. Literacy and military background were perhaps the most important factors in explaining why some benefited. The younger generation adapted better, while the elderly and those with a strong religious commitment were the most bitter; the most discontented had little option but to leave.

The changing face of resettlement: the Derg period

The settlements had already been through radical transformations in the six years prior to the downfall of the Derg. The most profound changes

---

[3] The Muslims held *wedaja* sessions, especially on Wednesdays, and the Christians *mehaber* on the more important saints' days.

resulted from the abandonment of collectivization and the move to liberalization of the economy, which occurred gradually in the last few years of the Derg regime.

The first important change was the removal of the cadres after three years. Control over the peasant associations was handed over to the Ministry of Agriculture. This represented a move away from political objectives towards stressing economic goals. For the settlers the new regime was much less irksome, though production was still run by outsiders, using tractors, and largely for collective purposes. Much of the surplus was sold at official prices and placed in an account in the settlers' names, but to which they never had access. Over a few years the number of government workers in agriculture and health was reduced and the privileged support the settlements received, disproportionately high compared with services provided to the local population, was reduced. A concern emerged to make settlements viable, to make them pay the costs of tractors and other services, and to make settlers pay taxes.

The change from a political to an economic regime was not unconnected with global changes, the decline of communism, and last-ditch attempts at reform by the Derg. At the end of 1988 the markets were liberalized, allowing traders to purchase grain in the settlements. This initially led to a rise in maize prices, making settlers reluctant to sell grain at the fixed prices connected with the Agricultural Marketing Corporation quotas, though these were abandoned by 1990.

Decollectivization led to even more dramatic changes. Already in 1987 the peasant association had allocated small private plots on the edge of communal fields, counting on settlers simultaneously protecting the collective fields from wild animals. In 1988 the Ministry of Agriculture decided to allocate half a hectare per household for private use, and in 1990 collectivization was completely abandoned.

These changes were welcomed by most settlers, who could produce more, and produce it for themselves. The return to smallholder production on a household basis was seen as both rational and natural, even though settlers were amazed by the sudden about-turn. However, those in leadership positions and the militia who lost certain privileges, as well as some of the vulnerable elderly, widows and disabled who had previously had their plots cultivated for them, were less enthusiastic. Nonetheless, until the overthrow of the Derg the privileges of the elite continued, and settlers still had to work two mornings a week to cultivate their plots.

The major constraint which limited the settlers' ability to benefit from the changes was the lack of oxen for ploughing. In Qeto, the small number of oxen donated by the non-governmental organization Concern were prone to trypanosomiasis, and the village had to work out ways of sharing their use and responsibilities for joint management. Most settlers were still too poor to purchase oxen and feared losing them to trypanosomiasis. Nonetheless, they had been quick to convert cash obtained from increased sales of maize into small stock, the numbers of which increased dramatically.[4]

By the end of the Derg period, settlers were back to a familiar way of life in which they had access to private land and control over what to grow when and how. There were some signs that the settlers were becoming established and were developing a different view of themselves. During a return visit I made in 1989, some of them confided that when they visited relatives in Wello, they had been upset at being stigmatized as settlers. Likewise in Wellegga they began to reject the label 'settler', which they considered derogatory. In their own eyes they had ceased to be simply immigrants and thought of themselves as tax-paying peasants living in Wellegga.[5]

The post-Derg period

With the change of government following the overthrow of the Derg the fate of the settlements seemed precarious, and observers expected the settlers to leave. In this section I seek to understand why the majority left, the problems they faced, why some remained and how they fared, why some who had left subsequently returned, and why new settlers were attracted to come.

Evidence from Village Three and other settlements (Wolde Sellassie 1997: 65; A. Pankhurst 1997: 549–50) suggests that the bulk of those who left did so at the time of the overthrow of the Derg and in the period of the ensuing power vacuum until the transitional government become established. Departure became much easier since the restrictions on travel were no longer in place. However, many of those who decided to leave had in fact adapted to life in Wellegga and were fairly well established. Some pointed out that they would not have left had they not feared for their lives in the prevailing conditions of uncertainty and insecurity surrounding the change of government.[6] Most had returned to Wello with great apprehension. Some recalled with bitterness that their land had been redistributed 'as if we were dead' and that they had not been received with much sympathy. Some sought refuge with relatives who were facing their own problems. Others became sharecroppers or worked as labourers during agricultural peak periods or as daily labourers in urban centres; yet others became petty traders. Estimates of figures of returnees from resettlement in Wello vary from about a hundred thousand to over two hundred and fifty thousand.[7] Some were able to regain land from relatives, and in certain areas peasant associations gave them small amounts of land. However, this

---

[4] There were eleven sheep in 1985 and 644 by March 1989.
[5] When I expressed surprise at this new-found identity, one man wryly noted that the word for settler, *sefari*, rhymes with *assafari*, meaning 'shameful' (A. Pankhurst 1992: 270).
[6] In particular, they feared that the area would be caught up in fighting or that the settlements would be attacked by the Oromo Liberation Front, as happened in some other areas.
[7] Wide variations in figures point to lack of reliable data. See A. Pankhurst 1997: 551.

was generally unwanted or less fertile land from which it was very difficult to earn a living. Even when they obtained land, former settlers had to gain access to oxen, since they had not accumulated enough capital to purchase them and therefore became dependent on sharecropping. Given the prevailing shortage of oxen and feed, this meant entering into unfavourable arrangements to borrow oxen in exchange for labour, generally not at the most advantageous time, when owners reserved the animals for themselves.

When it became clear that conditions of relative peace had been restored and that, at least in the Qeto area, most of the remaining settlers had remained unharmed, some former settlers who had left for Wello gradually decided to return. Those who came back included people who were unable to (re)gain access to enough land to become self-sufficient; many from the younger generation who believed that they had better opportunities in the settlements; and the emerging generation who had been born in Wellegga and had no recollections of, or strong attachments to, Wello.

Those who came back to the settlements found that the land they had been cultivating had been redistributed among those who had remained. They thus found themselves doubly landless: in the areas from which they came more than a decade earlier, and in the areas they had been living in for many years.

In the atmosphere of fear at the time of the change of government in 1991, those who had remained comprised die-hard settlers who decided to stay come what may or who could not leave because they were too old, disabled or had small children. The former included people who felt they had turned their backs on Wello for good; who knew that they would not get land back, and did not want to become labourers; and a few who had left due to feuds and did not feel they could return safely.

Those who had remained recalled vivid memories of their fears in living through the changes. In Village Three the transition period left them relatively unscathed. There had not been any severe fighting for control over the area, the eventual transfer of power had occurred fairly smoothly, and the settlers had been disarmed by the new government.[8] In the land redistribution, some of those who remained obtained two hectares or more. Later, when those who had left began to return, they were initially only given back their garden plots of about a thousand square metres. However, gradually and through negotiation most were able to obtain additional plots, often ending up with a total of at least a hectare. Nonetheless, this process resulted in differentiation in the village between those who had remained and those who came back subsequently. However, even for the latter this represented much more than they could have hoped for in Wello.

---

[8] There had been attacks earlier on settlements, especially by the OLF, and at the time of the transition Village 18 was burnt down. Further attacks took place in 1992–93 (Yimer Yesuf 1996: 39, 50).

In addition to the former settlers who returned, a new wave of people began to seek ways to join the settlements, notably during years of drought and famine in northern Ethiopia.[9] Some of these were 'sponsored' by existing settlers who had gone to visit relatives in Wello and brought back a younger sibling or other relative, often arranging marriages for them with young men or women already living in the village. I heard of several such marriages during my visit in 1991.

At one wedding I attended, in addition to the official gift of clothing to the bride, covert negotiations were held about the deal to be struck. The young man from Wello had been sponsored by a relative and undertook to hand over to the bride's family, with whom he was going to be living, the six sacks of grain he would earn by working for his sponsor that season. In exchange, the bride's father agreed to allow his son-in-law to use some of his land, with the tacit understanding that it would eventually be transferred to him. This process of incorporation through affinity enabled settlers with more land to overcome labour shortages and also circumvented official restrictions on non-settlers gaining access to land.

However, new settlers without sponsors willing to provide them with board and lodgings and a share of the harvest tended to face difficulties. Their main option was to become wage-labourers at low rates of pay. During my 1999 visit I met a young man who had left Wello a couple of years earlier after his parents had died. He was looking for a settler relative in Wellegga. He came with a sum of 700 *birr*, which had been stolen in Addis Ababa, but fellow passengers paid his fare to Gimbi, from where he walked to Qeto. He had been looking for relatives in vain and had been working as a daily labourer for 3 *birr* a day. He confided to me that he had decided to return to Wello.

Other arrivals have come from different resettlement areas. In Village Three these include incomers who have moved from other Qeto settlements, particularly through marriage, as well as people from other 'conventional' planned villages, notably several households from other parts of Wellegga, especially Assosa to the north-west and Anger Gutin to the north-east, as well as some from a few 'integrated' settlements. This could indicate that conditions in Qeto are relatively attractive.

By 1999, however, the settlements were much reduced in size. A quick survey suggested that Village Three had a total of 159 households, representing about a quarter of the initially settled population and a decrease of more than half in a decade, despite natural population growth. Over a third of the decrease had taken place in the first five years, notably in the first year, but the bulk of the departures occurred at the time of the change of government. Although figures for the total population are not available, average household sizes seemed larger, as families regrouped, single persons whose spouses had left remarried,[10] and more children were born.

---

[9] New migrant settlers were noted in other cases, some even coming with livestock (Wolde Sellassie 1997: 64).

[10] Over two thirds of households had had a spouse leave since the beginning of the settlement.

There were interesting differences in areas of origin among those who were now living in Village Three. A much larger proportion of those who had stayed on were from northern Wello (Lasta), whereas the greatest proportion of those who had left were from southern Wello (Dese Zurya and Qallu).[11] This can be explained by a combination of environmental, economic, social and political factors. First, northern Wello is characterized by more severe land shortages, degradation, and recurrent drought than southern Wello. Second, settlers from northern Wello tended to include more of the frustrated younger generation of single men and women from dispersed localities who had made individual decisions to leave in search of land and better opportunities. By contrast, settlers from southern Wello tended to come in groups on the basis of a common decision, group pressure to leave together also being stronger. Third, in certain parts of southern Wello returning settlers seemed to have been more welcome and were provided with at least some marginal land, whereas land shortages in northern Wello seem to have been so acute that returning settlers were marginalized into living as labourers, often in roadside towns.[12]

Production, exchange & differentiation

The changes from collectivization to smallholder agriculture and from dependence on tractors to ploughing with oxen meant a return to a mode of production with which the settlers were familiar and competent, resulting in an increase in their confidence, independence and self-reliance.

Gradually settlers were able to increase their access to land. During the Derg period they first had just garden plots measuring 25 metres by 40 metres for their private use. Then they obtained in addition a quarter of a hectare, later half a hectare, and finally three quarters. After the overthrow of the Derg, those who remained or who came back promptly were able to increase their holdings to at least two hectares, whereas those who came later gained access to at most one hectare, resulting in some differentiation. Some land is used for growing coffee in the riverine areas, though protecting rights to cultivated coffee is difficult. Although those who returned later have less land and new settlers have none, most, especially the younger generation, have more than they could have hoped for in Wello.

Labour is in shorter supply, especially for weeding, guarding crops against animals, and harvesting. This in part explains the presence of new

---

[11] The small number of settlers from Lalo Qile in Western Wellegga all remained, presumably because they had little to go back to. They were made to leave mainly because they were too poor to pay their taxes following production shortfalls after the rains first failed and were then excessive, and after cattle diseases decimated their livestock, preventing them from manuring the fields (A. Pankhurst 1992: 69).

[12] During a short visit to northern Wello in July 1998, I was able to witness and hear of a few villages in southern Wello where returnee settlers represent a significant proportion of the population.

settlers working as wage-labourers. At times of peak labour requirements, settlers often resort to traditional labour-sharing groups, either working in turn or hosting a meal in exchange for their labour. However, the only remaining institutionalized form of collective labour[13] is the joint herding of livestock.[14]

Arguably, though, oxen are the decisive factor of production, especially given the threat from trypanosomiasis. Settlers are keen to own a pair of oxen, but only a minority have succeeded in doing so. Since oxen are vital for ploughing and threshing, those with a pair can hire them out and obtain land and labour in exchange on favourable terms through sharecropping.

Settlers' produce is becoming more varied, despite a continued dependence on maize and sorghum.[15] Some have been growing cash crops, notably sesame, peppers and coffee. Although settlers are worried about fertilizer dependence and price increases, yields of grain crops do not seem to have decreased as much as some observers feared.[16] In the past settlers used to carry grain on an individual basis for sale in markets. Nowadays entrepreneurs among the settlers make deals with local traders.[17] Although a handful of settlers in Village Three have become more prosperous on the basis of the trade in grain and consumer goods, differences are still fairly minor compared to other villages, notably those in the Mechara area, where trade has been the main source of differentiation.[18] Generally most settlers still rely primarily on the sale of maize, and those who have been selling pepper, coffee and sesame in more than small quantities are exceptions. Settlers have no option but to sell their maize at the same time in the couple of months after harvest to meet immediate needs and to avoid the danger of losses due to weevils. This means that prices tend to fluctuate dramatically, increasing by up to 300 per cent between the pre- and post-harvest periods.[19]

---

[13] During the latest war with Eritrea, settlers were made to work on the land of two members of the militia who had gone to the front, as used to happen on a larger scale under the Derg.

[14] There are nine such groups in Village Three, corresponding roughly to the former teams based on living along one row. Each member guards the livestock for one day.

[15] However, the varieties have changed. Settlers seem to prefer a white sorghum which requires less labour rather than the short-stemmed red sorghum promoted by the Ministry of Agriculture, which is allegedly more susceptible to damage from birds. Likewise the settlers are not keen on the maize varieties promoted by the Ministry as a package with fertilizers. Settlers seemed to fear indebtedness and objected to the semi-compulsory nature of the purchases.

[16] Settlers said they harvested 40–60 kg of maize and 15–20 kg of sorghum per hectare.

[17] The trader may offer them money in advance and a small profit of about two *birr* per sack, and the settler collects the number required. Alternatively a settler who has the capital or trust of settlers will buy sacks on the understanding of paying the settlers later and will reach a deal whereby the trader will provide the transport and they will split the profits.

[18] It is suggested that the settlers who have managed to buy generators worth 5,000 *birr* or who have even joined up to buy a mill worth 28,000 *birr* did so mainly from trade in grain and sesame. Likewise some have built tin-roof houses in Mechara, each costing up to 7,000 *birr*, from the proceeds of trade.

[19] Prices of maize in 1998 were 45 *birr* per quintal after harvest, rising to 150 *birr* prior to the 1999 harvest.

Looking Back on the Projects of the Socialist State

There are a number of indicators that settlers are now relatively well off. The houses they have built are generally larger and have several grain stores. All settler households have some livestock, including chickens, sheep or goats, and some even have cattle. However, those with a pair of oxen and a cow are considered well off; a few of the most successful even have two pairs of oxen and more than a dozen sheep. The settlers' diet now includes more fruit, vegetables, pulses and oil seeds and occasional animal products,[20] and they are able to buy some luxury items. Sugar is now occasionally added to coffee instead of salt, and settlers buy sachets to mix with water for soft drinks.[21] Settlers are better clothed and the women wear jewellery. They also have more consumer goods. Radios, which used to be exceptional, are now quite common. The more prosperous settlers in some villages have bought bicycles and even sewing machines. In the Mechara area some settlers have been constructing houses with corrugated iron roofs near the roadside, one settler has bought a generator with which to show videos, and several have joined together to purchase grinding mills.

Generation, language & identities

During my visit in 1999, Meseret, one of the first children born in the settlement thirteen years earlier, was married to a young man from Village Six. Her father, who had earlier served in the peasant association administration, had named her Meseret, meaning 'Foundation', to symbolize his commitment to starting a new life. Meseret was her mother Alem's first child, and since then Alem had given birth to four other children. Meseret's wedding was a grand event; an ox was killed, and her mother brewed fourteen barrels of beer. Meseret was the second girl born in the village to be married off; so far none of the boys there have been married, which is indicative of the custom of the early marriage of girls, especially in northern Wello.

Revisiting the village a decade after my initial stay, I was struck by the contrast in meeting people of different generations. There were adults who knew me, many of whose faces I recognized, even if I could not remember their names; there were the former youngsters, many of whom had grown up so that I no longer recognized them; and there were the children, all surprised by my presence, not having even heard that a foreigner had lived in their village. Although detailed population figures are unavailable, the majority are clearly young adults and children who were either born in the

---

[20] This is mainly due to the greater availability of animal products (especially eggs, milk and yoghurt and occasionally meat), more types of grain (notably millet and even some of the prized cereal *teff*), and more fruit (mainly papaya and banana, but also mango and some guava), pulses (in particular beans and chickpeas, but also lentils and peas), vegetables (especially cabbage and beetroot) and oil seeds (notably sunflower, linseed and sesame).
[21] These come from South Africa (and are past their sell-by date).

village or who were too young when they came to have strong recollections of their homeland. Increasingly, the village is composed of a new generation whose experience is rooted in life in Wellegga and for whom Wello remains a distant and hazy land brought to life only through the talk of elders and the occasional visit of relatives.

A discussion I had with Yimer Yesuf, a settler who completed his BA on the history of Qeto settlement at Addis Ababa University, was particularly revealing when he told me about his family. His parents left Qeto settlement and returned to Wello, but his elder sister remained with her husband. When he went to visit his parents in Wello, his younger sister begged him to take her back to the settlement, which he agreed to do. She is now living with his elder sister and goes to a local school where education is in Oromo. Yimer believes that she stands better chances in the resettlement village than in Wello.

Linguistically, the settlers represent a small island largely of Amharic-speakers among the Oromo of western Wellegga.[22] Under the previous government they were privileged, since Amharic was the official language as well as their mother tongue, and they did well in the local schools, where Oromo children were studying in their second language. Indeed about ten young men from the Qeto resettlement complex of twenty villages, who were in the middle of their schooling when they were resettled, were able to complete secondary school in Wellegga, and six went on to higher education or teacher training colleges. Three of these joined Addis Ababa University.

However, with the regionalization policy after 1991 and the change to a curriculum in local languages, the roles were reversed and it was the settlers who found themselves at a disadvantage. Although some settlers stopped sending older children in particular to school, many continued to do so. Some villages raised funds to employ a teacher to teach their children Amharic.[23] For a few older Christian boys, church education, which is inextricably linked with Amharic and the liturgical Ge'ez, offered another option. Priests have been providing some basic literacy training, and prospective deacons in Village Three were studying in the evenings, at weekends and on holidays.[24] However, for most, education is unlikely to offer a means to employment, especially since jobs in government are now allocated on the basis of regional ethnic affiliation. Although prospects for settler children reaching high school do not seem bright, most seem to be benefiting from at least some exposure to primary education, which will

---

[22] In several of the other Qeto settlements, there were also some Tigrinya, Kambata and Hadiya speakers. In Village Three there were also some Oromo speakers, mainly from Qallu in Wello, but also from Lalo Qile in Wellegga (A. Pankhurst 1992: 134–5).

[23] Bulcha, an Oromo settler from Lalo Qile living in Village Three who had completed the twelfth grade under the previous system, is teaching basic Amharic to children from several villages.

[24] While I was there in 1991, a couple of boys were preparing to go to Nekemte to try and obtain credentials from the regional church administration.

presumably help them become more integrated and improve their Oromo second-language skills.²⁵

In Village Three, identities have shifted from a situation in which, in many contexts, people felt that their common identity as settlers was more important than their original backgrounds to one in which earlier regional and especially religious identities have become much more salient. This is visible even in the settlement pattern. Settlers from different areas were initially given household plots in mixed rows. Gradually, however, as some left, they began to move closer to relatives and friends. With the large-scale departures at the time of the change of government, a clearer distinction began to emerge. The upper part of the village is now inhabited mainly by people from Lasta in northern Wello, and the lower part by people from Desse in southern Wello. Rivalries and differences between them have also begun to emerge.²⁶ Moreover, religious identities have become more evident. The Christians have built a church with a corrugated iron roof and often refer to Village Three as Qeto Medhane Alem.²⁷ Many go to other villages for annual saints' days.²⁸ The Muslims have built a mosque and join in prayers together, especially on religious occasions.

Environment, regionalization & relations with local people

One of the prevailing criticisms of resettlement has been that the presence of settlers would have a devastating effect on the environment (Alemneh 1990a, 1990b; Dessalegn 1988). In the lowlands, sites were selected in areas assumed to be 'virgin', but which were often used by indigenous peoples for shifting cultivation, thus not overly exploiting the fragile soils. The land-use rights of local people were ignored and settlements were established without local consultation, consent or compensation for lost resources (A. Pankhurst 1997: 551–4). Sites were selected without land-use planning and were often abandoned after much investment and wasted energy, owing to a lack of water, or conversely water saturation, etc. (Alemneh 1990a: 107–8; A. Pankhurst 1992: 111–12). The local environment, top-soils, bush and forests were assumed to have been severely affected by the massive influx and excessive population concentration in the villagized settlements.

---

²⁵ Children living in the settlement village have few opportunities to practise Oromo except at the market. However, the few remaining settlers from Qallu in south Wello speak Oromo with only slight variations in dialect, and a few children with parents from Wello who are living close to Oromo settlers from Wellegga have picked up some Oromo.

²⁶ The divisions became clear to me when I bought a football and gave it to my friends from Desse, with whom I was staying in the lower half of the village. The children from the upper part of the village petitioned me to purchase another one for them.

²⁷ 'The Saviour of the World', i.e., Christ.

²⁸ While I was in Village Three in January 1999, large numbers of settlers went to Village Six for the annual St. George's Day celebrations there.

Surviving Resettlement in Wellegga: Qeto

However, fifteen years later, this scenario – to which many writers, including myself, subscribed[29] – may be worth reconsidering. Certainly the deforestation of the riverine belt is clearly visible, and I can recall the difference from a decade earlier and remember the accounts of the settlers about their fears of the forest and of wild beasts when they arrived.[30] Likewise, the initial clearing of fields by bulldozers and several years of tractor ploughing no doubt had a deleterious effect on the soil, especially on the slopes. In 1991, visiting some other villages, I also heard women complain for the first time about the distances they had to walk to collect firewood.

Nonetheless, there are a number of mitigating factors and even some positive developments. This is in part due to the substantial reduction in the population to about a quarter of the number of those initially resettled. The riverine forest has not been completely destroyed, and some trees have also been left in the fields, the larger ones often used for hanging up hives. The move from tractor to ox-plough cultivation has reduced the threat to the top soil, and the smaller size of the separate private plots in the fields results in less erosion than in the large collective fields.

In the settlements, the houses are now more spaced out and the homesteads surrounded by fruit trees, especially banana and papaya, but also mango and guava. Settlers have been planting trees for fuel and building purposes, although these are almost exclusively eucalyptus and some *gulo* (*Ricinus communis* or castor oil tree), whose stems are used for roofing supports and seeds for oiling cooking pans. However, tree planting is mainly restricted to household plots.

It might have been thought that the 1991 change of government would result in tensions between the settlers and the local people. However, fears of violence at the time of the change of government, which were common elsewhere, were not borne out in this area. There were a few incidents involving the villages,[31] but these related to OLF activities rather than to actions by the local Oromo peasants. Regionalization, with its logic of ethnic homogeneity, might also be thought to create a sense that settlers coming from other regions and cultural backgrounds were unwelcome. However, this does not seem to be the case, for several reasons. First, since the settlers have effectively been disarmed,[32] they no longer pose a

---

[29] For instance: 'Serious ecological problems have been created due to the destruction of forests for cultivation, construction, the sale of wood, charcoal and wooden products, the cultivation of fallow areas in the highlands and the threat to the bio-diversity and the river system. Thus, in addition to the immediate and direct threat to the livelihood of locals, the settlements represent a long-term environmental disaster' (A. Pankhurst 1997: 554).

[30] One settler even recalled that for several days they did not realize there was a river, so thick was the forest (A. Pankhurst 1989: 249–51).

[31] Even in the case of Village Eighteen, where the settlers had their houses burnt and were forced to leave by an OLF attack, there were no reports of massacres. However, it is said that the attitude towards the 'integrated' settlers, mainly from Tigray, was harsher.

[32] Their automatic weapons were confiscated. Some settlers still have old rifles, but bullets are scarce and settlers are prohibited from hunting game.

significant threat. Second, the local administration seems to have taken a pragmatic position in appreciating the direct revenue from the taxes they pay,[33] as well as the increase in production, trade, and markets they have fostered. Third, the settlers have good relations with local grain traders, many of whom have become prosperous, initially through the aid business and more recently through trade in grain and consumer goods. Fourth, settlers have entered informal land deals which have been beneficial to some of the local Oromo farmers. In the Mechara area, a small roadside town has grown up composed largely of settler houses with tin roofs built on land obtained from local farmers.[34] It thus seems that, contrary to expectations, several powerful sections of the local population see benefits in the settlers' presence.

However, most settlers in Village Three and the other Qeto settlements have had limited direct relations with most of the local Oromo except through market exchanges.[35] Intermarriage is most exceptional.[36] There is one case of an Oromo man who gained access to land in the village and has a sharecropping arrangement with a settler. The exceptions in Village Three are the Oromo settlers from Lalo Qile who have strong ties with the surrounding Oromo and have formed joint religious groups.[37] Former disputes over land for growing coffee and beehives seem to have subsided, as settlers have been planting coffee and hanging hives in areas allocated to them. In general though, the resettlement villages have stronger relations among themselves and with other 'integrated' villages. Thus intermarriages with other settler villages are fairly common, perhaps becoming more common than marriages within villages, and moving from one village to another is quite common.[38] Settlers also attend marriages and religious celebrations in other villages.[39]

---

[33] Between 20 and 80 *birr*, depending on the size of their land. In Village Nine a proportion of the income from wood sold to settlers from the village forest goes to the *wereda* in the form of taxes.

[34] The settlers bought land from local Oromo for 1,000 to 3,000 *birr* and have been building houses requiring an additional 3,000 to 4,000 *birr*.

[35] Although market exchanges are generally peaceful, occasional brawls do occur, since market-goers earn money and sometimes get drunk.

[36] There is only one case in Village Three, between a settler woman who came from an integrated settlement and an Oromo husband who died afterwards in the village. There is also a case of a settler from Wello who is still married to an Oromo woman from the Lalo Qile settlers. I also heard of a case of a settler marrying an Oromo woman in Village Twenty.

[37] Some have established a Protestant Mulu Wengel Church, a few have remained Orthodox, and others have even converted to Islam under the settlers' influence.

[38] I counted 23 cases of settlers from Village Three who moved to other Qeto settlement villages, and there must be many more.

[39] When I arrived, many of the Christian villagers were attending a marriage at the nearby integrated site of Chanqa Bururi, and others attended a wedding at Village Six. During my stay, villagers attended the annual celebrations at the Church of St. Giorgis in Village Six, and an annual feast to the Virgin Mary at the Church of St Mary, which settlers from 'integrated' sites had built with the local population at Chanqa Bururi.

Surviving Resettlement in Wellegga: Qeto

Conclusions

Relations between the settlers and the state have undergone radical transformations over the last decade and a half. Peasants who had been self-reliant smallholders using oxen for cultivation were uprooted and turned into wage-labourers on mechanized collectives. The experience of loss of independence and control over production was alienating and generally resented. However, this chapter has sought to show how responses varied. The few in leadership positions and the militia fared best, and the younger generation adapted better than women, the elderly or the devout, who found the experience difficult to come to terms with.

In the past decade, however, further changes have completely altered the circumstances in which settlers now live. Towards the end of the Derg period, the decollectivization and liberalization of the economy meant that settlers once again found themselves becoming smallholder producers dependent on oxen for cultivation.

At the time of the change of government, the majority of settlers departed, fearing fighting and attacks. Those who remained felt that they had nothing to return to or were unable to leave because of young, elderly or disabled dependants. In the event, in Qeto, at least, settlers were hardly affected. Those who remained redistributed the land of those who had left, obtaining much larger holdings and increased harvests as a result. Those who left hoped to reclaim their land in Wello, but a significant number returned disillusioned to the settlements.

Today, most settlers have enough land. Labour shortages are overcome through sharecropping, work groups, wage labour and the incorporation of new settlers. A shortage of oxen is the most serious constraint, given the threat from trypanosomiasis, as cattle losses and veterinary costs drain settlers' surpluses of grain converted into livestock. Nonetheless, most settlers in Qeto are by now relatively well off and would say that their living conditions are better than they would have been in Wello. Their diet has improved, since they are growing a wider range or crops, including fruit and vegetables, as well as some cash crops, notably sesame, peppers and coffee. However, maize and sorghum remain the main crop, and a few settlers and local traders have prospered through grain trading, which has been the main factor explaining the differentiation of wealth within the settlements. Some settlers can thus afford luxuries.

Gradually there have been changes in the settlers' sense of identity. Fifteen years on, in many ways settlers no longer see themselves as immigrants. Some have cut their ties with their former homeland, while others have sought to keep a foot in both places, and a few have become long-distance traders. A major change has been demographic, given that the younger generation were born in Wellegga or have no recollection or attachment to Wello. However, paradoxically, former regional and religious identities have become more salient within Village Three, as reflected in changes in settlement patterns and social organization.

Looking Back on the Projects of the Socialist State

Under the current government's regionalization policy, the settlers represent isolated pockets of mainly Amharic speakers, whose children are learning Oromo in schools. Some have sought to continue to teach their children Amharic by employing teachers or through church education. However, in contrast to several other areas, regionalization so far has not led to the settlers being made unwelcome. This is partly because the local administration is benefiting from the revenue from taxation that the settlers provide and partly because certain sections of the population, notably traders and farmers who have reached deals with settlers, have benefited from relations with them. The dire consequences for the environment do not appear to be as acute as was predicted, owing to a much-reduced population, the abandoning of tractor cultivation, a reduction in the size of private plots and some tree planting.

Despite the considerable problems the programme faced at the start (the abuses of human rights, the haste and the lack of proper preparation), resettlement from Wello to the Qeto area – which a decade ago seemed to offer little likelihood of sustainability – has, after fifteen years, not ended in complete failure. Although probably only about a quarter of the settlers have remained, they are now clearly self-sufficient and producing a small surplus. Those who have remained seem committed to staying in Wellegga if they are allowed to do so. They have invested in the land by building larger houses and religious buildings, as well as in social capital, by establishing relationships.

Finally, it should be stressed that these findings cannot be generalized, since conditions in Qeto are undoubtedly very different from other resettlement areas, which may well have faced more severe crises (compare, for example, references by Eisei Kurimoto and Wendy James to resettlement schemes in the far west and south-west, Chapters 12 and 14). Moreover, the untold human as well as material costs of resettlement have been immense, and resettlement probably had a minimal impact in alleviating population pressure in Wello.

III

The Promise of 1991
Re-shaping the Future & the Past

Introduction
Donald L. Donham

In this part we group together chapters that focus on the period of the EPRDF regime, from 1991 onwards. As historians often observe, it is inherently difficult to gain purchase on the present. We do not yet know how the story will end. What we do know is that the most dramatic aspects of the high modernist moment have ended in Ethiopia, at least at the national level. (For indications that it sometimes continued at a local level even during the 1990s see Chapter 10, Ren'ya Sato's account of the events among the Majangir, where evangelical Christians took up the project of villagization after the Derg had fallen.)

As they took control of the state in 1991, TPLF leaders gave up the Marxist metanarrative of socialist progress for a more qualified rhetoric of the devolution of power and economic liberalization, democracy and human rights. Compared to the Derg's overt project of *encadrement*, the Ethiopian state of the 1990s would seem to have made a significant retreat vis-à-vis local society.

Jon Abbink's chapter on the Suri demonstrates just how misleading such a conclusion would be. The EPRDF government, it is true, has less readily resorted to coercion. But this does not mean that local society has been left alone. Just the opposite: while extolling the right of self-determination, EPRDF cadre among the Suri campaigned against 'backward' customs such as lip plugs and ritual stick fights. In addition, they presided over elections to the new Surma Council that would administer the newly-drawn administrative district, the 'Surma *wereda*'. The young men 'elected' – not unlike the ex-Christians in Maale who were taken into the Workers' Party as described by Donham (1999) – were overwhelmingly those who, through a variety of culturally anomalous experiences, were able to speak Amharic. Ironically, then, in Chapter 8 Abbink shows that the potential for social and political change in Surma expanded after 1991. During the Derg, Suri age grades and traditional ritual leaders – like similar institutions described by Tadesse Wolde for the Hor – had continued to provide the local structure of decision-making. By the 1990s, the Council was poised to take over these functions – ironically, in the name of 'tradition'.

Abbink emphasizes the increasing openness of local society to regional changes – the flow of automatic weapons from the Sudan, the increased presence of missionaries and tourists. These themes are echoed in Chapter 9, Matsuda's study of the Muguji. Like other examples in East Africa, the Muguji in the nineteenth century were a submerged group of hunters and gatherers attached to an agropastoral people, the Kara. In *Southern Marches*, David Turton described a similar example of the Kwegu attached to the Mursi. Traditionally, the relationship between such groups was hierarchical and caste-like: the two did not intermarry or eat together.

Towards the end of the Workers' Party period, the Muguji fought their Kara overlords and had to flee to the protection of the Nyangatom. By that point, the Nyangatom had armed themselves with automatic rifles, obtained largely from their Sudanese neighbours, the Toposa who had been armed by the Sudanese government as a bulwark against the Sudan People's Liberation Army (SPLA). As the Muguji began to trickle back to their homes, the TPLF came into power in Ethiopia. By 1993, the Muguji were given their own *qebele* administrative unit oriented in the direction of the Nyangatom (rather than being bound in with the Kara). At that moment, one could say that the Muguji became an 'ethnic' group. And they did so in a way that symbolizes all the intersecting complexities of our postmodern world: Matsuda's own former research assistant, a Nyangatom man who had lived with him among the Muguji – and who by 1993 was a zone governor in the new order – played an important role in gaining government recognition for the Muguji. The new visibility that equal ethnic status conferred upon the Muguji allowed them to participate in the interregional gun trade in a new way. Matsuda demonstrates how the Muguji subsequently armed themselves with modern automatic rifles. Armed and 'ethnicized', the Muguji began to enjoy a new equality with their neighbours – but at the price of increased threat and volatility.

In the case of the Majangir, Ren'ya Sato shows how a new form of ethnic consciousness was connected with conversion to evangelical Christianity. As described by Jack Stauder in the 1960s, the Majangir were hardly touched by imperial administration. Villages often moved over abundant supplies of land, local deaths being the precipitating cause. The notion of preventing all outsiders from using one's 'own' land was a foreign concept. Sato shows how dramatically different Majangir life was by the late 1990s. Conversion to evangelical Christianity played a critical part, he argues, in the formation of a new consciousness, particularly among the

Introduction

young, that Majangir were both different from and equal to their neighbours. When the remapping of district boundaries by the EPRDF failed to match realities on the ground, Majangir grew upset. Some had been included in Oromiya, some in the region of the Southern Nations, Nationalities, and Peoples, and finally the bulk in Gambela (which the Majangir were to share with the Anywaa and Nuer). In frustration, a large group of armed Majangir attacked and murdered more than a hundred non-Majangir living near Tepi in April 1993.

Even if boundaries could be established in relative conformity to peoples' own sense of cultural identity – and this is made virtually impossible by the fact that many Ethiopian peasants have pursued intentional strategies of spreading their kin ties across social boundaries – in Chapter 11 Elizabeth Watson presents the inherent ambiguities of ethnic federalization among the Konso. The Konso were given the status of a special *wereda* with the right to 'self-determination up to secession'. For many Konso, this was greeted positively for it appeared to mean that they would now be able to resume the status they had at the end of the nineteenth century – before they were conquered and incorporated into imperial Ethiopia. Konso 'tradition' could now be restored.

Watson demonstrates why this has not occurred. With the status of a new special *wereda*, the Konso now had authority over their own Ministries of Health, Education and Agriculture. The work in all of these was carried out in Afa Xonso, the language of the Konso. This meant a 'one-hundred per cent' employment rate among educated Konso – something that could never have occurred in any previous order in which educated Konso had to compete (necessarily at a disadvantage) with elites from northern Ethiopia. But nearly all educated Konso had been produced by evangelical Christian missions and so had been influenced by mission ideas that equated Konso tradition with 'backwardness' and 'superstition'. Consequently, the project of retraditionalization – even though, ironically, promoted by some foreign NGOs working in the south in the 1990s – did not catch on in Konso.

In Chapter 12, Eisei Kurimoto is able to trace for the Anywaa processes that parallel what Watson details for the Konso, but to follow them well beyond the early 1990s 'honeymoon' period. Considerable historical depth of memory is also drawn on in this study, as the Anywaa come to reflect back on the former strength and grandeur of their kings and nobles. As among the Konso, the Anywaa – Anywaa men in particular – at first welcomed the changes brought by the EPRDF. When he interviewed Anywaa women in the mid-1990s, however, Kurimoto was shocked to discover their considerably more negative views. It is interesting to speculate that women – having been married into their husband's villages, some perhaps from across the boundaries that were being 'ethnicized' – were well placed to appreciate the coming contradictions of the new order. However that may be, the (men's) initial euphoria was over by the late 1990s, as population flows of Nuer into Gambela prompted by patterns of conflict in the Sudan threatened to swamp the Anywaa in the new game of 'ethnic' politics.

As the decade closed, the fragility of today's conditions and the social confusion threatened by 'ethnic' federalism to the Anywaa is reflected linguistically in how they described 1991 as compared to 1974. They used their own traditional word, *agem*, to refer to the, for them, more ordered and comprehensible changes of 1974. This does not mean, of course, that Anywaa necessarily recalled the 'revolution' of 1974 with affection, only that they felt they understood it, and could place it in terms of their own history of dynastic coups and power struggles. However, they borrowed the Amharic word *girrgirr* to refer to 1991 – with all of its connotations of riot and confusion.

Finally we come to Chapter 13, Cressida Marcus's study of the Amhara of Gondar. Marcus demonstrates the centrality of Orthodox Christianity, and particularly, the intense sociality of local church organization to Gondari Amharas' sense of identity and history. At the height of its project of *encadrement*, the Derg tended to see such local institutions only as obstacles. In that setting, church going and church building became forms of what Marcus calls a 'silent monumental resistance'. The religious efflorescence that began during Derg times, however, continued and took on new meanings under the EPRDF. Remembering the famous forty-four churches of Gondar became a way of nostalgically recalling a time in which Gondar and the Amhara constituted the very core of the empire. This process recalls various efforts at retraditionalization across Ethiopia in the 1990s. Among the Maale, for example, a ritual king was reinstalled at the end of 1994 (Donham 1999: 183) – a dramatic event after all reference to royalty had been suppressed by the Workers' Party for the previous decade and a half.

If, then, the Derg's explicit project of *encadrement* has been rejected at a national level, this does not mean that the Ethiopian state's fundamental project of 'capturing' its citizens has ended. Rather, this undertaking has only changed character in the 1990s. In converging with international expectations and rhetorics of democracy, it has come to mirror global discourses on rights, reparative group rights in particular. What this mirroring has allowed, in many cases, is a co-optation of local (male) elites. In this new context, political leaders – even at the village level – have become increasingly conscious of acting upon a stage in a kind of theatre of 'ethnicity'. Not only has the range of 'characters' brought into the play expanded, but the pace at which events occur has accelerated. It is in this context that we appreciate the 'unreality' of the world for people like Kurimoto's elderly female Anywaa informants: just at the point when they should have been able to preside over large households of coresident children and children of children, their progeny is scattered to foreign cities like Oslo and Dallas – not lost completely but now in only ephemeral and disembodied contact via letters and telephone calls.

Eight

Paradoxes of Power & Culture in an Old Periphery
Surma, 1974–98

JON G. ABBINK

This chapter discusses the dynamics of interaction between the Ethiopian political centre and the Surma or Suri people, a small group of independent agro-pastoralists located in the Maji border area of the south-west, from 1974 to 1998.[1] It will be argued that, in this period, the Suri – like neighbouring smaller groups in the Maji area such as the Dizi and the Me'en – provide an example of how allegedly 'marginal' populations were challenged, if not forced, to break out of their peripheral condition into one of engagement and co-optation that necessitated indigenous responses to an encroaching 'modernity'. Modernity is an ensemble of socio-economic and institutional conditions,[2] as well as a state of mind and a new conception of interpersonal relations. In the Ethiopian case, I see it as being manifested in an expanding bureaucratic and military state, politico-adminstrative incorporation of local communities, the import of new ideologies (codified law, Marxist socialism, ethnic self-determination) and the governance, or, more specifically, 'surveillance', of citizens.[3] Up to 1974, imperial policy towards the smaller and politically less relevant populations on the margins of the empire was marked by a situation of what one might call 'tolerated difference'. Under the two post-revolutionary

---

[1] I gratefully acknowledge support for fieldwork and further research provided by the Netherlands Organisation for Scientific Research in the Tropics (WOTRO, WR 52-610) and the African Studies Centre (Leiden). I also express my thanks to Dr Taddese Beyene, Professor Bahru Zewde and Dr Abdussamad H. Ahmed, successive directors of the Institute of Ethiopian Studies, Addis Ababa University, for their interest and support during the years 1992-97.

I would like to dedicate this chapter to the memory of a remarkable man, the Chai *komoru* Dollote (Ngatúlúl), who to my deep regret died in 2000 (see Plate 8.1).

[2] See, for example, Giddens 1990.

[3] The influence of missionizing and tourism can, especially in the case of peripheral groups like the Suri, also be reckoned part of a process of modernization, but this will not be treated as central here.

The Promise of 1991

regimes of the past twenty years, this situation came to an end.

The Ethiopian south-west is traditionally the habitat of a number of smaller ethno-linguistic groups that became part of the Ethiopian Empire under Emperor Menilek II in the late 1890s. This implied their being drawn into a hegemonic order, a 'national project' defined by the highland Ethiopian state, but in which they were deemed peripheral on account of their political, economic and cultural characteristics. They became part of the political economy of this state through indirect rule (via *balabbat*s or local chiefs), taxation, and the *gebbar* system (see Donham and James 1986). Ethno-cultural differences and indigenous social organization were largely tolerated in the imperial system of governance: there were no mass campaigns of forced conversion or abandonment of customary law, traditional socio-political organization, or 'harmful customs' as long as political loyalty was shown to the centre.

The position of these mostly Omotic, Cushitic and Nilo-Saharan speaking groups in southern Ethiopia has changed significantly in the last decade, because of new interdependencies between 'minorities' or 'nationalities' and wider Ethiopian society. Especially in the case of the Suri (who are central to this chapter), this process of change in the south-west must be interpreted as part of a gradual process of the redefining of a periphery.[4] Under the EPRDF-led federal republic after 1991, this process has decisively accelerated and gone in the direction of 'dissolving' the idea of periphery and marginality itself, by redefining ethnicity and ethno-regional identity as constitutive of a new federal Ethiopian state.[5] As this chapter will show, however, this process of the ethnic redefinition of the nation is fraught with problems and paradoxes.

The political context

It could be said that under the empire and the Derg, relations between the Suri and the state were reproduced as ones between core and periphery in a politico-economic and cultural sense: the Suri people (and the region) were seen as being quite separate in culture and mentality, as 'having no religion' and as being politically peripheral (if not inaccessible, since they lived in one of the remotest lowland areas), though their area was economically of use as a hinterland for cattle and natural products (coffee, hides and skins, ivory, and later some gold).

The underlying hypothesis of this chapter is that under the post-1991 regime installed by the Tigray People's Liberation Front, with its policy of recognizing the rights of 'nationalities', Suri society, now seen as 'different'

[4] The people under discussion call themselves *Suri* (when linking Chai and Tirmaga, the two sub-groups). As this chapter deals with the perception of 'periphery' I also speak of *Surma*, because this is the term by which they are known among outsiders (including government agents) and also indicates the area where they live (now the 'Surma *wereda*').
[5] Article 39 of the 1994 Ethiopian Constitution, which recognizes the right of the 'nations, nationalities and peoples' of the country to self-determination up to secession, illustrates this.

Paradoxes of Power & Culture: Surma

in a relevant (rather than irrelevant) way, is being co-opted into the political dynamics of a new, de-centred, 'ethno-federal' state project. The Suri are also being affected by new globalizing cultural forces such as Christian missionizing and tourism, both present in their area since the late 1980s.

It is interesting to examine the question of whether and how the peripheral status of the Suri community is being reduced. This could be argued in reference to changes in core practices like local 'leadership', ritual life and community organization, as well as in cultural values and identity. However, I argue that peripheral status, at times and in some respects, provided the Suri with leverage and privilege in maintaining their autonomy. Under the imperial regime and the Derg they were largely left to their own devices. Under EPRDF rule they can claim separate status as a 'minority' entitled to self-government under the 1995 Constitution and the new administrative arrangements, and they are now organized into a new administrative district (the 'Surma *wereda*'; see Map 1.3, p. 31). This has led the Suri into a new and challenging phase of their history, where an emphasis on their 'ethnic identity' may turn out to have a short-term beneficial role.

Despite the fact that the studies in *Southern Marches* (Donham and James 1986) paid attention to the politico-economic connections between the southern margins of Ethiopia and developments at the 'core', there is still ample room to develop more integrated historico-anthropological perspectives on 'centre-periphery relations', especially in the wake of the demise of the empire in 1974. This is especially true for the historical exploration of the nature and impact of political and cultural dominance and its long-term transformative impacts on local societies. This applies not only to 'ethnically different' societies in the west, south, and east of Ethiopia, but also to many rural societies in the highlands, which are just as politically marginal as the south, as a study of relatively isolated rural areas such as, say, Wollo Borena, Amhara Saynt, Wolqait, Quara, Raya-Azebo, Selale or Genda Beret would show (and see, for example, Chapters 4, 5, and 7 above). This shows that ultimately the notion of centre–periphery relations should not be based on geographical or cultural criteria, but primarily on a model of the structure and distribution of political power.

The 'unknown' Suri

The Suri are an agro-pastoral group of some 26,000 people, organized on the basis of an age-grade system with three ritual leaders. They have come to be known as 'Surma' but usually call themselves 'Chai' or 'Tirmaga'[6] (two territorial sections with different ritual leaders, and with a dialect difference), or in collective self-reference 'Suri'. Further, they reckon

---
[6] Formerly called in the literature 'Tirma'. Tirmaga is the correct self-name.

themselves to be members of named groups or 'clans' based on patrilineal descent. They are marked off from their neighbours – except the Mursi and Baale,[7] with whom they have much in common linguistically and culturally – by having their own language, ritual traditions and norms of kinship. The Suri live in a territory which has acquired significance for them in that it forms their 'meaningful area of living', with historical, cultural and economic dimensions. Their cultural space is a combination of territory with the ritual and other socio-cultural means to sustain and sanctify it. The latter refers to specific practices, like chiefly ritual blessing and reconciliation ceremonies, initiations, divination with reference to intestines, and ritual places like *komoru* (ritual chief) burial and initiation grounds. The perimeters of this cultural space were not defined or influenced by *state* referents: Suri territory was a domain outside state rule or regulation. It was not clearly bounded, as reflected in the (overlapping) territorial arrangement of named local Suri groups as well as neighbouring peoples, with whom the Suri shared water holes, pasture and cultural arrangements.

The state as an (external) hierarchical politico-administrative structure bent on exploiting the resources (both natural and taxable) generated by the local society and establishing a monopoly in the use of armed force made its presence felt only in the late 1950s, after the imperial regime had made the Suri tributary to the state for tax purposes. This lasted until 1969. In the Derg period and under EPRDF rule to late 1999, the Suri still did not pay taxes, although about eighty Suri youths were forcibly recruited into the Derg army. In the late 1970s, the government started a few primary schools in the Suri area (the last one was abandoned in 1988). The only other service given by the state was periodical inoculation of cattle by a mobile clinic, which reached only a small number of Suri. Primary health care posts were located in highland villages outside the Suri territory and were frequented only sporadically.

For local society, the Ethiopian state continued to figure as an external, impermeable framework of power and hegemony, led by elites located outside local arenas of interaction. In the decades of contact between Suri and state agents, the conflictual dynamics of Suri–state relations in their various forms were shaped by the unresolved competition of cultural models of the nature and construction of political power, space and sociality. State representatives, northern in-migrating settlers and local (Suri) society, all had different notions about the specific hierarchical power structure in Ethiopia.

Among the Suri, political power relates to elements like the status equality of individuals, collective authority, the absence of coercive force being used on members of society, and political deliberations in public

---

[7] The Baale (formerly known as 'Zilmamo') are a group of some 8,000 agro-pastoralists living in the Ethio-Sudan border area, partly in the Boma Plateau, northwest of the Suri. They speak a Southwest Surmic language, related to that of the Suri. See Dimmendaal and Last 1998 and Bader 2000.

until 'unanimous' decisions are reached. The state usually worked in opposite ways to these. For the Suri, the unarticulated concept of sociality (being human, and being socially connected) implied, amongst other things, the idea of equality among persons, the observation of balanced reciprocity, and the idea of open access to resources by all who knew how to use them (i.e., for making a living, not for personal exploitation). The approach of state agents, by contrast, was marked by ideas of boundary, territory, privileged access, and political and cultural hierarchy.

The exchange of material goods and persons which emerged in the Maji area after the turn of the century (see below) did not create a common sphere of understanding or a voluntarily entertained, durable relationship between the state and local societies; on the contrary, emerging perceptions (and practices) of rivalry, 'competition over resources', the unpredictable and often arbitrary use of force, and inequality between groups often exacerbated difference and distance.

Suri space in the context of the imperial state, 1898-1974

After 1898, the imperial state and its agents tried to include the Suri in a 'tributary' economy, which attempted to establish political hegemony while being indifferent to local culture. Northern settlers in the highland villages traded cloth, iron tools and later guns and bullets with the Suri for cattle, grain and game products (ivory, hides and skins), and also occasionally raided them. They did not incorporate the Suri into the *gebbar* system (as was done with the Dizi; see Garretson 1986).

In this situation, the role for the state in exercising coercive power and establishing an administration was limited. Until the Italian occupation in 1936, it rarely intervened to stop local conflicts and inter-group violence (e.g., cattle-raiding), and indeed may even have stimulated them in order to gain access to a supply of war prisoners taken as slaves and retainers to the north. Traditional cultural mechanisms of retaliation and reconciliation were usually observed among the various local peoples in times of conflict: state suppression was not the norm. On the side of the settlers, who deemed themselves culturally superior, or at least more developed, in terms of religious identity and political tradition, there was no forceful effort to assimilate the Suri to northern, Christian culture. The aim was to enforce political loyalty and the economic incorporation of local peoples into the wider Ethiopian politico-economic framework. The outlet for local products from Maji (however limited in volume) was the Ethiopian market (via Jimma to Addis Ababa), and the people in Maji (including the Suri) worked with only Ethiopian currency before and after World War Two.[8]

[8] There was also, however, trade with groups living in the Sudan, especially the Baale and Murle (beads, ornaments, some cattle and grain). The Italian interlude from 1936 to 1941

The Promise of 1991

Plate 8.1   The Komoru *of Chai-Suri (son of Dollote IV), 1996* [J. Abbink]

After liberation from Italian occupation in 1941, Emperor Haile Selassie attempted to forge a more unified and developed country, proposing to introduce the state and its civilizational project among the various local groups in the periphery of the highlands. He tried to establish law and order as well as taxation. The latter was to replace the tribute extracted, often by force, by the state-linked elite in pre-war days. As said, from the early 1950s to 1969, the Suri paid taxes in kind (a monetary value but converted into heads of cattle).[9] The Chai-Suri *komoru* Dollote IV (Wolekorro) had nominally been appointed as a local 'chief' or *balabbat* (his son appears in Plate 8.1). The Emperor also tried to introduce other elements of highland civilization, for example by providing the Suri – predictably – with clothes, tools and improved seeds, urging them to start plough agriculture.

Apart from the above-mentioned *komoru*, the Suri were not involved in local administration, which remained marginal at best throughout the entire period. The Suri took no interest in this partly because hardly any Suri knew Amharic, the lingua franca (although several northerners trading and living in the area had learned the Suri language). The Suri continued to see themselves as a different politico-cultural unit. In terms of their segmentary political ideology, they differentiated themselves not only

---

[8] (cont.) did not leave a lasting impact on the Suri, who were largely left to their own devices. While trade with Italian soldiers (in grain, meat, coffee, game products and ivory) was permitted and their elders and *komoru* gathered for consultations with officers, there was no attempt to draw the Suri into the administration or to develop the infrastructure of the area for their benefit.

[9] An effort to find the records in Jimma and Maji was unsuccessful. In Maji I found a ramshackle wooden storage room where most of the old archival papers from the Haile Selassie era had either not been sorted out or had been destroyed by rats and rain.

from neighbouring peoples like the Nyangatom, Anuak (or Anywaa), Toposa or Dizi, but also from the highland Ethiopians in general, collectively called *golach*, seeing their own initiated elders (a 'ruling' age grade) as structurally equivalent to the Ethiopian central government. Finally, the political economy of land and labour exploitation after 1941, while not feudalist like the pre-war *gebbar* system (cf. Garretson 1986), remained predatory and hierarchical, traders and district officials dominating the scene by illicitly augmenting their incomes. The Suri were usually left alone and kept in check militarily if necessary. In cases of persistent disputes or violent incidents (for example, cattle-raiding) between Suri and non-Suri, mediation talks were organized under the auspices of the government with village chiefs, sometimes successfully, sometimes not. But the indigenous and state political traditions did not confront each other head on, and Suri traditional leaders were not captured by a state structure.

The Suri cultural model of 'political' authority was thus maintained throughout the Haile Selassie period. It was based on: a) the three ritual figureheads (the *komorus*), 'servants' of the people, agents of reconciliation and mediation symbolizing a normative unity of the Suri moral community; and b) the reigning age-grade or *rora*, the normative forum through which community decisions were reached in public debates (see Abbink 1999; also Turton 1978, 1992, for the comparable Mursi situation). Suri identity (or better, Chai and Tirmaga identities), the idea of their forming a distinct politico-ritual unit, remained anchored in a culturally and materially 'significant' territory; while a cultural aesthetic of 'difference' (consisting of ritual, song and dance, and bodily decoration) was also self-consciously maintained.

The revolutionary period 1974–91: the Surma as a 'nationality'

After 1974 a socialist ideology gradually became the basis of governance. The Suri were depicted as a 'primitive communalist' society, the lowest stage on the 'evolutionary ladder'. As such they presented an ideological and developmental challenge to a regime committed to collectivist–socialist development (and to the overthrow of the 'ruling classes' throughout the country).

In contrast to the imperial regime, the Derg introduced a policy of the positive recognition of the existence of ethnic groups or 'nationalities'. The founding of a political research bureau, the Institute for the Study of Ethiopian Nationalities (ISEN), in the early 1980s was meant to give expression to it, being based on Marxist–Leninist and student movement views of the 'national question'. The rights of nationalities to equality and development were mentioned in Article 2 of the 1987 Constitution. However, in the eyes of the Derg the development of nationalities had to be seen in terms of a 'progression toward socialism'. This implied a sustained attack on traditional elites based on the control of land.

The Promise of 1991

Plate 8.2   EPRDF troops presenting themselves to the Suri, November 1991
[J. Abbink]

The Suri, like many other peoples, were subjected to a revolutionary ideological campaign (*zemecha*) by cadres and students in 1976. A small group of six to eight students came via Maji to the Kibish area, near the old missionary-built airstrip, to meet elders of the Tirmaga sub-group. With the help of some government soldiers stationed there, they called a meeting and raised issues like inequality in land tenure, oppression, and also 'bad or harmful' customs (in the context of religion and ritual). The students were hosted by the *wereda* people in the Kibish area, who relied on the Suri to supply them with extra food. The 'conversation' with the Suri, however, did not go well. The local Dizi translators could not always convey what the students wanted to say and made mistakes in the translation of the students' neo-Marxist arguments. Suri realities did not answer to the model of an unequal, oppressive, feudalist society. No oppressive landowning stratum could be identified, for example: there simply were no 'landowning chiefs' in Suri society; Suri elders and *komoru*s were not very different from average Suri, except perhaps in numbers of cattle. The cadres thus came to focus on an ideological, 'developmental' offensive, suggesting, for example, that the Suri start wearing clothes, settle in one place and practise plough agriculture, stop cattle sacrifices, and abandon the wearing of large lip-plates and ear-discs by women ('harmful customs', a concept later enshrined in Article 23 of the 1987 Constitution).

But the Suri did not take the visitors very seriously. They knew at the time that the Emperor had been deposed, but had also seen that in the subsequent political turmoil no new legitimate leadership of Ethiopia had yet formed. They did not see a worthy equivalent to their own elders and *komorus* in the group of young cadres and students, and they could not deal with them on an equal basis. The impact of the suggestions of the *zemecha* visitors was virtually nil. Scepticism, subtle ridicule, and sometimes outright rejection remained a characteristic response vis-à-vis all subsequent local administrators who ruled from Maji town, located in the Dizi area.

Although the Derg made an effort to implement its new programmes, few had any impact, due either to lack of funds, a draining of manpower and finances to the war front in the north, or misguided authoritarian policy. In the Suri area their efforts to organize people into peasant associations – a new form of collectivist unit of rural producers – or herding associations (for pastoralists) failed: these plans remained paper constructions, existing only in the Maji government offices. As mentioned above, in only two locations in the Suri area were primary schools set up, and the mobile veterinary service for Suri cattle served for only a few years. Local officials attempted to re-instate tax collection (which had collapsed in 1969), but were not successful due to difficult logistics, non-cooperation by Suri elders and *komorus* and persistent difficulties in estimating wealth or pinning down the 'responsible people'. Neither was much headway made with the 'cultural offensive'. The Suri lowlands were hated by administrators and soldiers, and in the later years of the Derg the two remaining primary schools, three soldier posts and a sub-district (*mikitil wereda*) in Suri territory were abandoned (1989-90), giving the Suri back their de facto autonomy.

Reviewing the Derg's approach to local society in the south-western periphery, it can be said that in its radical drive towards modernization, it succeeded in many areas in removing 'traditional' chiefs from the political arena and replacing them with peasant-association chairmen, a new type of politicized and dependent local leadership. This often proceeded in a highly traumatic manner, in which the hereditary chiefs (*balabbat*s) and ritual specialists of the various groups were delegitimized, humiliated, robbed of their insignia and cultural objects, and often also physically eliminated by the revolutionary authorities, as among the Me'en and Dizi. As a result, the remaining headmen and chiefs 'retreated' to the cultural domain, where their survival was deemed harmless. However, the elders and *komorus* of the Suri, being institutionally and geographically elusive in the Maji lowlands, escaped this campaign, losing neither access to their land nor their visible 'chiefly' insignia and objects. As we saw, their authority was constructed in non-material domains and continued to be recognized by the great majority of the Suri population.

Crucial developments in the area took place in the period after 1989 (just before the 1991 changes). In that year, the Derg government lost the monopoly of the means of violence when the Suri armed themselves with contraband rifles, mainly from the Sudan. This sudden influx of modern automatic rifles (imported in the context of worsening conflicts in southern Sudan and bordering areas of Ethiopia) was a factor that unexpectedly changed the entire political situation in the Maji area, undermining government authority and local patterns of peace and co-operation. Other groups also obtained access to more weapons, but in far smaller quantities (compare the account in the next chapter by Hiroshi Matsuda for the Muguji). In the last years of the Derg, skirmishes and incidents had indeed become very frequent, due to the atmosphere of government failure and growing lawlessness, with fatal casualties in inter-group clashes (ambushes, raiding, robbing) an almost daily occurrence.

In addition, the Suri political system itself also came under threat. To understand this development, it is useful to recall Chai Suri political organization.[10] Most prestige and influence is traditionally accorded to the 'reigning' set of elders of one age-grade (*rora*), already mentioned, and to a less important one of senior or 'retired' elders (*bara*). These together form the backbone of political society, to be respected and honoured on occasion by the junior set of (uninitiated) men called *tegay*.[11] Each set of *rora* elders (a status not determined strictly by biological generation and age) has a collective name, marking them off from their predecessors. Until December 1994, the reigning set was that of Neebi (the 'Buffaloes'). Since the early 1980s the Suri had been in violent conflict with the 13,000 or so Nyangatom or 'Bume', their southern agro-pastoralist neighbours. Until the late 1980s the Suri dominated them, but after the influx of large quantities of modern automatic rifles (which reached the Nyangatom first) and an emerging alliance between the Nyangatom and Toposa, the Chai and Tirmaga Suri have been raided, defeated and chased from territories they formerly held. The Toposa and Nyangatom (whose grazing areas and resource territories are contiguous) had always recognized their historical and cultural affinities, but closer ties emerged in the context of the rekindled Sudanese civil war after 1983. The Toposa were armed by the Sudan government, who turned them into a 'tribal militia', which meant more arms, ammunition and food aid. This was noticed by their Nyangatom brethren, who then acquired new Kalashnikov, M-16 and other assault rifles from them. From the mid-1980s, this enabled the Nyangatom effectively to occupy and incorporate essential pastures and lands of the Suri both in Sudan and in Ethiopia and to push them out of their strongholds around Mount Shulugui (called Naita on most maps).

The alliance of the Nyangatom with the 65,000 or so Toposa (whom they also call 'Bume') is still overwhelming for the Suri, even though the latter are now also well-armed with automatic rifles (mostly possessed and handled by young men of the junior age-grade *tegay*). The Suri have instead expanded towards the north and east, thus encroaching on Dizi territory. Violence has increasingly been directed towards the Dizi and other highlanders instead of the Nyangatom. The situation in the Maji area thus stimulated a process whereby the younger Suri of the *tegay* grade started taking violent initiatives of their own, beyond what the *rora* elders 'expected' or could 'tolerate' (in the context of the age-grade system and its

[10] For a fuller account, see Abbink 1994; 1998b.
[11] Among the closely related Mursi, this grade is called *teru* (Turton 1973: 125).

structural tensions). The pattern of often unprovoked violent attacks against Dizi went on, unsanctioned by the Chai elders.[12] Partly as a result of this, the Neebi elders kept on stalling the preparation for the new initiation ceremony for the *tegay*, which was long overdue. They tended to perceive the *tegay* as too uncontrolled to take on *rora* responsibility and appeared to think that *tegay* were endangering the social order of Chai society. This was the situation at the time of the EPRDF takeover in May 1991 (see Plate 8.2, p. 162).

Suri identity in federal Ethiopia: a 'periphery' on the wane?

The EPRDF government, which came to power in May 1991, instituted a new discourse on ethnic relations in Ethiopia and translated it into a policy of ethnic regionalization. Ethnic groups ('nationalities') have to realize self-determination in the management of their own affairs, including education, language use and political organization. This policy potentially has great implications for the 'peripheries'.

Compared to all previous types of government, the impact of the current EPRDF administration on the Suri seems to have been greatest. Combined with other forces emanating from an increasingly 'globalizing' world (see below), the new regime is posing the final challenge to the political and cultural autonomy the Suri were able to sustain under previous political regimes. The paradox is that this is happening under a regime which has proclaimed local ethnic identity and 'self-determination' as core defining elements of political participation and group identification.

I have suggested that, for the Suri so far, the state was never a normative agent with which a mutually beneficial relationship could be established. This still holds for the majority of Suri. But there were changes after 1994. The new state authorities started an ethnic co-optation policy intended to empower the Suri, engaging them in local-level government, self-administration and education (as well as in making them give up 'harmful customs', though there is unresolved confusion over what these are and over who decides what these are). It was hoped that this policy would also reduce the tendency of the Suri to resort to violence in dealings with other groups.

Even if the ethnic federal model in Ethiopia is fraught with problems and may not live up to its expectations[13] – essential power is still retained at the centre and new ethnic tensions are being evoked in some areas – the

---

[12] This process frequently occurs in age-graded pastoral societies. For a fascinating example, see the case of the Turkana Ruru age-set (Lamphear 1992: 227), which emerged as a semi-independent fighting force in Turkana society after 1917, mainly as a result of an earlier round of firearms accumulation supplied by Ethiopian highland traders (ibid.: 144–15).

[13] For critical reviews of the Ethiopian ethno-federal experience so far, see, Cayla 1997, Clapham 1996b: 249, Kidane 1997, Ottaway 1995, Vestal 1996.

exercise, both rhetorically and practically, is crucial in reshaping perceptions of the importance of culture difference, redefining group relations and creating new forms of collective self-consciousness, whether these are based 'on the facts' or not.

After the establishment of the new regime and the instalment of regional and zonal administrations, group tensions largely inherited from the Derg past did not abate. Indeed, in the years after 1991 the Suri have had some of their most violent confrontations with both Dizi and Anywaa (Anuak) and the state. Whether there is a causal link between renewed state interference and violence cannot be ascertained. The Suri's conflicts with other groups in their vicinity were in part a continuation of the same structural problems indicated in the previous section: persistent economic and ecological pressures, increasing 'resource competition', and the new power position of the youngsters due to the spread of weapons in earlier years. But new 'ethnic borders' and ethnic party formation have tended to lead people to translate all problems in terms of ethnicity (instead of in terms of, for example, socio-economic inequality, educational shortcomings, environmental problems, faulty administration, or a lack of fair justice).[14]

The new authorities have frequently tried to mediate and negotiate local differences and disputes, but have not been very successful (cf. Abbink 2000b). There was also local criticism of their 'half-hearted measures' and style of mediation, which often alternated – in nontransparent ways – with violent means of enforcing order.[15]

A series of incidents in 1993 in which Suri attacked Dizi (in Kolu, Dami and Adikyaz) and, more importantly, killed **EPRDF** government soldiers in the Omo National Park finally led to retaliatory action by the latter. In a confrontation in late October 1993, in which hand grenades were used, it is estimated that a few hundred Suri died, mainly women and children. After this event,[16] the Chai elders from various villages (among them the ritual leader or *komoru*, and those responsible for age-grade initiation) called a big public meeting in Makara, the village of the *komoru* (see Abbink 1998b: 340–1). This was one of the more important Chai gatherings in recent years. Elders of the reigning age grade (Neebi) reviewed the situation of the Chai, called upon those present to search for the reasons for the escalating violence, and implicitly castigated the *tegay*, the young men, for their record of excessive violence. Shortly after this meeting the elders decided to organize the new *rora* initiation, in order to force social adulthood upon the youngsters who had been responsible for this violence.

[14] Especially in zones grouping together several smaller populations, a common policy to address shared regional problems would be perhaps be more conducive to development and local popular participation.

[15] In 1994, some Me'en people, using a familiar local cattle metaphor, said: 'The government troops behave just like bulls', meaning: they have power but do not use their brains.

[16] I was not in the Suri area at that time. Information was collected from local healthworkers, Surma informants, a university scholar visiting the area at the time, and local Dizi people in Maji and Kolu areas.

Paradoxes of Power & Culture: Surma

This was done in late 1994. Thus military action by the state resulted in precipitating a major Suri ceremony and in a 'reaffirmation' of its cultural value.

A year later, the government installed a *Surma Council*, a kind of local authority for the newly designed 'Surma *wereda*', with only Suri members, to operate under the guidance of the zonal administration.[17] With the installation of the eleven-member council, a new phase of state penetration of Suri political space was initiated. The programme of the new national government was, in a sense, to 'recapture' the Suri people, who, because of their strategic position along eighty kilometres of the Ethiopian-Sudanese border and the growing economic relevance of their area (the gold trade, game resources, and the potential tourism in national parks), became more important. One strategy was to elicit a new indigenous but loyal leadership stratum. Ethnic groups had to be represented through their own ethnic party (founded under the auspices of the EPRDF), their own people in the reorganized local administrations (in the zones and the regional-state governments) and in the national parliament in Addis Ababa (the House of People's Representatives).[18]

The members of the Surma Council were nominally elected by the Suri people, under close supervision by the zonal authorities.[19] As a rule, members had to be men who knew some Amharic (the national working language), in order to be able to communicate with government representatives at the zonal level. Thus most members were young, ex-soldiers formerly in the Derg army and marginal youths who had grown up with an Amharic-speaking soldier or trader family in the area. However, this policy of selecting only on linguistic ability led also to incompetent and even 'criminal elements' being included.[20] In the first council there were three monolingual Suri (both Tirmaga and Chai) and for a short while even the Chai *komoru* was a member (see Plate 8.1).[21] The composition of the council has shown numerous changes over the past five years, members frequently being removed after 'elections' and 'evaluation sessions'. Council members received what, by local standards, was substantial government remuneration and 'fringe benefits' (this opened the door for embezzlement, allegedly practised by several members). The Council's contribution

---

[17] Bench-Maji Zone, with its head office in Mizan Teferi.
[18] The Suri have one representative in the House, a young man in his late twenties and educated by an Amhara man in Mizan Teferi. He speaks good Amharic and some Dizi (being married to a Dizi woman). In May 2000 he was re-elected by the local Suri to the Ethiopian parliament, the House of Peoples' Representatives, remarkably as an independent candidate.
[19] There are also at least two non-Suri government party cadres active in the council, though they are not formally members.
[20] According to local townspeople and Dizi, several council members were known to have been involved in multiple homicide. Council members were therefore not screened by the administration as regards their past record, a mistake which undermined the confidence and willingness of non-Suri to co-operate with the council.
[21] He resigned after eight months, probably convinced that the job was not suitable for him as a ritual, non-political leader.

The Promise of 1991

to constructive Suri self-government or to stimulating development has been marginal, and its authority among the Suri is limited.

However, the role of the Surma Council as a new local political body may eventually bypass the traditional arena of political decision-making formed by Suri assemblies or public debates held under the auspices of age-grade elders and the *komoru*. Against this, the state has its own programme that it wants implemented, and in its view 'democratization' means primarily ethnic representation and working through ethnic elites connected to or co-opted at the regional and national level, and not grass-roots decision-making. The Surma Council is thus also – predictably – used as a conduit for implementing national policy (as the previous regime used the districts and peasant associations).

One interesting future aspect of social transformation will be 'Suri culture': social organization and economic and cultural practices will very likely be a 'target' of planned change. One example has been the declaration by the zonal authorities that the popular ceremonial stick-duelling (*thagine*) of the Suri, a major cultural event, should be prohibited, or at least toned down, because it is seen as 'too violent'.

Even though the Suri may be co-opted into a state structure where they have little real influence, they now do have some voice at the higher echelons of the state. They are formally represented in the local and regional administration and in the national parliament on the basis of the ethnic quota system. In this sense, their peripheral position as an 'ethnic minority' or 'nationality' has now become a privilege, because other local people – for instance the dispersed descendants of northern settlers – are not politically represented.

Whether this new position will break the Suri's ingrained perception of the encroaching Ethiopian state as an imposition and a threat remains to be seen. The Suri will perhaps remain a dissatisfied and unstable element in the region if local problems with neighbouring groups are not solved, if participatory local administration is not established, if they detect local ethnic favouritism biased against them, and if public debate, consensus-building and the ritual confirmation of decisions within local society neglected by the new authorities.

Suri identity vis-à-vis tourists & missionaries

Apart from the political arena, there are other spheres of change in the Maji periphery which affect Suri society and identity, and will continue to do so in the future: namely the encounter with tourists and the impact of a slowly encroaching evangelical Christian mission. Both are developments that emerged in the 1990s, and they form a major challenge to the Suri, decisively invading not only their territory and economic system, but also their cultural and political space. These two factors will connect them to a global discourse on religion and identity construction. The encounter with

Plate 8.3 *Missionary teacher Mike Bryant with Daniel Kibo (a Christian convert) and a non-converted Suri (left). November 1999, Tulgi* [J. Abbink]

tourists (Europeans, Americans and Japanese) has so far mostly evoked irritation and anger, because Suri see them as not observing the rules of reciprocity and as showing an imposing, exploitative attitude. Mutual contacts have been aggressive and occasionally violent. The Suri have developed a disdain for tourists that is only thinly veiled, and in their encounter the 'boundary' between them and the white foreigner is reinforced. This is not based on any inherent negative attitude towards white foreigners as such – they had known the Italians in the 1930s, as well as some well-liked missionaries and development workers in the 1960s – but purely on the stunted nature of the current social interaction with tourists (cf. Abbink 2000a).

The foreign missionaries who have resided in the Suri (Tirmaga) area for about eight years are Evangelical Christians or Presbyterians associated with the Lutheran World Federation, working together with the local Mekane Yesus Church and the Qale-Hiywot Church (see Plate 8.3).[22] They are engaged in infrastructural and agricultural work, in setting up a clinic, a literacy programme, and a church building together with local people. They are also trying to introduce new crops (fruits) and cultivation among the Suri. One couple is carrying out linguistic research (partly funded by the Wycliffe Bible Translators) in order to prepare primers for the school and to translate the Bible (mostly the New Testament) into the local language. The programme started in 1989 in the district capital of

---

[22] In the late 1960s there was a Presbyterian mission station in Merdur, a remote spot in the Tirma area, also staffed by some Americans. They were forced to leave in 1977 because of logistical difficulties and Derg pressure.

The Promise of 1991

Tum, where water supply and medical services were provided. The present mission station is in Tulgi, a locality at the northern fringe of the Tirmaga area.

This effort, which is so far not unfavourably received by the Suri, shows another face of global culture: next to their valuable modern medical and eductional assistance, used as an entry point, the missionaries formulate a new appeal to a transcultural religious ethic and worldview that is held to be universal but may tend to bypass central tenets of the Suri way of life (for example, gender relations, traditional religious and ritual ideas, healing and divination, alcohol use in the form of native beer consumed in work parties, decorative body culture). This is regardless of the sincere and open attitude they have toward Suri history and culture. The missionaries have found some response because of the Christian message itself, but also because of the internal disarray in which Suri society finds itself in the last decade (due to excess violence and persistent threats of drought and famine). It can be expected that, by creating a Christianized Suri group which may come to agitate against the premisses of Suri culture, the missionaries may inadvertently contribute to an internal 'power struggle' in local society and thus link the Suri ideologically to a translocal modernity formulated outside their own domain, and also beyond that of Ethiopian Orthodox Christianity.[23] This factor, together with the political challenges initiated by the new federal government, will constitute much of the future dynamics of Suri society. The Suri will thus become involved in a national discourse on political ethnicity and a global discourse on religious and cultural affiliation. Both processes will undermine Suri local autonomy, redefine their value system, and transform their group integrity and cultural identity.

Conclusions

Since the demise of the imperial regime, the Suri have been increasingly drawn into the wider Ethiopian political system. Although they frequently resorted to excessive force when they saw their political and cultural space being invaded by enemy groups and state agents, they could not stop this process, nor foresee its unintended consequences. Because of persistent inter-group conflict, they were also drawing more of the attention of the reconstituted, post-1991 Ethiopian state, which saw them as a challenge to its policy of ethnic devolution and 'empowerment' in an increasingly sensitive border area.

A divisive process has been noticeable in the Suri community in recent years. We have seen that internally the Suri have not been able to contain the contradictions in authority relations of the age-grade system, due

---

[23] In mid-2000, the mission in the area reported that there were at least ninety potential Surma converts enrolled in Bible classes. At least a dozen others had already been converted and baptized earlier.

Paradoxes of Power & Culture: Surma

especially to the usurpation of firearms by the members of the younger age-grade. The metaphor of 'age' as an organizing principle in their society (see Turton 1995) may be on the wane.

While the new presence of automatic weapons and large amounts of ammunition gave the Suri – the younger age-grade – a false illusion of power, they did not solve any of their long-standing conflicts with their Nyangatom, Anywaa, or Dizi neighbours (see Gerdesmeier 1995), conflicts which were partly about guaranteeing free access to economically important territory (pasture, fields, gold-panning sites) and partly retaliations for past killings. Nor have the Suri been able to reoccupy their core area near Mount Shulugui. Violence has also bred rivalry and conflict vis-à-vis local groups *within* Suri society itself.

The presence of the state in both a military and an administrative sense has inevitably grown, exploiting internal social divisions between young people who had served in the former national army (now becoming leaders in the Surma Council) and those not exposed to outside life in the wider society. In addition, schooling and missionary influence may exacerbate social and family divisions further and break the normative and ideological unity of Suri culture.

One of the underlying arguments of this chapter has been that, in this part of Africa too, a conception of 'ethnic groups' based on geographical boundaries is becoming less and less viable as an explanatory element in understanding cultural and historical developments in a globalizing world. Seen in terms of the flow and exchange of commodities, images and persons, boundaries are becoming more and more permeable. They are flexible and manipulated, despite the rhetorical counterclaims by some governments (see Clapham 1996b).

Thus, in the study of a complex society like Ethiopia, the focus of analysis may have to shift from how supposed 'ethnic' groups interact with the state to how individuals come to 'represent' such groups within the arena of national politics. It is difficult to speak of ethnic groups as collective, acting agents. The critical factor in the new political space for ethnicity created by the federal Ethiopian state are local elites and individuals, people emerging as political and cultural agents in the name of 'ethnic groups' which are presumed to exist and to have the right to express themselves collectively. These agents of change may be brokers in the classic sense, crossing boundaries, making use of differential access to 'resources' (including an identity and legitimacy derived from mandatory, ascribed ethnicity), bringing together formerly separate spheres of interest and carving out a power base that is not located entirely within the local society. The new federal structures and the system of ethnic representation (in parties, zones and the national parliament), as well as emerging missionary church structures, are thus creating different opportunities for social action for people who have long remained in the margins.

These facts in themselves construct new modes of both political communication (transcending and partly invalidating the old age-grade

The Promise of 1991

structure and the idea of a *komoru*) and cultural communication (through global religious connections and the tourist trade that is commodifying 'Suri culture'). The Suri will thus be part of the new wider Ethiopian political structure and of global cultural and identity discourses. At the same time, as the highly significant example of the efforts to end Suri ceremonial duelling mentioned above make clear, there is a distinct possibility that they will also gradually lose much of their specific culture under the impact of a continued project of 'reform' and 'development'. These projects envisage sedentarization, more agricultural activity, a halt to cattle-raiding, controlled grazing, disarmament, the disempowerment of traditional authorities such as elders and ritual leaders, and the abandonment of 'harmful customs' (as determined by the national government and its 'experts') – customary law, initiation ceremonies, ritual stick-duelling, animal sacrifice, intestine divination, body-piercing, scarification, etc. It is likely that, despite the current national rhetoric of all groups 'developing their culture and language', the material basis of the cultural distinctiveness of groups like the Suri will be de-emphasized. While the rhetoric of identity may increase, its substance will become elusive but tightly controlled within the new political framework. Perhaps the Suri will gradually become part of a general Ethiopian rural 'underclass', as has been happening with other groups. That is, their dependence on outside forces and on non-Suri political agents may increase and their local autonomy in decision-making decline. This scenario may perhaps seem paradoxical if not sad in an era in which ethnic and cultural rights to self-expression have been proclaimed in Ethiopia as a duty, but it is nonetheless a likely outcome of the current process of co-opting local elites.

Nine

Political Visibility & Automatic Rifles
The Muguji in the 1990s

HIROSHI MATSUDA

> *Yadik[1] came and talked to us.*
> *They came, so we all welcomed them;*
> *Meles Zenawi gave a qebele to the Koegu.*
> Muguji song in praise of the prime minister Meles Zenawi, 1996

When I first visited the Lower Omo area in 1986, the Muguji (one of the marginal groups widely labelled Koegu or Kwegu) were a minor and little-known population speaking a Surmic language. People had a few goats and sheep, but no cattle in their camp. They formed a subordinate class in a political unit called Karo, which consisted of a majority of Omotic-speaking, cattle-herding Kara. Neighbouring peoples would say that the Muguji were a poor and baboon-like people depending on bush products, while to outsiders like highland Ethiopians they were simply an 'invisible' people. During the last decade, however, the most drastic change for the Muguji is that they have come to be 'visible' to everybody. Whenever I revisited them, their political position revealed itself more clearly in the area. Although their economic situation has not changed, they are recognized today as an autonomous group by both their neighbours and the present Federal Government of Ethiopia.

The increase in the number of rifles among the Muguji, with which I am concerned in this chapter, has followed this recognition. This is not to say that the rifles have been playing an important role on the battlefield or that they obtained autonomy as a result of modern armament. I would rather emphasize that it is a recent transformation of the centre–periphery relationship that has existed in the Omo valley in a variety of forms since the turn of this century that made possible a new inflow of automatic and semi-automatic rifles into Muguji society. Having become 'visible', it was

---

[1] *Yadik* is from the acronym IHADEG (Ye Ityop'iya Hizb Abyotawi Demokrasiyawi Ginbar), that is, the EPRDF.

possible for them to take advantage of their position as intermediaries in a spreading regional and international pattern in the flow of arms, specifically of modern rifles deriving on the one hand from the demobilization of the Ethiopian army in 1991 and of ammunition from the intensifying Sudanese civil war on the other.

The influence of firearms on Ethiopian history has been an object of study for a long time (Caulk 1972, Merid 1980, Chapple 1990, R. Pankhurst 1990, 1997). Most of these authors have analysed the impact of firearms on the progress of battles from the historical perspective. Garretson refers to centre–periphery relations through the arms trade in the Maji area in the first half of the twentieth century (1986). Although anthropologists have paid most attention to 'ethnic' conflicts (Fukui and Turton 1979; Fukui and Markakis 1994), the study of modern weapons and their impact on local societies have not been examined so far. Abbink has suggested the importance of examining this topic (Abbink 1993; and compare his analysis in Chapter 8 of this book). Given the recent rapid increase of arms on the African continent, their influence cannot be compared with what the situation of ten years ago. We now need a wider, more in-depth perspective to study the importance of arms in Africa.

Turning now to research methods, I obtained most of the information for this chapter during fieldwork carried out in the Lower Omo Valley since 1986, carrying out research specifically among the Muguji in 1988 to 1990, 1994, 1996, 1997 and 1999. I have only limited information on the relationship of the Muguji with the Ethiopian centre before the Derg, the socialist period in Ethiopia from 1974 to 1991. Donham and James's volume *Southern Marches* (1986) gives some general indication of the position of the Muguji at that time. Like other 'Koegu' populations who appeared quite outside the reach of the imperial state, they were very probably drawn into its orbit through the indirect effects of the ivory trade at least since the beginning of the century (see especially Turton 1986). This chapter continues the theme of the Muguji's struggle for autonomy up to 1990 which I introduced in an earlier paper (Matsuda 1994). The main data on rifles in this chapter were collected in 1997 and 1999.

The Muguji & the Lower Omo

The Lower Omo valley, located in the south-western corner of present-day Ethiopian territory, is a very peripheral area in both the geographical and political senses. The south of the valley extends into the northern semi-desert of Kenya, and the west is contiguous with the southern plains of the Sudan. In terms of physical and cultural features, the peoples living there have more in common with those living in adjacent areas across national boundaries than with those in the Ethiopian highlands. On the Sudanese border in particular, there is a daily traffic though the Ilemi Triangle on the Sudanese side, which is now under Kenyan control. It would therefore not be true to say that this area has been isolated, even though there was little communication with the Ethiopian centre. What I wish to stress here is that, although a periphery, this area now relates to plural centres rather than a single one and that it has links across national boundaries with global phenomena (see Map 1.3, p. 31).

Before turning to the main subject of this chapter, I would like to explain the name of the group with which I have been concerned over the last decade, the Muguji. Until recently, I used the name 'Koegu' for the same people. Muguji is the name that the surrounding peoples used to refer to them, while Koegu is their own name for themselves. However, peoples calling themselves by a name of this kind have been known to ethnographers and linguists as at least three separate groups, with a total population of about 1000, all living along the lower Omo, speaking a Surmic language and being engaged in riverine agriculture and foraging. The first group are the Kwegu associated with the Mursi as described by Turton in 1986, in which he says that they are called 'Nyidi' (singular Nyidini) by the Mursi (Turton 1986: 149). The second group are those the Bodi call Idinitt (Fukui 1994: 39). The third are the group I am discussing here, formerly associated with the Kara, who as mentioned are known to them and others as Muguji. They number around 500, mostly living at the junction of the Omo and Mago rivers, making them the southernmost of the three groups.

Apart from these, there are a few comparable groups, called Bacha or Baacha by neighbouring agriculturalists, who have a similar way of life. For example, there is one group along the Omo in the Gofa area and another along the Shorum river, a tributary of the Omo. Clan names like Koigi and Kwoigir are also found among the Dizi and the Majangir (Deguchi 1996: 124-5; Ren'ya Sato, personal communication). Hence the name Koegu and its variants have a range of different applications according to context. This is why I have adopted the term Muguji for the group described here.

The Muguji mainly live on agricultural products like sorghum and cowpeas. According to my own survey, in the three-month period just after the harvest, 67 per cent of the daily food intake is obtained from agriculture. Fishing, which accounts for 11 per cent, is important in the diet as a side dish, as well as providing food out of season. Compared to these two activities, hunting is insignificant (4.8 per cent). Some types of traps were set formerly, but today only rifles are used for hunting. Honey-gathering is crucial in a different sense, because it is the Muguji's special product, used not only as a gift for bond-partners inside and outside the Muguji group but also as the only commodity exchangeable for cash in the markets.

It is worth mentioning that most Muguji do not practise pastoralism as a daily activity apart from some 150 people living in the southernmost village called Galgida, who raise goats and sheep through familiarity with the Nyangatom, an agro-pastoral people speaking an Eastern Nilotic language who border them on the south-west along the Sudanese border.

The Muguji do not raise their own cattle but keep the few they possess with the herds of their Banna bond-partners to the east. Although the Muguji are generally recognized as poor by their neighbours because they have virtually no domestic animals, their recent rapid acquisition of arms has transformed their position.

The social position of the Muguji before & after 1990

'Muguji' was not a well-known name either to local people or to the government. For example, in 1988, when I visited the chief administrator of Hamar *wereda*, who had been born in Gofa, he knew nothing about this group inhabiting his district. This can be explained by the fact that, when a Muguji is asked where he comes from, he usually answers that he is from Karo. This is one reason why they were entirely invisible to outsiders. The two groups were united through a bond-partnership called *bel* in Kara, *belmo* or *eeda* in Muguji, but the Muguji were actually in a submerged position in Karo society. Intermarriage and drinking local beer together were tabooed between them, even though they lived in the same village. Under the pretence of *bel* a Kara man would use a Muguji man as a labourer to clear and cultivate the land. The Muguji explained that a long time ago the Kara moved into territory along the Omo then occupied by the Muguji, and took over their riverine farms by force. The relationship between the two groups, until very recently, was partly similar to that between the Mursi and the Kwegu, which Turton described as a patron–client one (1986: 154).

In 1989, the Muguji started fighting against the Kara majority within their shared administrative unit, Karo. A summary of events before 1990 is as follows (for further details see Matsuda 1994). After some skirmishes with the Kara, the Muguji all left the homeland temporarily and spent some months from the beginning of 1990 in Nyangatom territory to the south-west. One informant told me that it took almost a year before everyone came back. Thereafter, the Muguji, who had fewer than ten old rifles at that time, succeeded in separating from the better-armed Kara thanks to strong support from the Nyangatom, militarily the most powerful group in the Lower Omo valley, armed with automatic and semi-automatic rifles. There is no doubt that the separation between Muguji and Kara could not have been achieved without the Nyangatom's rifles. Gezahegn, who studied Karo at the beginning of the 1990s, mentions nothing about the conflict in his thesis (Gezahegn 1994). I can only suppose that the Kara were not prepared to admit the change of the situation between them. We may consider the era after the separation as a new regime of autonomy for the Muguji, while the era before 1990 might be regarded as an old regime of dependency on the Kara. Since then, as far as I know, there has been no major battle between them.

Political Visibility & Automatic Rifles: Muguji

The change in the political situation after the fall of the socialist government in May 1991 was regarded as working in the Muguji's favour. They hoped to have their own *qebele*, that is, the minimum unit in the Ethiopian administration system. In the old regime before 1990, all the Muguji belonged to one of two Karo *qebele*s in Hamar *wereda*, named after the Kara villages of Dus and Labuk along the Omo. The Muguji did not have their own *qebele*. One informant told me that they had never received the food and clothes delivered to each *qebele* by the government and aid organizations, because they used to be forcibly deprived of all goods by the Kara. I saw the people pleased at having the first food in their history distributed by the Swedish Philadelphia Church Mission (SPCM) through the Relief and Rehabilitation Commission (RRC) in 1994.

Their wishes were fulfilled with the change of district division carried out by the Transitional Government of Ethiopia in 1993. The Government approved the Kuchur *qebele*, which most of the Muguji would belong to, and its affiliation at *wereda* level was changed from Hamar to Kurraz. This meant that the Muguji's status within the administrative structure was officially recognized and, at the same time, the new alliance with the Nyangatom was also confirmed at local government level. In 1996, I heard that a young Muguji man was working as a local officer in Omo-rate town in Kurraz *wereda*. I also met a Muguji man at Jinka, the capital of South Omo Zone, who had been nominated as the official leader of the Muguji under the present government and had been summoned to attend a meeting (although he did not speak Amharic) to discuss the plan and the budget for the coming year. Significantly, the zone governor at the time was a Nyangatom man who had stayed in a Muguji village for a few years as my research assistant. Although he definitely played a role in promoting the position of the Muguji in the administrative structure, the recognition of the Muguji as an 'ethnic group in their own right' could have not been achieved without their own struggle to redress the stigma associated with their identity. The achievement was a cause for some celebration; see the song in praise of the prime minister quoted at the head of this chapter.

A short history of rifles among the Muguji

Not much time has passed since the Muguji first acquired rifles. The situation before I started research among them in 1988 could be re-constructed through interviews. In 1988, the rifles which some men carried on their shoulders were old single- or double-loaders. They said that most of these rifles had been brought into Ethiopia by the Italian army about fifty years earlier. These are models called Mauser, Belgic or Dimotfer (DM-4) in the Ethiopian highlands. The oldest rifle that the Muguji acquired, which they call *sodon*, is still in the hands of an elderly man, who told me that it had been used in the time of Menilek II. As for these old rifles, some parts of the body, for example the checkering and the stock, are made of wood, the bullets are quite large, and the recoil fierce. Although they may cause great damage if they hit a target, handling and control of these weapons is difficult and misfiring is common. I estimate that fewer than twenty out of one hundred adult men in the village at that time possessed rifles.

However, the rifles were highly valued by the people, not only as a tool for hunting wild animals but also as a weapon against neighbouring enemies, and particularly as an item of bridewealth. The Muguji obtained these rifles by bartering with the Banna living in the eastern mountains and the Nyangatom in the southwestern plains. A rifle was exchanged for more than ten head of cattle. An elder told me an interesting story: in former days a Kara man would lend a rifle to his Muguji bond-partner to hunt an elephant. He was given the rifle after giving the tusks to his bond-partner (cf. Turton 1986). Since it was apparently difficult for the Muguji, who had no cattle, to obtain rifles, this might be a major means for them to do so. It is also said that the imperial government distributed rifles and shells to the lowland peoples so that they could guard the national boundaries. However, the Muguji were not able to obtain any because of their relationship of dependency on the Kara.

When the Mago and Omo National Parks were created in the Lower Omo valley in the 1970s, a ban on the ivory trade also prevented the inflow of rifles to the Muguji. At the same time the socialist government stopped providing the lowland peoples with rifles. On the other hand, a new inflow route opened up across the Sudanese border, with the ending of the first Sudanese civil war in 1972 and of the Ugandan civil war in 1979. In the mid-1980s rifles coming by way of the Toposa were popular among the Nyangatom (Turton 1996: 103), though they were still expensive to the Muguji.

Let us move on to 1988. The Nyangatom owned many automatic rifles, mostly AK-47s (Kalashnikovs), a type commonly used by Sudanese government forces (Abbink 1993: 220). I noticed also many M-14s in their hands, which were used in the SPLA, the Sudan People's Liberation Army. These arms drastically changed power relations among the various groups in the Lower Omo valley. The events mentioned below owe much to these arms: the Nyangatom pushed the Dassanetch to the eastern bank of the Omo, the Mursi and the Suri (Surma; see Chapter 8) to the north, and all the Kara to the left bank of the Omo. The Muguji did not have automatic rifles, nor did their erstwhile patrons and present 'enemy', the Kara. Back in the 1970s the Kara had equal power with the Nyangatom, but by the end of 1980s the difference in power was clear. We may assume that, analysing the balance of the situation, the Muguji chose to change sides and tried to establish a new alliance with the Nyangatom.

When I left them in March 1990, no Muguji yet had an automatic rifle. On my next visit to them in February 1994, they had three automatics, although the total number of rifles did not seem to have increased. I did not identify the models, but they said that two of them were given to the government chief or *rikamambar* and sub-chief by the local administrator in Jinka. As far as the third rifle was concerned, a young man told me an interesting story of it being given to him by the SPLA. He went on foot up towards Juba in southern Sudan (still controlled by the Khartoum government) with two friends and took part in military training outside that city for some months. The trip was not easy, one boy dying on the way and another returning to the village.

By August 1996, at the time of my fourth visit, a Muguji elder told me that the number of automatic rifles had increased to forty-one. The number was surprising, given their economic situation. Most of those who had rifles said that they had acquired them from the Banna by bartering the cattle deposited with them. I knew that some cattle of the Muguji were kept among the herds of Banna bond-partners, but I was surprised at the number of these weapons and their rapid action. And the fact that almost half of the Muguji men now owned automatic weapons gave them a new freedom; instead of living in one place for security reasons, to protect themselves against attacks by the Kara and the Mursi, they could return to their original way of life scattered among the riverine forests of the Omo. Here they were in more frequent communication with the Nyangatom than before, and could operate more independently in their links with all sets of neighbours.

The inflow of arms in the 1990s

Assuming that the Muguji's rifles came to them from many directions, such as the Ethiopian highlands, southern Sudan, Uganda and Somalia, we can say that 'centres' for the Muguji are now not one but many. Their distinctive and newly 'visible' identity is being influenced more directly by the political situations in neighbouring countries than before. I therefore decided to make a detailed list of the guns they possessed, particularly the automatic rifles.

In 1997 I interviewed 27 owners of automatic rifles about the name of their guns, when and from whom they had bought them, and how much they paid for them. I was not able to interview twelve owners directly, but the necessary information was acquired from their age-mates. Actually they were well informed about the histories of their friends' rifles. I recorded all the signs, marks and letters seen on the rifle bodies. The total number of the automatic rifles owned by the Muguji in September 1997 was 39. I also collected such information on the older types of rifles, most of which had been imported into Ethiopia from Europe or used by the Italian army before World War II. I took photos of 40 rifles, including 27 automatic rifles, and showed them to an ex-soldier of the Derg army living in Addis Ababa, a gun enthusiast. He gave me some information about the names of the models, countries in which they had been produced, the original users of these weapons, and so on.

The Promise of 1991

I followed up this research in February and March 1999, but could not conduct direct interviews this time, because the number of those possessing automatic rifles had rapidly increased. Altogether, I collected some information on 101 automatic rifles (including four semi-automatics) and thirteen old-style rifles as a result. Eight automatic rifles were in the process of being traded, so that we may say that the Muguji owned 93 automatic rifles in March 1999. I shall describe these in detail.

First, I recorded nine names of automatic rifles among the Muguji (although these names did not originate with them), namely *natolobok, chicha, parko, nyad'ed'eya, riyada, neethe, parko-ka-neethe, muchacha,* and *tingul*. These names seem to be common among peoples along the national border. All of them are so-called 'Kalashnikov (AK type)' made in China, the former USSR, DDR, Yugoslavia and Bulgaria (Table 9.1). The Muguji call the automatic rifle *klash*. This is one of the most popular rifles used in disturbances around the world, because it is strong and easy to take apart on the battlefield (Togo 1982: 134). It is likely that the Muguji's rifles had been used by Ethiopian forces in the civil war. Some of them may have come from Sudan and Somalia via the Borana on the Kenyan–Ethiopian border. I did not find any mark on the rifles by which the original users could be identified. It proved to be a false expectation on my part that the Muguji would have M-14 assault rifles made in the USA, which were popular with their neighbouring allies, the Nyangatom. This is the model used by the SPLA in the civil war in the southern Sudan and it is called *ngilamalai* by the Muguji. I noted only one M-14 among them in 1994, the one that had been brought back from the SPLA training camp.

Table 9.1 Types of Kalashnikov held by the Muguji, 1999

Muguji name	Type	Country of manufacture	Number
natolobok	Type 56	China	59
chicha	Type 56 (SKS carbine)	China	4
parko	Type 56-1	China	3
nyad'ed'eya	AKM	USSR	6
riyada	AK47	USSR	5
neethe	MpiKM	DDR	10
parko ka neethe	MpiKmS	DDR	1
muchacha	M64	Yugoslavia	3
tingul	AK47	Bulgaria	2
TOTAL			93

Second, Figure 9.1 and Table 9.2 show the route whereby the Muguji obtained rifles. I found out that 64 rifles had come from the Banna, which means that they came via an internal route within Ethiopia. The rifles

Political Visibility & Automatic Rifles: Muguji

Table 9.2 Inflow routes

Source (mainly by named group)	Number of rifles
Banna	64
Kara	11
Muguji	11
Hamar	5
Dassanetch	3
Nyangatom	2
Aari	1
Bachada	1
Mursi	1
Merchant in Jinka	1
no data	1

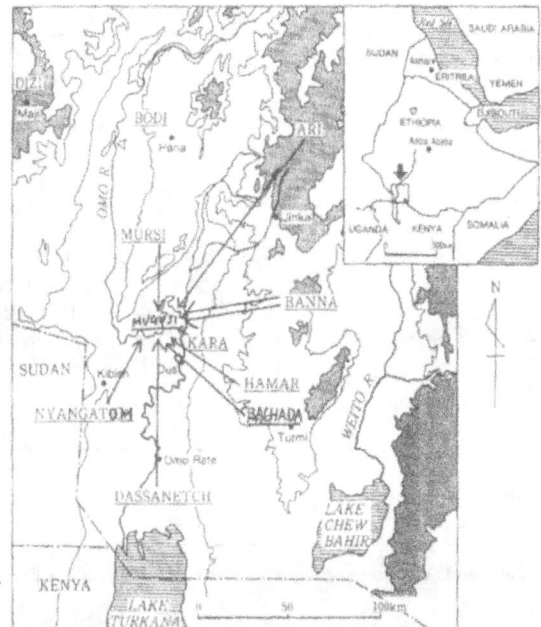

Figure 9.1 Origins of Muguji rifles

from the Dassanetch and the Hamar are also supposed to be of internal origin. The rifles from the Nyangatom are unexpectedly few in number. They say that rifles from the Nyangatom are expensive. It is strange that eleven rifles came from their immediate 'enemy', the Kara. Added to this, I may note that the Muguji bought all the bullets from the Nyangatom and therefore from the Sudanese side, since they were cheaper than those from the Ethiopian highlands.

The Muguji economy is basically self-sufficient, though some sorts of goods, such as cow-hides and pots, were commonly obtained through the bond-partner, the *belmo* or *eeda*. As far as the rifles are concerned, however, only seventeen of the 101 originated from bond-partners, eleven from other Muguji, and the majority (at least 60) were purchased from people with whom there was no special relationship. This indicates that rifles are different from other commodities for the Muguji.

Third, Table 9.3 shows when they bought the rifles and how much they paid for them. Twenty-one people bought their rifles in 1994 and 1995, but there was still a difference between prices when converted into numbers of cattle. More than half of the adult men obtained their rifles in 1998 and 1999 (up to March). The price was also gradually falling, and was steady at four cattle in 1998 and 1999.

Let us now look in detail at how the Muguji make money to buy rifles. They usually exchange honey, sorghum, cash, and particularly bullets for cattle, which they ask their bond-partners in Banna to keep in their own herds. The exchange rate in 1997 was one head of cattle for 700 or 800

The Promise of 1991

Table 9.3 Automatic rifles purchased from 1990 to March 1999

Price in Cattle*	1990	1991	1992	1993	1994	1995	1996	1997	1998	1999 (Jan. to March)
11										
10		1								
9										
8		1		1						
7			2		2	1				
6		1		2	2					
5			1	1	2	2+(1)	1+(2)	(1)		
4		1			4	4		4	31	24
3					(1)		(1)	1	2	1
2					(2)					
1										
TOTAL rifles purchased	0	4	3	4	11	10	4	6	33	25

* Numbers in brackets represent cattle equivalents for purchases actually made with cash

birr (1 US dollar = about 6 Ethiopian birr) depending on size, or 3 calabash containers of honey (about 60 kg), or 8 sacks of sorghum (about 160 kg). Recently they have frequently gone to Nyangatom to buy large quantities of bullets at 2 birr each, then selling them to the Banna at 3 birr apiece. Now (in 1999) they can buy a cow for 100 bullets and an ox for 80 bullets. This is why there has been a sudden increase in the number of the rifles among the Muguji since 1998.

It may be useful to sum up briefly the points in this section. The first is that both the variety of rifle models and the importation routes are fewer in number than I expected. In short, they mainly obtained Kalashnikovs, formerly used by Ethiopian forces, through the Banna by way of highland traders. The bullets, however, come from the Sudan via the Nyangatom. This suggests that the Muguji take advantage of their geographical position as intermediaries in the Lower Omo valley.

The second point is that the Muguji did not purchase the rifles in order to win their battles with their former patrons, the Kara. Although the number of rifles that the Muguji have been buying started to increase in 1998, no major battle has occurred between the two groups since 1990. The Muguji's hopes of acquiring a separate identity from the Kara had already been fulfilled in 1991 to 1993 under the Transitional Government.

I shall outline only briefly here the social impact of the increasing number of automatic rifles on Muguji society. First, it should be mentioned that the neighbouring groups, the Banna and the Aari in particular, have started visiting the Muguji more frequently than before. Furthermore the Muguji often visit the Nyangatom to buy bullets and the Banna to sell them. It is surprising, in addition, that trade between the Muguji and the Kara recovered in 1994 (Table 9.3 shows the eleven rifles purchased from their former patrons, who always used to be the source of arms to the Muguji). These frequent visits also provide material satisfaction in the form of coffee or iron tools, as well as confirming interrelations between them and their neighbours. When I was with the Muguji in 1999, I met an Arbore man (ie., 'Hor', see Tadesse Wolde's account in Chapter 2) who had been staying with them since he visited the Mago plain to hunt buffaloes more than a year before. He would sometimes speak at meetings or beer parties and insist that the Muguji and the Arbore had been one people historically. No Muguji openly contradicted him. Viewed in this light, the same may no doubt be said of the Banna, Aari, and Nyangatom.

Second, cases of murder are increasing among the Muguji. I recorded three cases in 1997, in all of which automatic rifles were used as lethal weapons. A drunken quarrel and impulsiveness are also common elements in these accidents. The details are not important here, but it is fair to say that these accidents have weakened the unity of the Muguji as a single group. This is because the family of the murderer usually leaves the village of the injured party and they do not return for some years, until the ceremony of reconciliation is carried out.

Third, the spread of automatic rifles has brought with it the modern concept of commoditization. The rifles are not goods that are exchanged with the bond-partners but must be bought at a fixed price. The bullets are used like currency on any occasion and people become accustomed to evaluating things in a consistent standard. Yet the phenomenon has only just been observed in Muguji society, so that much study still remains to be done.

Finally, I must mention the decrease in the number of wild animals in the Lower Omo plain. Although Muguji do not actually live in Mago National Park, they go there and to the adjacent area daily to hunt. In fact, they are well aware that buffalo numbers are rapidly decreasing because of the rifles. I often heard of shots being exchanged between the local people and the Park guards. Not only the Muguji but also other groups living in the area are equally responsible for this, of course. It is not likely that the decrease will have a direct influence on their lives for the time being, because because they are not so dependent on hunting for food, but because of poaching they may come to be accused of illegal ownership of the rifles.

From ivory to arms

Today, no other people in Africa looks more 'primitive' than the inhabitants in the Lower Omo valley, particularly the Muguji. Being far from a market town, their economic life remains self-sufficient in terms of both daily food and tools, outside the range of education, medical care or social welfare. Their way of living resembles that of the primitive tribe described in old anthropological textbooks. Johnson (1986) and Turton (1986) showed, however, that the south-western borderland of Ethiopia had already been incorporated into the world ivory market by at least the early twentieth century. The ivory trade made clear the existence of a centre–periphery relationship in modern Ethiopia.

On the other hand, the ivory trade had been declining in the 1920s throughout Ethiopia (Garretson 1986: 211). I have no definite information on the ivory trade since then, but it seems that the total amount of ivory traded under the Derg era (1974–1991) was trivial because of the official ban and the global movement to protect wild life. In fact, I have never heard the Muguji say that they made a profit through the ivory trade since the start of my research in 1988. It thus seems that while the centre–periphery relationship under the Derg regime did not entirely collapse, it was fairly inactive as a result of the fall in commodity trade.

Under the Transitional Government, however, and the end of the civil war, itself linked to the major change in the world political situation, relations between the Ethiopian centre and the south-western periphery were re-activated. The Muguji were officially recognized as an independent 'nationality' by the Government, which claimed to stand for the rights of local communities to self-determination. As a consequence, the Muguji were marked out as a newly visible part of this modern periphery, and it is in this context that we must understand the part played by the Muguji in trading and acquiring automatic rifles and ammunition.

What difference has the shift from ivory to automatic rifles as a trade commodity brought into the centre–periphery relationship? Ivory was a kind of local, vernacular commodity used by the Muguji in exchange for tax payments or protection against violence (Turton 1986: 167), whereas the automatic rifle is a foreign, imported commodity coming to them from long-distance and even anonymous outside sources. The honey, sorghum, and cattle used in exchange for rifles are 'exported' from Muguji, but consumed within a fairly restricted area. As regards ivory, although people still found it possible under the old pattern to break the link by refusing to hunt elephants, under present conditions it is virtually impossible to exert any control over the circulation of arms. Passivity seems to be their sole attitude against the system at the present.

We may therefore reasonably conclude that the Muguji did not arm themselves in order to win a local battle for independence. However, as a result of being newly recognized as a peripheral 'ethnic identity' within the state, they were in a position to buy into the regional trade in arms stimulated by the end of the Derg era and the continuing war in neighbouring Sudan. At the end of the twentieth century the Muguji are certainly included within the world system again, and as a direct result have been able to acquire modern weapons themselves.

Ten

Evangelical Christianity & Ethnic Consciousness in Majangir

REN'YA SATO

In 1993, when I was conducting fieldwork among the Majangir in Godare *wereda* (where the Southern Nations, Nationalities and Peoples Region abuts Gambela: see Maps 1.3 and 1.4, pp. 31 and 32), I made a trip on foot to the small settlement of Jaman with a Majang friend. During the trip, I found that surprising changes had taken place which made me seriously re-think my understanding of their society and of their relations with other locally distinguished groups and the state.

First, in Jaman, I found a new villagized settlement being constructed. This was a great surprise to me, as the former Derg policy of villagization, which had supposedly been extremely unpopular among Ethiopian peasants, had already been abolished. Here in Jaman, however, people were constructing a new village of their own free will. I also discovered that the construction of a church was central to this new, voluntary villagization project. Then, on the way back from Jaman, we came across Majang women fleeing into the forest. Soon we realized that EPRDF forces had been attacking Majang villages. This occurred because there had been an incident some months earlier, in which a huge group of Majangir armed with automatic rifles attacked several towns, including Tepi and Meti, occupying them for a couple of days until they were retaken by EPRDF forces.

These violent events came as a shock to me, as the Majangir were known to me, as well as to their neighbours, as a 'shy' and vulnerable people without centralized political institutions, living scattered in the forests. On this occasion, however, it appeared that they were not only well organized but also bold enough to carry out a military operation in which they successfully occupied some strategic towns.

In this chapter, I shall aim at an understanding of the changes I observed, with a special focus on Christianization and villagization, which the Majangir have experienced during and since the Derg period. The

The Promise of 1991

Majangir, who engage in shifting cultivation, bee-keeping and hunting in the thick rainforest of southwestern Ethiopia, have been regarded by neighbouring groups as a people with little property and neither livestock nor a complex political system of their own. They are also allegedly always drinking local beer and honey-wine, which causes disputes, sometimes even homicides, among them.

Nowadays, however, many of them, especially the younger generation, never drink or smoke, and they live in sedentary villages where homicide and fighting rarely occur. Villagers worship the Christian God (Waqoyo) regularly three to four times a week in churches in the sedentary villages. What kind of factors have led them to choose such a stoical and moralistic alternative? In this chapter, by describing the process of Christianization and villagization, I examine the social and cultural background which compelled them to adopt these choices.

I conducted fieldwork in Majangirland on several occasions between 1992 and 1999. The total length of my stay in Ethiopia amounts to thirty months. Most of my time in the field was spent in a sedentary village called Kumi, where 523 Majang people live (Central Statistical Authority 1995), approximately six to seven hours' walk to the west of Meti town. In the village, I had opportunities to observe religious and political activities, as well as conduct daily conversations and interviews with people of different generations.[1] In addition, I sometimes went on field trips on foot with my Majang friends to peripheral areas in the forest around Bebeka (Birhani), Dembidolo, Gambela and Kabo, where small traditional settlements are found. My field observations, however, may be somewhat biased towards the younger generation because I usually conducted my work with their help. Taking this limitation into consideration, this chapter represents a preliminary report which will be extended in the future. At present, however, it can offer recent information about the Majangir, about whom ethnographic reports these days are quite rare.

Evans-Pritchard, travelling in Anywaa (Anuak) country in 1935 and again in 1940, described the Majangir (termed by him Masango) as 'shy' (Evans-Pritchard 1947: 73). According to Stauder, who conducted the first intensive ethnographic research among the Majangir (1970, 1971, 1972), there are no organized, corporate, local or territorial groups in their society, and formal political and legal administration is absent (1972: 155–6). He argues that the Majang political system is based on 'underlying anarchy'. However, they had traditional ritual experts called *tapa* (singular *tapa'd*). *Tapa* are supposed to have the power to divine the future, confer protection and good luck, and diagnose and cure sicknesses of various sorts (Stauder 1972). Most *tapa* are from the dominant Meelanir clan, whose ancestors are said to have come from the south, around Maji. Stauder also

---

[1] During my fieldwork in Majangirland, I used Majang as my sole means of communication. As for translations of speeches and interviews, I did this with the help of several Majang friends through the Majang language. I would like to thank especially Redat Gebrekidan in Meti town and Ismael Yohanes in Kumi village for their help in this respect.

suggests that in the 1960s, at the time of his fieldwork, interference by the government in Majang affairs was minimal. No administrative or police post existed in the main areas of Majang settlements. It should be noted that it was during those days that rapid social change started among them, as will be explained in the following section.

First, I shall outline the main topics of social change the Majangir have experienced since the 1960s, namely their contacts with foreign missionaries and the government's general implementation of villagization, which had a great impact on their society. It seems clear that the 1960s can be regarded as a prelude to 'Majangir modern history', when the American Presbyterian Mission started its work among the Majangir. It should be noted, however, that much deeper contacts with other peoples, including Ethiopian central and local governments, started after the Derg took over power. Here I would like to discuss the process with special reference to the following three topics: the arrival of 'Odola' and the beginnings of missionary work; the Derg and the villagization process; and the new phase of competition after the Derg. Then I shall return to the scene I briefly sketched above, in which a young Majang leader persuaded local people to make a sedentary village and accept Christianity. Through this description, I shall argue that the Majangir's acceptance of Christianity is inseparable from the condition in which they had been put by the Derg, whose policies had a great influence on their lives.

Odola & the Godare missionary post

It was in 1964 that the American Presbyterian Mission started its project in Majangirland. An American missionary, Mr Harvey Hoekstra, who is known to the Majangir as 'Odola', and his family constructed a large airstrip, a school and a clinic at Godare, in the core area of Majangir forest. His nickname, 'Odola' (which is probably an Anywaa name), has been one of the most famous foreign names known to them up to this time.

The missionaries obtained permission to continue their activities from a powerful ritual expert (*tapa'd*) in Godare and began promoting their missionary work. Until then, the medical care of people suffering from various kind of diseases, such as smallpox, malaria or tropical ulcers, had been in the hands of ritual experts or *tapa*. Odola wrote in his autobiography that Balti, a powerful *tapa'd* living near the missionary post, at first discouraged his followers from going to the clinic (Hoekstra 1995: 271). People living nearby, however, gradually came to know that the missionaries' medicines were very effective against such diseases, and later Balti himself went to the clinic to be vaccinated against smallpox.[2]

---

[2] People's expectations of the effects of these medicines can also be seen from my observations in the villages. When I asked them about their memories of Odola, many of them referred to the medical care they had experienced. They would go to the clinic in Godare from quite remote areas, which may take a few days' walk.

The Promise of 1991

The main factor that initially interested them and brought them towards the missionaries and Christianity may have been their great expectations of the effectiveness of the medicine, as well as the friendly and devoted attitude of the missionaries. The narratives of Muse Adirman, who was a student of Odola's and an important leader of villagization during the Derg period, show the importance of these aspects:[3]

Text 1. Muse's narrative on Odola's coming and his work

I know Odola. He came in 1957 (E.C.). Was it 27 years ago? He brought the true words about God (*peeni waqoyotongk sino eekerk*). In the old days Majang people worshipped stones put around *tapa*. While we lived such a life, he came and said, 'Don't pray to stones on river banks: God exists.' It was Odola who brought words of God. Then he lived with us. We listened to the words of God, and he taught us. It was Odola who brought it first.

What Odola did was to teach us how to do various kinds of work. How to make fields. And various things, for example, he taught us that it was bad to kill elephants, while he taught us how to pray, and gave us medical care. He told sick people not to go to see the *tapa'd* but to come to his clinic. And he taught us about Sundays, on which we must not work and must go to church. And we discovered our past mistakes. He taught such things.

Muse's insistence that they accepted these new customs not only because the missionary persuaded them is noteworthy. The impact which the missionaries had concerned material aspects, not spiritual life. He first notes that what Odola did was to show them 'how to make fields' and that he provided 'medical care'.

Odola tried to persuade them to accept Jesus Christ at the same time, using cassette tape recorders through which people listened to the gospel in Majang. At the time this was an innovative means of preaching to people who could not read the Bible. Many people around Godare more or less felt the power of the Christian God and the affection of foreigners towards them through the missionaries' persistent efforts. Odola wrote about the impact of cassette recorders as follows:

I wish you could have been there to see their reaction. Frequently someone would hold his or her face tightly, turning the head from one side to the other to hear more clearly. Some spoke back to the voice coming from the 'box that talks', exclaiming, 'Tia' ('I hear you'), 'Moko nyun' ('It's no lie') or 'Yang jet' ('It's sweet'). It was marvelous to behold.

---

[3] Texts 1, 2 and 3 are from a series of interviews I conducted between 1994 and 1995 with Muse Adirman. The interviews were recorded with a cassette player and in total amounted to more than five hours. Here, I give some extracts concerning his explanations about and evaluation of recent changes in Majang life. For further information on these texts, see Sato 1997.

Evangelical Christianity & Ethnic Consciousness: Majangir

They were hearing the saving message of God's salvation.
(Hoekstra 1995: 248).[4]

Odola and his family left the Godare Missionary Post in 1976, when the influence of revolutionary policies was predominant around Tepi. Then Majang people lost contact with foreign missionaries, as well as their medical care. Interestingly, however, Christianization among the Majangir became widespread under the Derg, despite the absence of foreign missionaries.

Villagization under revolutionary policy & Christianization

The socialist revolution occurred in 1974 and the Emperor was deposed. Among the main policies of the Derg, the villagization programme had the greatest influence on Majang society. It began at the end of the 1970s, when an administrator called Tamuru took up his post and implemented the policy. On the Majang side, chairmen (*lekanbali*, a term borrowed from Amharic) were assigned to *qebeles* or peasant associations. Most of these village heads were relatively young men who were often the sons of local *tapa*.[5] Meetings of the villages were led by several young men who assumed positions of leadership.

Villagization radically changed the Majang way of life. Before its implementation people would move their settlements with their agricultural fields every a few to tens of years. This often happened when conflicts happened between clans or neighbours, or when someone died in the settlement. When a powerful *tapa'd* died especially, his large settlement used to be completely abolished, and no one would return until people had put the memory of his death out of their mind. Now that people live in sedentary villages, they seem to have abandoned the custom. They have a graveyard in the village, and when someone dies, the person is buried there. Many of them build their huts along the main roads of villages, and leave them in place even when they clear new places for their fields, which are now relatively remote from the village centres.

During the period that villagization was going ahead, Christianization was promoted by several leaders, many of whom had been taught by Odola in the missionary school. They were still young, in their twenties, and they preached to people living in distant places where new villages were to be built. They did this using the cassette players which Odola left behind, and persuaded the people to accept 'the great God' (Waqoyo Sin

---

[4] Odola later developed this idea and founded the Audio Scripture International, of which he is currently president, to provide the Bible on cassette.
[5] In many cases they did not hold the post for many years: the post was often temporary and was taken over by another person after a few years. In Kumi village, the total number of *lekanbali* appointed was twelve in almost thirteen years of Derg rule.

189

The Promise of 1991

Obik) and his word (*peeni waqoyotongk*) instead of local spirits (*waqoyo tapa'dongk, walde*, etc.) with which *tapa* were able to make contact. They encouraged local people, especially the younger generation, to construct new villages with churches at the centre. Muse explains the reason why they chose Christianity and its associated customs, such as 'giving up drinking and smoking', as follows:

Text 2: On the difference between spirits in the old days and the Christian God
I know *saloy* (place for worshipping spirits). For example, they dig a hole there while a house of a *tapa'd* is over there. According to him, he puts stones and *eemuy* (a plant name), so that Waqoyo lives there, and he also puts sorghum beer, meat of wild animals, or chicken's blood. All this is because people found a wrong thing. Probably in the old days Satan would drink the blood. At the sorghum beer parties, also, people at first would go to a *saloy* place and pour some of it out there. Then people would drink. They lived such a life.

Gods of *saloy* (*waqoyo saloyongk*) and God of church (*waqoyo so gode sambateyongk*) are entirely different. Gods of *saloy* are small. Why is this? They are called '*kayii*' or '*walde*' and what they do is to possess people. The God of the church lives in the sky. We can say that. Why is this? Gods of *saloy* help everyday, usual things. For example, they are good for such things like making good tea, but they never help with big problems. They help only with tiny things. It is like that. But they don't know about future things. They are troublesome and always want sorghum beer and possess people. They live with some people and tempt them, and bring them to some places and have them drink. They say you will die if you don't drink. And you really die, without beer.

Text 3: On giving up smoking and drinking
In the words of God, he says, 'Don't drink beer; don't drink honey-wine.' If you are drunk, you cannot recognize your brother. Medical doctors say that you can drink a little but you should not drink so much that you have a headache, while God says 'Don't drink any more.' Why is this? It is because you could not recognize who I was. And nowadays we are taught so. If we, the Majangir, continued to drink, we would not have lived together as now. We, the Majangir, would have lived in scattered places, away from each other. This is because people will kill others when they drink if they live together. People otherwise smoke. How many people died because of intoxication in those days! Now we have automatic rifles, so if there is only one person drunk, he will hurt many people. In the old days we had a rifle called *meniseeri*, load one bullet and kill one person. Then people would escape. Nowadays, however, if a drinker has an automatic rifle, he will kill many. So in the old days if people found drinkers, they would keep away from them. It was very bad. And they wouldn't work.

It is apparent from these texts that Muse consistently emphasizes that their acceptance of Christianity is so monumental that their life and customs today are entirely different from those of the old days. Proclaiming the contrast between the old days and now, he evaluates the conversion as a radical change.

Muse clearly states the difference between *waqoyo* of the old days and *waqoyo* of these days. In his own words, '*Waqoyo so gode sambateyongk* is in the sky, while *waqoyo saloyongk* exists nearby and helps us for only small things, such as tea-making.' Moreover, '*waqoyo saloyongk* wants beer very much and often possesses people.' In everyday conversation, I often hear young people say, 'Waqoyo whom we regarded as God was actually Satan (*seytan*). We didn't know it in those days.' Thus the acceptance of Christianity was a turning point. The Majangir abolished several previous customs and chose a new style of life. Giving up drinking and smoking is one of the most important and symbolic choices.

It is notable that the main promoters of this movement were from the Meelanir clan; they were the sons of *tapa* or, if not, at least sons of traditional leaders among the Majangir. It can be said that their attitude towards villagization and Christianization was well received, mainly among the younger generation, partly because of the authoritative power of their descent. For example, there was a very famous leader from the Meelanir named Gebrekidan in a settlement near Godare. Odola describes him as 'the stocky and obviously intelligent Mesengo [Majangir] chief' (Hoekstra 1995: 253), and in fact his name was broadly known among the Majangir from Godare *wereda* to northern settlements near Dembidolo.[6] His two sons became important leaders of the Christianization and villagization movement, one of whom will be referred to in the following sections. In Kumi village, as an additional example, five grandsons of a powerful *tapa'd* in that area are still holding leadership positions in political issues, one of whom is currently a member of the *wereda* council in Meti town, and another has been *lekanbali* of the village for four years.

After the Derg: the new period of disputes over 'ethnic boundaries'

After the collapse of the Derg in 1991, the Majangir faced a new phase of political crisis. Under the new federal structure, the Majangir are regarded as one of the 'representative ethnic groups' of Gambela Region, where they are the third group in numerical terms next to the Anywaa and Nuer.

[6] I once made a long journey to the northern fringe of Majangirland around Mogi in Oromiya Region with a son of Gebrekidan. During it, I saw an elderly woman cry, deeply moved at seeing the son of Gebrekidan, and saying, 'Aah, you are the son of Gebrekidan, that big and clever man! Those days I would live near his land and take honey-wine to him! He was a very very big and good man!' She continued crying for several minutes, repeating the name of Gebrekidan. I was deeply impressed to learn that the great names of *tapa* were known in such a distant place, where daily connections with the Godare area do not seem to exist.

The Promise of 1991

The problem is, however, that their habitat actually extends over the boundaries of three Regions: Gambela, Southern Region and Oromiya. There have been a series of disputes concerning regional boundaries around Tepi and Bebeka, the Majang insisting that they be included in Gambela, though they are ultimately striving to establish a Majangir Zone.

There are three factors about present political conditions around Tepi which are increasing the Majang sense of distrust of neighbouring peoples. First, Tepi had traditionally been inhabited by the Majangir before Amhara pioneer merchants came to settle down there and open a market.[7] Later more and more settlers came to this area to grow crops such as coffee, turmeric and Madagascar cardamom (*kororima*), while several Majang settlements remained around Yeki *wereda*, which surrounds Tepi Ketema. Second, there is a traditional sense of antagonism between the Majangir and Shakacho, who inhabit the north of this area and recently came to assert their rights to extract political and economical benefits in Tepi and Yeki *wereda*.[8] According to Amharic pioneers and Majangir, Shakacho people have rarely lived in this area, but in 1993 the presidency of Yeki *wereda* administration was occupied by Shakacho, and the group took the political initiative in the administration of Yeki *wereda*. This situation seriously angered Majang leaders. Third, it is possible to assume that there was some connection between the Majang leaders, the SPLA and Surmic-speaking peoples around Gurafarda and Dima. It was said that these often attacked the EPRDF and neighbouring highlanders after the collapse of the Derg, though there was no local proof of this, despite the EPRDF's claims. At the very least, such resistance by their neighbours might have stimulated the Majang leaders into reacting aggressively to what they saw as an oppressive situation.

The incident I sketched in the introduction occurred in such a situation. In April 1993, a large number of Majang men armed with guns were mobilized. They systematically and successfully placed Tepi and Meti under their control, blocked traffic in and around the towns, and kept control over them for a few days, until EPRDF soldiers came to mediate. More than a hundred people who were regarded as political opponents of the Majangir – Majang people called them Donjiyer, which meant Shakacho in their language – were shot.

In May, when the EPRDF arrived in the area, several meetings were held among representatives of the EPRDF together with Yeki, Mizan Teferi, Gambela and Majang villagers. As a result, the administrators of Yeki *wereda* were recalled and a Majang man appointed as vice-president, while an Amhara man became president. Later, however, several meetings were held in September and October, at which several Majang leaders

---

[7] One of the first Amharic settlers gave evidence of the process in which they cleared forest where Majang people were living: the town's name, Tepi, is the name of a Majang man who had lived near the present market place.

[8] On the tradition of antagonistic relations between Majangir and Shakacho in highland Illubabor, see Yasin Mohammed 1990.

Evangelical Christianity & Ethnic Consciousness: Majangir

were blamed on suspicion of conspiracy with the SPLA.[9] Shortly after, more than 150 Majang were arrested around Tepi, Meti and Bebeka, including the then President of Godare *wereda*, and accused as suspects in the incident.[10] EPRDF soldiers went to Majang villages and held meetings, at which they blamed the incident on the SPLA and their sympathizers.

This incident shows that the Majangir were facing a problem which they had never experienced in the past. In their traditional concept, land for them to live on is abundant anyway, and disputes over land were totally unknown among them. If outsiders came to live on their land, they would leave for another uninhabited place in the forest. Since the early 1990s, however, wood-felling by several private companies has been taking place in the forests around Meti,[11] and Majang leaders have begun to feel that their habitat and resources were no longer inexhaustible. It is clear that the new policy of the federal government assuring them regional autonomy has encouraged them to stand up for their own rights in alliance with neighbours like the Anywaa.[12]

Constructing a new sedentary village: the discourse of promoters

As I briefly described in the introduction, in 1993, together with my Majang friend, Redat Gebrekidan, I made a trip to a peripheral area one and a half day's walk into the northern forest from Kumi village. At our destination, a small settlement called Jaman, a plan to construct a new sedentary village was being implemented.[13] I actually went there to carry out research on the relationship between Majangir and Shabo, who are hunter-gatherers living around there mixed with the Majangir, but unexpectedly what I observed was a new village being constructed.

Redat Gebrekidan, my research assistant, then in his thirties, was a Christian as well as a promoter of various development plans among the

---

[9] None of the Majang leaders who were arrested, however, accepted the accusations of conspiracy with the SPLA, despite EPRDF's claim.

[10] Most of them were put in prison in Masha and Minzan Teferi for two to five years and were released before the end of 1998.

[11] This information comes mainly from Godare *wereda* administration offce, especially the Natural Resource Department, and several Majang people. They say an influential local leader in Meti is concerned in this illegal wood-felling, and that no one can stop it.

[12] See Kurimoto 1994. The situation of ethnic relationships, of course, seems quite complicated. The apparent alliance with Anywaa does not seem that solid because the Majangir very often complain about their behaviour, including exclusive possession of public property. In addition, they often seem to regard Anywaa as traditional enemies who used to raid them.

[13] This villagization programme was first proposed by Godare *wereda* administration, though almost all the actual implementation activities within the village were left in local people's hands. Despite this, they made the project administrators hesitate because the Jaman area was very difficult to access, lacking a clear path from Meti. As this case shows, 'villagization' is still continuing as an extension of the Derg's policies, though implementation is often found difficult without enough support from the local administration.

The Promise of 1991

Majangir, including villagization and agricultural development. His father Gebrekidan was a famous local leader around Godare from the Meelanir clan, as described earlier. When I later visited various places in Majangirland, I realized that Redat was so enthusiastic that he never failed to preach and persuade people, especially the young generation, to construct sedentary villages and to go to school. The younger generation of Majang who are assuming the leadership of religious activities around Godare area often behave like him, and many of them are from the Meelanir clan.

Below I give excerpts from a speech that Redat gave in the evening for the purpose of encouraging village construction, in front of some dozens of people who had gathered for the meeting. The place they gathered, a hilltop, was to be the central meeting place of the village in the future, and he started with an Amharic Bible in his hand. The speech continued for more than forty minutes, and I will give some important and repeated expressions from it in the following texts.

Text 4: On the universality of Jesus Christ
Our fathers believed and worshipped different things. Our fathers thus lived disconnectedly from each other. Now we are different from what we were. Shakacho people (Donjiyer) worship Jesus Christ (Yese Kirisiye). Highlanders (Gaaleer) worship Jesus Christ. Foreigners (Farenji) worship Jesus Christ. Majangir also worship Jesus Christ. Bencho (Meriyer) worship Jesus Christ. Those who remain are only us. Anywaa ('Beriyer') worship Jesus Christ. Why do you remain alone? Today those who worship Jesus are good. Others who don't accept Him and remain in small places will not be able to go towards the good way. Those who construct in the same place will thus say: 'We worship the same God, and there is no difference between highlanders and Majangir.' All of us go the same way.

Text 5: On the new future life after the village has been constructed
Now we have contacts with the government. We sometimes find mistakes in what they say. You work here in the new village. What do you want? What do you do? You will plant lots of cassava. What do you do? You will plant lots of sugar cane. What do you do? You will make large fields of sweet potato. And what do you do? You will make a church. You will do it here. What do you do? You will worship God. Here it is. And what will you see in the future? You will see lots of relatives (*tekanir*) here. ... What causes that? We owe it to the church. Have we ever benefited so much in the past? Have we ever learned in school up to now? Now we can open the door towards the outside. ... We plant sugar cane, cassava, sweet potato, and American taro (*meti*, *Xanthosoma sagittifolium*), all over, up to that river. Villagers in distant places also work like this. We make a village like this and make lots of food. And there will be a church over there. And we will find guests here. We will invite school teachers. ... Elders and youngsters don't follow the old customs. ... How did our fathers get along? They drank

Evangelical Christianity & Ethnic Consciousness: Majangir

and danced, and killed others. We, however, shall construct a new village, and enjoy life there. But we shall never drink.

At several points, Redat's speech has the same emphasis as Muse's narratives cited in the previous section. One of them is their common perception of Christianity as being instrumental for material development; another is the contrast between the old customs and present-day customs. Even in everyday conversations these expressions are heard quite often: 'In the old days we Majangir were like those who drink every day and kill others. Nowadays we know the true God.' Christianity and material development, which prove and symbolize 'the new Majangir', are inseparable. We should note that, in his speech, Redat emphasizes the universality and 'ethnic equality' of Christianity. He uses the phrase, 'Equality among the Majangir and neighbouring peoples' to encourage them to accept Christianity, and relates the narrative to the construction of the new village. Christian churches in their sedentary villages may represent a monument symbolizing the new, materially affluent life. Young people in Jaman made apparently quite positive responses to Redat's suggestion. There were already several new huts and a church built for the future in the central square of the village. These were all the products of the younger generation. The elders, on the other hand, kept quiet.

However, I must also mention the ultimate outcome of this project: villagization in Jaman was later abandoned. Before I visited Jaman again after two years in November 1995, I had expected the plans to have seen further development. Contrary to my expectations, the village site on the hilltop was ruined and deserted. The reason for the abolition, according to the villagers' explanation, was simple. In 1994, a middle-aged woman cursed a young man, following which four people died of disease within a few months. They said that one night, many people saw candles floating in the air. As a result of these incidents, everyone left the hilltop for fear of possession by evil spirits.

Meanwhile executive officers from Godare *wereda* became hesitant to continue the project because the Jaman area was too far away to be administered as an independent *qebele*. It is said that the Jaman villagization project sometimes become an issue for administrators in Meti, but it had still not been realized as of January 1999. This episode may suggest that, despite my argument regarding the Majangir's positive view of Christianity, their traditional notion of spirits is still alive in their minds.

Conclusion

Odola and his colleagues played quite an important role in Christianization at the start. In the preceding sections, we have seen how Odola's activities were influential in shaping their images of Christianity. Had the missionaries not been in some ways acceptable, the later move to Christianization might not have been so successful. A more important point is that the missionaries were able to get the *tapa* and their followers to accept the healing power of modern medicine. *Tapa* had to do this because people's primary concern was health, and the authority of the tapa rested largely on their healing powers.

In a sense, however, the activities of Odola and other missionaries merely provided 'an opportunity'. This is shown by the fact that rapid progress with Christianization was made rather through the villagization process, during which there was no contact with any missionary or foreigner. The villagization policy greatly changed the situation of the Majang way of life, in terms of both internal communal structure and relations with other societies and the central government. Sedentary villages are knotting points which tie up people of 'underlying anarchy' with lower reaches of the Ethiopian central government. Villagization brought great changes which they had never experienced in the past.

It is significant to note that Christianity ideologically fits the new sedentary village life in several respects. A typical example is the relation between the traditional death beliefs and the pattern of settlement transfer. Traditionally the Majangir would abandon their settlement and move to a new place when a famous *tapa'd* died, through fear of evil spirits. After villagization they solved this problem by adopting the Christian mode of burial and Christian death beliefs. In a sedentary village, there is a fixed graveyard where all the village members are to be buried.[14] The changes in settlement pattern and religion are thus directly connected. Drinking customs are another example. One of the reasons why this custom has been given up is that, as Muse says, it is dangerous for village life in causing disputes among themselves. A traditional function of drinking customs, namely to gather people together for labour co-operation, has been replaced by other forms of reciprocal association in the village following villagization.

It is quite important to note, on the one hand, that many promoters of Christianization are from younger generations of the Meelanir clan. It would have been very difficult for ordinary Majang people to accept Christianity if promoters of the movement had not been members of the Meelanir clan, the clan of their ritual leaders. In traditional Majang communities, the special status of *tapa* was a focus of centripetal force. In a sense, the religious and political leadership of the Meelanir clan continued irrespective of the change of religion. These leaders presumably believed that Christianity is also appropriate in solving their chronic problems, notably the necessity of political and cultural negotiation with the central government and neighbouring peoples. These leaders and their ideas were comparatively well accepted by ordinary Majangir.

---

[14] Stauder predicted that his research site (Gelese settlement) would be abandoned in the near future because extremely high population pressure had been making people's lives critical (1971: 183–4). Gelese settlement, however, has never been abandoned to date in spite of the death of the then *tapa'd* (Yodn) in the mid-1980s. He was buried in the village graveyard like other dead people.

As for external relations, on the other hand, Christianity became a symbol of 'ethnic equality' in Majang consciousness. As we saw in Redat's speech, young leaders stress the equality of different ethnic groups as all followers of the Christian God. Such discourses are also quite often heard at the ordinary sermons and prayers in village churches. I assume that there is a strong correlation between Christianization and traditional intercommunal relationships in which the Majangir have been discriminated against as 'poor forest people with hardly any property'. They themselves often admit that they have traditionally been powerless. They think that their conversion to Christianity is likely to transform such discriminatory treatment. An awareness that Christianity was spreading among other groups was used by its protagonists, who argued that conversion would ensure that the Majangir would not be 'left out' but would become equal with other peoples, all being Christians. In a word, they are making a statement, to themselves as much as to others, that they now have the same rights to resources and benefits that historically were the preserve of others, especially highlanders. Thus it can be said that their conversion has been promoted as part of a contest in power politics among various groups and their relations with the central government.

Eleven

Capturing a Local Elite
The Konso Honeymoon

ELIZABETH WATSON

> *The government of Haile Selassie did some bad things to Konso people. They took all our livestock and the people passed this time as slaves. Mengistu's government was the same. What was different was that they took our children instead and sent them to the war front. Some died, others returned handicapped; others, we don't know what happened to them, whether they are alive or dead.*
>
> *This new government of EPRDF is now using our language and our culture officially. If we have a problem we can go and meet with people and talk to them. Before we had to buy translation. Our children have also been given jobs, which has never been before.*
>
> (Konso elder: preliminary statement to blessing at a political rally, 5/12/95)

This paper is concerned with the impact of the post-1991 Ethiopian government's radical federalization policy at the local level. By taking one example, that of Konso special *wereda* in south-west Ethiopia, it examines what the federalization process has meant for a relatively small, previously marginalized group of people. First, the paper describes some of the hopes and expectations which were raised with the establishment of the special *wereda*. It then goes on to describe some of the challenges and constraints which have limited the fulfilment of these hopes by examining two matters in particular: the process of standardizing the previously unwritten language of Konso for official use by the special *wereda*; and the nature and form of the local state structure itself. In places, comparisons are made with parallel processes in the neighbouring region of Borana.

Hopes & ideals: Konso special *wereda*

In 1995 to 1996, at the time of my fieldwork, the Konso were experiencing something of a honeymoon period in their relationship with the new

Capturing a Local Elite: Konso

administrative government of Ethiopia. After the 1991 change of government, Konso, an area with a population of 157,000 (1992 Census) in a remote part of southern Ethiopia (see Map 1.3, p. 31), was designated a special *wereda* (or *liyyu*) and became a self-governing semi-autonomous unit. This gave it the status of a kind of mini-state: it existed as a state within a state, a region with a right to 'self-determination up to secession' (Ethiopian Constitution 1995). Whereas under previous regimes it had always been a small, often marginalized part of the larger Ethiopia, in administrative terms it now had an importance and integrity of its own. Konso had received the status of special *wereda* because it was considered ethnically too distinct to be grouped together with any of its neighbours.[1] Important in this process must have been the high visibility of Konso 'culture', namely the Konso dress of home-spun cotton for men and women, and the well-known Konso agriculture with its terraces. But its geographical boundedness, located as it is on the Konso highlands, which rise out of the Rift Valley, also facilitated its designation as a unique area.[2] It thus became a semi-autonomous, self-governing part of Regional State 7, the Southern Nations, Nationalities, and Peoples; its dealings with the larger government are through their headquarters in Awasa. An official of the new Konso local administration, whom I shall simply call K, explained the situation as follows:

> We work mostly with Awasa but according to the formation of the state, the power is given down, the power is more in the bottom than with the state. More power is given to Konso; even the people themselves have the power to decide.
>
> (K. 1996).

Across the country, the process of federalization has given rise to many practical problems. Some administrative units are large, sometimes even spatially dispersed (for example, Oromiya) and thus difficult to manage. Others have been made up of culturally various groups combined together for ease of administration (for example, South Omo Zone). By contrast, in Konso the boundaries of the new special *wereda* defined an area and a population which was not too large to handle nor too small to be ineffective. It was an area of a people with a strong sense of their own uniqueness and identity. Konso was where it seemed that the new dream of an Ethiopia decentralized on 'ethnic' lines could be realized, one where people could re-create a new form of administration which would be appropriate to them and which would serve their needs. It was a very exciting time, and the new administrators were kept busy designing the form and nature of their new Konso, empowered as they had been by the state.[3]

---

[1] Other special *weredas* included Gidole, Janjero and Majangir.

[2] This is despite the fact that German scholars have sometimes included Konso in part of a larger group: for example the 'Burji-Konso cluster' (Amborn 1989) or the Konsoid group (Sasse 1986).

[3] The information presented in this paper is based on twenty-one months' fieldwork in Konso in 1995 and 1996 and a subsequent brief visit in 1999. During this time I worked with diverse

The Promise of 1991

The implications of the designation as a special *wereda* were immediately evident: the main junction town of Bekawle, which had grown up earlier this century, first as a garrison town and administrative outpost of the northern empire and more recently as the Derg headquarters, was renamed Karate, after the Konso name for the area in which it is situated. The administrative offices, which earlier had largely been staffed by officials from other regions of Ethiopia, were locked up and deserted. These had been situated in one part of Bekawle known by the Amharic term for the old garrison, Arogé Ketema. New offices were opened in another part of town and staffed by as many educated Konso as could be found. Through this renaming and relocation, the new local administration reclaimed Konso linguistically and spatially, at the same time distancing itself from its colonial history.

Konso now had its own Ministry of Health, Ministry of Education, Ministry of Agriculture, Ministry of Culture and Sport, and so on. It had its own administrator, its own court and judges, the ability to make its own laws, and its own budget, which it could spend as it saw fit. The new administration needed personnel, and there was nearly 100 per cent employment for Konso people with education. There were not many Konso people who had received tertiary education, but those who had were placed in influential positions; others with Grade 12 schooling were placed in supporting positions. This was in real contrast to the previous regimes, where those with education had found few employment opportunities.

The official working language of the administration was immediately changed from Amharic to the local language. This language had previously been unwritten and was known in Amharic as Konsinya. The new administration called it by its Konso name, Afa Xonso or Afa Karate. Preparations were made to standardize the writing of the language and to introduce it into primary schools as the main language for teaching. Those in the final stages of their own education who could speak Afa Xonso were encouraged to go to the teacher training institute in nearby Arba Minch to meet the needs for new teachers. The quotation at the head of this chapter reflects much of the feeling of the moment. An old man had been asked to give a Konso blessing at a political rally, which was an attempt to bring

---

[3] (cont.) groups of people, men, women, farmers, administrators and ritual leaders in different parts of Konso. With the exception of one taped interview with an administrative officer which is quoted here (K. 1996), I rarely discussed the political processes taking place or interviewed people explicitly about them, as political issues were sensitive, and many interviewees appeared to feel uncomfortable and to believe that such questions were inappropriate for me. However, issues concerning the recent political process could not help arising in my research, and also simply because I was living there, I was privy to different discussions and meetings and became aware of popular feelings. Different groups have different feelings towards the developments in the post-1991 period, and I try to include some of these in this chapter, while at the same time giving an overall impression of the broader trends which are taking place. The majority of interviews quoted here took place in Karate town and were carried out in the English language (K; EB; G). RP9 was interviewed in his house, some 5 kilometres from Karate town in the Konso countryside. This interview was carried out in Afa Xonso. It was taped and transcribed with assistance.

together Konso culture and the new administration. He expressed what appeared to be genuine happiness on the part of many Konso, namely that if they had a problem they would no longer have to 'buy translation', that their children were receiving some benefits from their education in the form of employment. This quotation cannot necessarily be considered representative, nor can a political rally be considered a place where a person can speak his or her feelings freely. However, I do think that it reflected something of the spirit of the moment, the hope that an administration made up of Konso people would both understand and represent the needs of local people.

The reversals in the make-up and form of the administration were part and parcel of a reversal in attitudes to local culture. The regimes of the pre-1974 imperial and post-1974 socialist governments were both characterized by views that denigrated local cultures and saw people in the south as backward. The imperial regime encouraged conversion to the Ethiopian Orthodox Church and assimilation to Amhara culture (Donham 1986a). The post-1974 socialist government saw local religion as hindering the emancipation of the people, and also often as extractive and exploitative (Donham 1999). In contrast, the 1994 constitution states that 'Every Nation, Nationality and People in Ethiopia has the right to speak, to write and to develop its own language; to express and to promote its culture; and to preserve its history' (Article 39, 1995: 18). It refers to 'rich and proud cultural legacies', and 'requires full respect of individual and people's fundamental freedom and rights, to live together on the basis of equality and without any sexual, religious or cultural discrimination' (Ethiopian Govt. *Federal Negarit Gazeta*, August 1995: 13ff).

These developments have also taken place in the context of a wider change in attitudes to development and administration at the global level. Whereas local institutions were previously seen as antiquated and inefficient, now they have been re-appraised as flexible, adaptable and in touch with the local people. This reversal in attitude has become mainstream in approaches to agricultural practice, with the heralding of indigenous agriculture and indigenous knowledge as a means through which sustainable development can be achieved (Adams 1992; Richards 1985). Similar principles are being developed in the realm of organization and administration at various levels, from that of local governance to that of the community, for example, of water-users' groups. Modern, externally imposed forms of governance and organization in developing countries have been seen as corrupt and inefficient; in many places the committees set up by outsiders (either state bureaucrats or NGOs) to co-ordinate local projects or natural resource management have often broken down altogether. In contrast, customary or 'traditional' forms of organization have been seen as effective and sustainable; they include incentives for participation, and are efficient at solving conflicts (Adams 1992; Bassett and Crummey 1993). These reappraisals have been incorporated into practice recently in the form of promoting community-based development projects, which have mushroomed in many parts of Africa. Government structures and other development and administrative agents are turning to indigenous institutions as a means through which grass-roots development with the participation of local people can be achieved.

In Ethiopia, the architects of the present decentralized state may not have drawn explicitly on these ideas, but they have no doubt influenced the developments that have been taking place today. The relevance of these ideas is particularly strong in a special *wereda* like Konso: there is the idea that a new Konso political society can be created, which will use the Konso language and build on Konso culture and institutions, being more appropriate and more in harmony with the needs of the people. It will therefore be more successful and efficient: it will serve the needs of, and empower, the people it represents.

In 1995 and 1996, therefore, Konso was exciting both for those involved in the reconstruction of the Konso mini-state, and for me watching these processes in action. The opportunities provided by federalization were understood by those involved as a chance to return to a golden age of Konso, to a pre-colonial Konso which was strong both in work and culture. This was after nearly a hundred years of rule by outsiders, who were now seen as people who had oppressed and exploited the Konso. When I asked the Konso administration official I had come to know, an educated man speaking in English, how things had changed since the establishment of the Konso special *wereda*, he explained:

> I may answer this by referring back to Konso of the 1890s. Before that time Konso was an autonomous area and leaning on its own cultural dynamics and administrative situation. But with the invasion of the Menilek's troops, Konso was invaded by an exotic tribe and it was forced to lose its culture and self-ruling administrative situation. The previous culture was basically very strong. Theft was very forbidden in the community. Everyone was hardworking and agricultural technology was very highly developed. But after Menilek's troops' invasion, the culture and the rights of this nationality were distorted. Through Haile Selassie's government, and the Derg as well, many problems were faced by this nationality.
>
> (K. 1996)

He viewed this period of outside rule and exploitation as coming from all non-Konso people, not just one Ethiopian group in particular:

> In the past it is not only the Amhara nationality that oppressed the Konso, but everybody coming through, whether Oromo or Gamo or whoever, came here as an exploiter because the state system was doing that, and everyone was doing the exploitations on Konso.
>
> (K. 1996)

He understood the process of the construction of Konso special *wereda* as a rejection of outside influence and a return to an authentic and integral

Plate 11.1 Konso: listening to the new message. The flag reads in Amharic 'Peace and democracy we gained through struggle: through struggle we will strengthen them'
[E. Watson]

Plate 11.2 Flag waving at the Konso rally [E. Watson]

The Promise of 1991

Plate 11.3 *Acclaim for the new policies, Konso. Two men are dancing holding candles: they are acting out the importance of encouraging people to come to school. The candles are symbols of 'modernity' and the light brought by education*
[E. Watson]

Konso, to Konso culture and Konso organization. These were ideas that I also often heard from different people at the time.

Needless to say, this was a more difficult process than might have been envisaged at first. When I visited Konso again in 1999, I found that many of the initiatives of the new administration were floundering: it seemed that it had failed to fulfil its potential. In what follows I examine in more detail two areas of the process of rebuilding this new Konso, and particularly some of the problems that have limited the realization of the hopes and expectations of its people. First I look at processes involved in implementing the use of the local language: this was a lynch-pin in the conception and working of the new mini-state. Secondly I examine the new administration's approach to working with indigenous institutions, which was also crucial to the viability of the project.

Difficulties & disappointments: reclaiming the voice of 'Afa Xonso'

Dr Johnson described languages as the 'pedigree of nations' (Hill and Powell 1950: 225). The decision to use Afa Xonso as the official language of the new special *wereda* was instrumental in the reconceptualization of Konso (both by Konso people and others) as a nation with its own identity and independence: it was no longer an insignificant group marginalized along with a host of others. The standardization and introduction of the Konso language that took place was thus a practical matter and also an important symbolic event. In practice, it opened up the local administrative structures to the local people. When the Konso people had to carry out business in another language, they were in effect prevented from expressing much of their experience, or forced to translate their experience into the words and phrases of that other language. A Konso person who

Capturing a Local Elite: Konso

wanted to take a dispute to the local courts, for example, was now able to discuss the matter directly and to use the exact words to refer to a Konso institution or a relationship or a transgression. This was empowering for the local people and heralded a new era in which the relationship between the people and the administration had changed: they were now able to talk directly to each other.

Symbolically, when Afa Xonso was made the official working language of the special *wereda* and the intention to use it in Konso primary schools was announced, it was given the status of other languages in Ethiopia, including Amharic. In schools, this meant that Afa Xonso was seen as being as good a medium of education as any other language. The introduction of Afa Xonso was both a symbol and a metaphor for the rejection of externally imposed mediums and measures: it facilitated a resurgence of Konso self-esteem and was crucial to the process of the reinvention of Konso in the context of federalization. In practice, however, the introduction of Afa Xonso was riddled with difficulties. As an unwritten lowland Cushitic language (Uusitalo 1989), a script had first to be selected and the orthography standardized. This proved to be an extremely complicated and also highly political process (as shown by the work of Gideon Cohen covering many other areas of the Region: Cohen 2000).

Many of the decisions regarding the language took place at a three-day meeting in Karate called by the Konso Ministry of Education in 1996. The purpose of this meeting was to make some key decisions on how Afa Xonso should be written. The Ministry brought together all the people who they thought would have something to contribute to this matter, including many of the educated people of Konso who were working in the administration, as well as a Finnish linguist called Mirjami Uusitalo. She lived in Arba Minch and had been working on the language for five years as a translator for the Wycliffe Bible School, translating the Gospels into Afa Xonso and also writing some notes on the language. The administration also sent a motorbike to the village where I was staying and fetched me, so I was able to attend. Although I could speak some Afa Xonso, my technical knowledge was limited. Other participants included Yohannes Hadaya and the late Gemechu Gedeno, who were the authors of a Konso *Agricultural Dictionary* written for Farm Africa, a British NGO active in Konso. This text (Yohannes and Gemechu 1996) was used as a starting point for the meeting.[4]

The first matter of concern was the script in which the language should be written. There was not much discussion of this, as the matter seemed already to have been decided: Afa Xonso would be written in the Latin script, not the Sabean (Amharic) script. Here, the Konso *Agricultural Dictionary* was influential. The Latin script had been chosen here for several

---

[4] The two authors are Konso men, but neither were trained linguists. This short dictionary represents a huge amount of work and a significant effort towards the standardization of Konso.

reasons: it was felt to be more suitable for writing Afa Xonso, as it is made up of 'several different sounds or mixtures of sounds' (ibid.: 6). This is of particular importance because of the length of different sounds and the doubling of both consonants and vowels. In the meeting, the authors gave the following example, which also illustrated the nature and difficulty of Afa Xonso:

Afa Xonso		English
Kateeta	–	act of selling
kateetta	–	to drop price
katteeta	–	to fetch water/drop something into a well
kaateeta	–	to be watchful
kaateta	–	to tie two feet together

The capital K is used to illustrate an implosive 'K'.

The *Agricultural Dictionary* also gives four other reasons for choosing the Latin script for writing Konso. First, there are fewer symbols in the alphabet and therefore it is easier to learn. Second, the alphabet is common to many other local and foreign languages, and so easier for non-Konso language speakers to learn. Third, the common alphabet will help Afa Xonso speakers to learn other languages. Fourth, writing Latin script is easier on typewriters and computers than Sabean script (ibid.: 8).

The choice of the Latin script meant that the Konso administration was standing squarely on the side of the southern peoples of Ethiopia, particularly the Oromo, who have historically used the Latin script to write their language, in contrast to the northern peoples, who use the Sabean script. The Oromo language is close to the Konso language: according to Hallpike, Afa Xonso has some 46 per cent of cognates with Oromo (Hallpike 1972: 3). The Oromo have a history of power struggles with the northern rulers of Ethiopia, under whom they were subjugated for most of this century (Mekuria Bulcha 1996, Mohammed Hassan 1996). By choosing the Latin script, the Konso people were stating that they have more in common in terms of identity and history with these southern people than with the Ethiopians in the north.

However, the Konso rejected the Oromo way of spelling, particularly the practice of putting an 'h' after a consonant to indicate that it is an implosive or a fricative. Instead they decided to use capital letters to indicate these sounds. Possibly this reflected a reservation about their identification with the Oromo. Thus, although the Konso were acknowledging some degree of shared identity and history with the rest of southern Ethiopia, by designing and implementing their own way of writing different sounds – that is, in their choice of orthography – they were expressing their political independence and their own unique identity.

The extent of the task ahead for the Ministry of Education was still huge. This was illustrated by the fact that the meeting started to try and find Konso terms to translate technical linguistic terms. The task was not just to find a way to write Afa Xonso, but also to create a whole new technical language for writing text books and to create new words for subject matter not previously discussed in Afa Xonso. The meeting spent several hours discussing this and trying to find a Konso term for the glottal stop. Finally, at least in the meeting, the term *nessa qapa* was decided on, literally 'to have a breath'.

It became clear in the meeting that we were only touching the tip of the iceberg of the problems that were involved in this process. The Konso language varies greatly from one area of Konso to another. Although mutually intelligible, there are variations in both words and pronunciations. Most Konso will recognize that Afa Xonso refers to the language spoken throughout Konso, as it literally means 'the mouth of Konso'. In general parlance, however, the Konso people use different words to refer to the language as spoken only some twenty kilometres apart, for example, Afa Kenna to describe speech in the Fasha (or Kenna) region, or Afa Karate to describe speech in the Karate region.

An unwritten language has a great deal more flexibility and variation than one which is written, where the constant process of writing it serves to reinforce matters which have, at some previous stage, been standardized or accepted. For example, the problems become particularly evident when it comes to writing down something as apparently straightforward as the days of the week. Here the problem of variation across regions became particularly acute. The days of the week in Konso take their names from the sites where markets take place on that day. There are many markets throughout Konso, and thus the names used for the days of the week vary depending on the nearest market. Time is understood in terms of these places. The Konso dictionary has seven different possibilities for Tuesday[5], three for Friday, and two for Saturday (and this excludes several other forms). The writing down of the language therefore involved some standardization of the culture and also the experience of being Konso.

Thus a certain degree of ethnographic knowledge is needed in standardizing the language, but ethnographic sources on Konso are limited. This was further illustrated in the long discussion that took place in the meeting about the Konso months. The Konso have twelve months in a year and a month has thirty days. They are described as fifteen dark days and fifteen light days, in reference to the situation of the moon at night. In this way they are like the Amharic months, except that the Amhara have a thirteenth month called Pagumé, which has only five or six days and makes the year up to 365 or 366 days. The Konso have no thirteenth month, and the days of the Konso year add up to only 360 days. The issue in the meeting was that, if this was the case, the months must slowly move out of synchronicity with the rains and the pattern of the year. But this does not happen: instead, certain months correspond with different times

---

[5] Acaco Dipapa, Aaacaco Parayta, Onpoko Dipapa, Onpoko Parayta, Dipapata Onpoqo, Dipapata Acaco and Lankaya.

of the year. For example, Oypa corresponds roughly with January and is a dry month, a time known as *pona*; this is followed by Sakaanokama, the month when rain usually starts and of sowing; Murano, Pelalo and Hari follow, the period for weeding, then Tola, when the heads of sorghum are tied together to ripen, then Olxolashe, usually the time of the harvest, and so on.[6] The best part of one day was spent discussing what mechanism the Konso calendar has to prevent the months from shifting, and the day ended with the participants leaving to ask their elders for an explanation.

The explanation that was accepted by the meeting the following day was that every so often the months are found to have moved out of step and the rains do appear in the 'wrong' month. When this effect is sustained, then slowly the people come to the conclusion that they have got the month wrong themselves, and although they thought it was Oypa (for example) it is evidently Sakaanokama. There is no grand proclamation or mechanism: the people simply make what they consider to be a self-evidently necessary adjustment.[7] It should be added that this is facilitated by the fact that few people at a village level will know what month it is if you ask them, and asking usually provokes a long and heated discussion. They simply do not need to mark time in this way; they will rather infer what month it is from the current agricultural process.

These examples illustrate some of the difficulties in the standardization of Afa Xonso. This is before the preparation of text books for schools has even been considered (if Afa Xonso is really to be used in schools) or enough teachers who can speak Afa Xonso have been found. It is still not clear whether the schools will succeed in implementing Afa Konso as the main language for education. The use of an indigenous language is therefore laudable in principle but fraught with difficulty and extremely time-consuming. Writing and standardizing an unwritten language is a task of enormous complexity, something which only becomes clear to those involved as each different problem is encountered. The demands on the new administrators are so many and varied that they were only able to put aside three days to devote exclusively to the language. This only enabled them to skim the surface of the difficulties involved. At present Konso does not have the technical expertise or resources to overcome these difficulties, and the requirement of the new special *wereda* in needing to have a standardized language to use now is at odds with the time needed to carry out this process.

---

[6] The remaining months are Sessaisha, Partupta, Kisha, Olindela and Poorinka (Yohannes and Gemechu 1996: 9).

[7] See Turton and Ruggles (1978) for review of anthropological approaches to such calendrical problems, and for a comparison with Mursi 'measurements of duration'. The Mursi approaches to their *bergu* time measurement are similar to Konso attitudes to their months (*lė*).

Capturing a Local Elite: Konso

Working with indigenous institutions: contrasting cases

The critique of some Third World and Western scholars asserts a people's right to their own culture, history and world view, and argues for a new form of development, one that is ... a creative synthesis of tradition and modernity, drawing on local knowledge and culture.

(Parpart 1995: 254)

As discussed above, one way of creating a new form of administrative structure appropriate to the locality and the people is to incorporate existing indigenous institutions. This facilitates the participation of local people, and enables local conflicts and disputes to be dealt with by familiar methods. The formation of Konso special *wereda* would seem to provide such an opportunity for integrating indigenous institutions and building a bridge between modern state structures and the local people.

Such ideas are being put into practice in other parts of Ethiopia: in Borana zone, for example, an area neighbouring Konso to the south-east, the local government[8] in partnership with the German technical co-operation group, GTZ, are turning to indigenous Borana institutions to achieve forms of range-land management and development with the participation and co-operation of the Borana people. Through this it is hoped that a return can be made to a Borana in which the range-lands were 'exceptionally good' (Helland 1996: 134; 1998: 54), with effective mechanisms for safeguarding access to resources and maintaining security, and with indigenous coping mechanisms for times of ecological and other stress. It is hoped that this will counteract some of the problems that have resulted from large-scale development initiatives and state interventions, which have largely resulted in environmental degradation in the form of overgrazing and bush-encroachment rather than improving the lives of the Borana (Gufu 1998; Helland 1998).

The local institutions with which the GTZ and local government are working are all male traditional leaders: they include the *abba olla*, who is head of the residential and livestock management unit; the *abba deheeda*, who is responsible mainly for access to land; and the *abba gada*, who is head of the generation-grade system in Borana (Sorra Adi 1998). The *abba gada* is particularly important as head of the generation-grade system of which all Borana men are members. This generation-grade system, the *gada*, has been seen to embody egalitarianism and to be a local form of democratic organization (Baxter and Almagor 1978; Mekuria Bulcha 1996). It is therefore particularly suitable for incorporation into an administrative structure seeking to represent the people.[9] In contrast, the programme did

---

[8] This is mainly the Ministry of Agriculture of Borana Zonal Administration. This information on Borana was collected during a visit to Borana in 1999, where I was able to interview local government officials and development workers.

[9] The question of the representation of women through the use of these institutions is problematic, but cannot be considered here.

209

not want to work with another grassroots institution, the peasant associations set up in Derg times, as they saw these as by-passing customary decision-making institutions and therefore as much a part of the problem as the solution (Sorra Adi 1998).

In 1999, I was able to attend a meeting and see this process in action. A GTZ officer, together with local Ministry of Agriculture officials, met about twenty Boran men in Arero to discuss their plans to work together with the community. The GTZ officer explained to them that they wanted to incorporate local people into their decision-making processes and to learn from them what their problems and priorities were. They wanted the people to design their own solutions to these problems and promised that the GTZ would help them to achieve these aims. The questions that the GTZ officer asked illustrates this general approach to development. He said, 'How can we reconcile old practices and modern times?' Initially, the Borana men did not respond to this directly. They appeared disillusioned and sceptical. They complained that they had already been discussing their opinions with other development workers. One group spent four days asking them what they thought.

Later in the meeting they warmed to the discussion, and several men spoke up, agreeing that it was a useful initiative. Present was the man who next year will take up the office of *abba gada*. Though a young man, he stood out, as he spoke with finality and authority, and he appeared to be a spokesman for the others. He drew together some conclusions, indicating that those present at the meeting accepted the validity of the GTZ initiative to involve local people in the decision-making process, saying 'only the wearer knows where the shoe pinches; what the cow needs only the cowherd knows'. The discussion then moved on to practical questions of how to implement this form of participatory development.

There are some parallels between the situations in Borana and Konso. Both are areas which were marginalized under previous regimes. Under federalization, both are now hoping to build a new form of administration which is not externally imposed but essentially Borana or Konso. In both places there is some harking back to a golden age and a sense that there should be a return to the situation that existed before the distortions caused by outside interference. But in Konso no serious overtures have been made to indigenous institutions, and there is no intention to incorporate them into the administrative structure or use them as a bridge between the administration and the grassroots, that is, between the modern and the traditional.

This lack of action demonstrates the contradictions that exist in Konso special *wereda*, namely that it is perceived as being based on Konso culture, though at the same time only certain aspects of that culture are deemed relevant by the serving administration. It also demonstrates the complexity of working with indigenous institutions. These have themselves been changing, have been shaped by historical events, and are embedded in different sets of local and national power relations.

Capturing a Local Elite: Konso

As in Borana, there are many different indigenous institutions in Konso. Three are appropriate for discussion here, the *apa timba*, the *xela* and the *poqalla*. These were all mentioned by the Konso administrative officer quoted above when he described the post-1991 period in Konso as a return to self-rule and explained this with reference to the pre-colonial Konso situation, when 'Konso was an autonomous area and leaning on its own cultural dynamics and administrative situation' (K. 1996). When I asked him what this administrative situation was, he explained it as follows:

> The indigenous institutions were based on the rounded villages. Each rounded village has its own internal law, and these laws are generated by the *apa timba*, the man carrying the drums.... Each rounded village is governed by its own *apa timba* and *xela*. *Xela* is a group of people aged between 18 and 40, who are the administrators.... According to that Konso people were being administered on generation-grade system. The eldest advised and killed the lions. The youth administered and decided things: we should go this way and act this way.
>
> (K. 1996).

Prior to the 1974 revolution, the *apa timba* was a cyclical position with an adjudicating function.[10] The name literally means 'father of the drum', and two drums moved between the men, indicating who was holding the office. The position was generally hereditary, though it could also be purchased. The position moved between prominent families in a fixed cycle, with each family holding the drums, usually for between one and three years.[11] For his time in office, the *apa timba* had to keep certain taboos, including not cutting his hair and never sleeping outside his village. Each village had one *apa timba* in office, but *apa timba* also existed at the regional level who resolved conflicts between villages or conflicts which had not been resolved in a village.

The *xela* is the name given to the lowest generation grade in Konso. The *xela* are frequently described as the messengers of the *apa timba*, and they keep order and protect the village. Their responsibility is to carry out the orders of the *apa timba*, which includes the capacity to detain people and to fine them. During much of their time as *xela*, the men are in a state known as *farayta*. This means that they are either forbidden or unable to have children. They may form liaisons, but any child that results from a relationship during this time is considered illegitimate and must be aborted (Hallpike 1972).[12] This element of the generation-grade system has influenced the way in which it has been considered in recent years.

---

[10] Like the administrative official, it is possible for historians to describe certain general elements of the roles, practices and principles of these institutions, which, though not unchanging, were prominent and endured for some of the period prior to 1974. A more detailed analysis of the history of changes in these institutions can be found in Watson 1998.

[11] In the village of Buso the whole cycle took 36 years (see Hallpike 1972; Watson 1998).

[12] Hallpike writes *farida* and describes it as a generation grade of its own (1972: 181, 188). The generation-grade system is hardly operating in Konso at present and this makes research into it difficult, but I believe Hallpike to be wrong on this point: *farayta* is not a generation grade, but a condition related to *xela*.

The Promise of 1991

During the initiation of an *apa timba* or a new *xela*, they are told to '*baleta toyiti!*' ('look after the village!') or '*baleta umbani!*' which means 'to hold the village', in the way someone would rock a baby. In this latter case they are being likened to children being left to care for their younger siblings. There are obviously some parallels between these generation-grade institutions and the *abba gada* that the local government and GTZ were working with in Borana.

The other indigenous institution to be described here is the *poqalla*. These are male hereditary leaders who are many and who exist at varying levels of importance. They are religious, political and economic leaders, and they are clan and lineage heads. The administration official explained his understanding of the position:

> The *poqalla*'s place in Konso culture is just being the source of the clan. According to the settlement of Konso people in this area, the first people who settled here put fire into the area and then the place the fire covers is their place and they use it for their descendants. ... As the *poqallas* are the first people to settle, they administer those who come after them, and the late-comers say that we are from that family and therefore we have to obey that *poqalla*. As time passed people became very many and the *poqallas* become the rulers and the leaders of many people.
>
> (K. 1996)

In Buso village in 1996 there were 23 *poqallas*, which means that *poqalla* households make up approximately three per cent of the population. Traditionally, the *poqallas* perform ceremonies for their descendants, control and cultivate large amounts of land, and also have large stores of grain. They provide a resource for their clan and for others who will come and ask for their help, borrowing a field or some grain when they are in need. In the past and today, acting *poqallas* are believed to have special powers which are associated with vague notions about God (*Waqa*) and the world of spirits. It is these powers that are believed to enable the *poqalla* to bless and bring health and fertility to his descendants, and also to know the truth and therefore to be able to resolve disputes. If a person lies to a *poqalla*, then he puts himself in great danger. As one young man explained to me:

> If you and I have a dispute, and I am in the wrong, but I deny it, then you say, 'Let us go to the *poqalla*'s house, and enter that house, and we will see who is really in the wrong and who is in the right.' The one who is telling falsehoods, when he goes into the *poqalla*'s house, he will die.
>
> (G. 6/96)

The *poqallas* give rulings in disputes both among their descendants and between others, and they also have the capacity to fine. They can rule individually, or more than one *poqalla* can come together and decide the matter. If for some reason a *poqalla* or *poqallas* fail to resolve a dispute, then the matter is referred to the *apa timba*. The situation regarding *poqallas* has changed greatly over the last hundred years, and there is much variation between the roles and statuses of different *poqallas*. The above is a summary of their main roles, and several *poqallas* were still acting in this way in 1995–6 (for more details, see Watson 1998).

These three institutions, the *apa timba*, *xela* and *poqalla*, could have been used as the building blocks for the new Konso. The generation-grade positions (*apa timba* and *xela*) can be seen as unifying and the *poqallas* can be seen as easily accessible to all. However, the processes that have taken place over the last thirty years have made it impossible for these institutions even to be considered for integration into the new mini-state. These processes in Konso are not dissimilar to those described by Donham (1999) for Maale. In the late 1970s, the double-pronged attack which resulted from a combination of the movements of the Derg campaign and Protestant Christianity acting at the same time had a powerful effect, destroying the *apa timba* and the *xela*, and radically transforming both the position of the *poqalla* and people's perception of the *poqallas* and of Konso culture in general. This latter process in particular has prevented the use of *poqallas* and many other aspects of the culture in the state-building process.

In brief, after the Derg revolution the *apa timba*'s drums were either confiscated by the revolutionary campaigners (*zemecha*) or hidden and therefore prevented from circulating. For example, in Buso village, the drums were stuck in one man's house (a man named Balti), where they stayed for the whole of the Derg period. At this time the peasant associations were also set up in the villages and took over much of the work of the *apa timba*. After the 1991 change of government, the people of Buso started to circulate the drums again and to re-instate the position of *apa timba*. This attempt soon faltered, as one *apa timba* refused to accept the drums when it was his turn. This man was already acting as the head of the *xela*. He was not a Protestant Christian himself, but many members of his immediate family were. They exerted pressure on him not to be involved in practices that they viewed as traditional and suspicious. While this man refused to accept the drums, the person who had had the office the previous year became frustrated and refused to continue to act as *apa timba*. He cut his hair (a taboo for *apa timba*) and said, 'I have done my duty and I will not tolerate these head-lice any longer.' The circulating of drums, which also marks the passing of time of the generation grade, stopped altogether.

The association of the generation-grade system with the condition of *farayta*, when it was forbidden to have a child, also made it a target for abolition from Derg campaigners and Protestant Christians. The extent of *farayta* was not an issue, but it was appropriated by both Protestant Christians and revolutionaries as a key symbol in their rhetoric, namely as evidence that local culture was 'backward' and that its eradication and transformation would make people free. These processes combined to put an end to the generation grades. Although they may exist in some places in theory, and occasional ceremonies are carried out, they simply do not provide a resource to be built on in this post-1991 period.

The *poqallas* have fared slightly better: many still function, but they have been considered unsuitable for incorporation into the new state on three counts. First, they are perceived by the educated elite, who form the administration, as old landlords and exploiters of the people. Here the earlier history of the *poqallas* is important. Many, but by no means all, served the pre-1974 government of Ethiopia as *balabbats*. As such, they were designated landowners, they collected taxes and rents on behalf of the state, and communicated the government's orders to the people. Although this undoubtedly transformed the nature and power of the position of some *poqallas*, it should not be forgotten that in some respects the *poqallas* were already pre-eminent among other Konso because they controlled access to land.[13] What is of relevance here is that from pre-1974 times up to the present, *poqallas* have been considered an economic elite and are thus disapproved of by the administration, which is still committed to building an equitable society in Konso. In this way it is evident that, although the Derg may have fallen, many of the modernist, anti-tradition elements of the socialist ideologies which they advocated still continue to prevail. These are incompatible with working with indigenous institutions such as *poqallas*, which are hierarchical and non-egalitarian.

Second, the position of the *poqallas* as religious leaders means that many members of the local administration do not consider them suitable for incorporation into the new state. This is at least in part owing to the impact of Protestant Christianity, which first came to Konso in the 1950s. According to Messeret Lejebo (1990), 27 per cent of Konso were Protestant Christian in 1987, mostly members of the Ethiopian Evangelical Church Mekane Yesus. There is a strong relationship between education and religious outlook, as many schools have some Protestant involvement or influence. Many of the educated elite who make up the administration, whether or not they are practising Protestants, have themselves been through local schools and are strongly influenced by ideas and associations promoted by Protestant Christianity. Some are also Orthodox Christians, and these are linked up in alternative networks of power (particularly with northern traders) and are more tolerant in their views of local culture. The dynamics of these processes are influential, but beyond the scope of this chapter.

The way in which Protestantism has developed in Konso is in opposition to local culture. In the Protestant Christian discourse, local culture is constructed as dangerous, evil and intolerable. In this, *poqallas*, who

---

[13] It is generally assumed both by local people and scholars that the position of the *poqallas* was transformed by their role in the northern regime and that they became a class. This is along similar lines to the process described by Donham for Maale, when Maale chiefs became a 'distinct class' (1986b: 76). Although in some ways I believe *poqallas* can be considered a class, I do not think this has been a result only of their incorporation into the northern regime, but rather is a development that took place long before northern intervention, being connected with intensive agriculture in Konso (Watson 1998).

perform ceremonies and are associated with a world of ill-defined spirits, are particularly despised and frequently described as 'devil worshippers'. For example, one influential pastor, whose grandfather had been a very important *poqalla* (and who himself would have inherited the position), explained his beliefs to me as follows:

> I believe one hundred per cent that cultural worship is not clear, it is not detached. There isn't any cultural life which is not mixed up with worship of the Devil. I remember that whenever my grandfather accused people, they died. Whenever he said bad words against them, because he was a *poqalla*, he was heard by his spirits.
>
> (EB, 1996)

Such ideas are widespread, and generally accepted by the Protestant Christian community. The impact of Christianity has divided people into different groups with different ways of living and different beliefs. This divide has been made stronger by the fact that it has a spatial dimension: the junction town is where many educated people and Protestants live. Elsewhere, although there are a few Christians, to some extent the Konso have been able to 'close ranks' in their walled villages and to resist the march of modernity and its agents.[14] Although to give the impression of a complete dichotomy between town and village would be an oversimplification, the Protestant Christianity–local culture distinction has led to 'culture' in many forms being viewed negatively. Not all people feel like this, but these associations represent a strong and powerful trend.[15] They led two educated Konso men who are Protestant Christians employed in the new mini-state to tell me that they remained living in their villages, despite the daily walk to work, exactly to try and break down these spatially organized perceptions and associations. These negative connotations and associations have meant that Konso institutions such as *poqalla*s could not be incorporated into the new state and that many other cultural practices are viewed with some disapproval.

It is also possible that the local administration would not involve these local leaders in the structure of the state for a third reason: it would involve a real devolution of power away from their new centre in Karate. This at least was the opinion of one *poqalla*, who talked to me about the attitude of the administration towards him:

> There are some people [in government] who say, 'Why are they still doing these things [the work of *poqallas*]? The *poqallas* are becoming powerful and stronger and stronger, they are like government, they are becoming bigger than us.' So they say, 'Do not do these things.' Still

---

[14] Here I am also referring to Amborn's work on the pre-1974 history of Konso (1984). Amborn concludes that the Konso were able to resist the impact of much of the northern state during this period by 'closing their ranks' (1984: 757) and that this was partly because it was difficult for the northerners to penetrate the densely packed, walled villages.

[15] This generalization also overlooks a good deal of the variation between different villages and does not take into account other differences or groups, for example, Orthodox Christians. Nonetheless these associations are powerful and are shaping Konso today.

The Promise of 1991

there are people like this and they advise people not to co-operate with us [the *poqallas*].

(RP9. 1996)

And also:

The people are not keen on the present government. They don't say this to their faces, but secretly this is how they feel. They see the administrators just looking for their own salary and their own lives, so people don't like them. They are destroying the culture. They want to be greater than us who are followed by the people in culture and that's why they hate us. Everyone pays lip service to the government but they don't like them. If the people follow the culture, the government will have no authority.

(RP9. 1996)

This showed an unusual degree of political awareness and consciousness and was not representative of general opinion. It is a dimension that should not be ignored, however, and shows again that although Konso special *wereda* might call for a celebration of Konso identity and therefore culture, the feelings about culture on the part of those in government are highly ambiguous.

The combined impact of these political and religious influences has meant that there is little chance of incorporating any of these indigenous institutions into the new Konso state. Here, the processes of, and values associated with, Protestant Christianity, which are not often considered to be overtly political, can be seen as having a strong influence on the nature and structure of the new political system. The rejection of much of Konso tradition has left a vacuum in which only certain kinds of Konso culture will be considered legitimate and suitable for promotion. The people who are deciding what is and is not legitimate are themselves removed from many other people by education, Christianity and often residence.

Conclusion

I have tried to capture some of the dynamics of the complex process of devolution that is taking place at this critical moment in Konso history, and is working itself out in different ways across the country, as I suggested in my brief comparison with Borana. Although there has been a great deal of optimism about a self-governing Konso, I have shown how realization of these hopes has been severely hampered by practical problems. I have discussed the problems facing the administration in the standardization of the language, but I have not touched on many of the other practical problems which have frequently overwhelmed the new local government. In my interview with the administration official, he stressed the lack of skilled personnel to meet the challenges faced by this new opportunity for Konso. Amongst other things, this meant that, although some of the educated elite who make up the administration wanted to do something innovative, they were forced to rely on the few people with education and experience, many of whom had served with previous regimes and become rather set in their ways. The turbulent political history of Ethiopia also means that many of these people have been on opposite sides of ideological and/or violent struggles in the past, but now have to work together. As a result, in 1999 the new administration has been characterized in practice by petty rivalries, in-fighting, political intrigue and accusations of corruption. Dismissals and job re-shuffles are extremely common, and this has greatly limited officials' ability to meet the needs of the new Konso.

Although I said earlier that people were excited by the new developments, it is necessary to qualify this and say that, in the villages at least, the non-literates, as an administrative officer would no doubt have called them, were somewhat disillusioned by the new government. They were definitely relieved by the change of government, but not optimistic. The government official was surprisingly frank:

> The Konso people are suspicious of government. They say, 'They are exploiters and they will deprive us of our rights, take us away from the culture and not lead us in a good way.' According to my understanding, the works that the people are doing with us, they are doing with us because they think, 'They have guns and money.' They say, 'Government is a means of cheating', because this is what they have experienced from Menilek to the Derg, and so they are still suspicious.
>
> (K. 1996)

The local people's fears were confirmed when the government announced their plans to confiscate private guns (of which there are many in the villages). This was considered highly unreasonable and unnecessary by the majority of people.

In the villages, the people's main interest in the policies of the new government was limited to the principle of democracy, which was embraced. At village level, however, this was taken to mean that people could do whatever they liked, that no one should force anyone to do anything against their wishes: it did not include any notion of responsibility for shaping the new state themselves, nor for self-governance. The special *wereda* has failed to galvanize the support of the majority of people behind its project.

I have shown that one contributing reason for this has been the failure of the administration to work with indigenous institutions, together with its perception of these institutions as a political or religious threat. In addition, the ambivalent feelings about Konso culture and tradition have led to contradictions and gaps in the rhetoric about the new mini-state. Rather than a new, young, healthy centre drawing the local people together and encouraging their participation in decision-making processes, there is a contrived unit which is likely to collapse under the weight of ambiguities, contradictions and a lack of skilled personnel.

The Promise of 1991

It is true that the honeymoon period is over and that there are considerable problems. There are also questions which have remained unaddressed by the new administration. How have attempts been made to incorporate and empower women in the new regime? What has changed in Konso special *wereda* since the boundaries were redrawn to include the neighbouring areas of Gawwada and Gommaidi, and what is the experience of non-Konso people living there?

But the problems should not necessarily lead to despondency. I have likened the relationship between Konso special *wereda* and the wider Ethiopia to any developing relationship. A honeymoon period is characterized by idealism and a belief that things will be perfect. After this, the partners are frequently disillusioned and have to abandon their hopes that everything will be perfect. At this stage the relationship can either fail, or those involved can become more realistic and mature, finding ways of working through the problems.

This moment in Konso is one of uncertainty. At such a time it is easy to focus on the difficulties. What should be emphasized is that, despite the problems, there have also been achievements: Konso special *wereda* as it exists today is the result of a great deal of hard work on the part of the educated elite, despite their lack of specialized training. The start that has been made on standardizing the language has improved its status. The institutionalization of Afa Xonso in the local government has allowed people to talk directly to the administration. Those with education and others have obtained employment. But there has been a more important positive development: the fact that Konso, which has been ruled by others for decades, is now being run by Konso people themselves. This has had significant repercussions in terms of Konso identity and improvements in self-esteem, though these are difficult to measure. This was illustrated in the interview (on which I have drawn greatly) with the Konso administrative official. When I asked, 'What is different in Konso since the 1991 change of government?', he gave this simple, but crucial and heartening answer: 'We are free. Despite our weakness, we are free' (K. 1996).

Twelve

Fear & Anger
Female versus Male Narratives among the Anywaa
EISEI KURIMOTO

Now, like this old woman, we are left here alone. Our children are not here to stay with us. They became angry and went to the bush [unknown place, referring to the young people who fled to Kenya, and then to America and Canada], *or to working places like Dambala* [for gold panning] *to get something to help us. But they got nothing. The problem became very serious... .*

(an Anywaa woman, February 1993)

The reason for this change, is it from you [addressing the author as one of the 'white men'], *or from where? It is not known by people. Anywaa never created a problem that could not be solved. But now Anywaa have become fearful. We have been left behind. We died behind. We have no relatives ... Now my son* [addressing the author], *we cannot do anything... .*

(another Anywaa woman, February 1993)

For those who are interested in and concerned with Ethiopia, the image and understanding of the Derg regime and its successor is shaped not only by reading written sources but also by listening to oral testimonies of those who have direct experience. Often these narratives are striking, and sometimes even distressing.[1] As we have seen in this volume, *encadrement* by the Workers' Party state affected virtually every sector of Ethiopian society. We have heard narratives from various kinds of people both in urban and rural areas and from different social classes. While the Party was still in power, these stories were told in the corners of bars, private houses, or on the road while walking, in low voices. There was a constant fear of the security forces.

At the same time, we have heard voices with affirmative tones. There were, of course, 'official' narratives in praise of the achievements of the revolution in such fields as education, agricultural development, mass

---
[1] See for instance, the life history of a young Oromo woman (Hawani and Aneesa 1996).

The Promise of 1991

Plate 12.1 Dance of an Anywaa Youth Association on May Day, 1990, at Abwobo. Note one girl at the centre with a wooden AK47 [E. Kurimoto]

organization, and indoctrination of revolutionary ideology. More importantly, we have also heard affirmative narratives from local peasants and pastoralists who were powerless minorities, conquered and subjugated by the Amhara-dominated empire. For them, the Derg *did* bring a certain 'liberation' (see Plate 12.1). The land alienated by the *neftennya*, armed settler-colonists, was returned; equality among different 'nationalities' was proclaimed and new opportunities were opened for people (especially men) to climb the social ladder and become government employees, professionals, or officials.

Interestingly, the same person can tell both types of narrative; in addition, he or she may change the valence of the same narrative over time, according to setting and context. These narratives, both negative and affirmative, are essential for the writing of the ethnography and historiography of the Ethiopian revolution. Attempting such a narrative was a highly sensitive matter until 1991, and we still have very few efforts, Donham's *Marxist Modern* (1999) being the first comprehensive study of its kind. For historiography the era is too recent to become an easy subject of study. It is obvious, however, that local narratives of various kinds will provide the vital sources for future work.

In this chapter, I present some narratives collected between 1989 and 1998 among the Anywaa (Anuak) of Gambela region in western Ethiopia. Their experience of the Workers' Party and the EPRDF regime was shaped by conditions specific to Anywaa society and to the Gambela region. The Anywaa may be a modestly-sized population in a small region

Fear & Anger: Anywaa

Plate 12.2 Anywaa petty traders at Pinyudo, 1989. Most of the goods on sale originated as relief items for the Sudanese refugees at Pinyudo (Fugnido) camp.
[E. Kurimoto]

(see Map 1.4, p. 32).[2] Nonetheless, Anywaa narratives share many themes in common with those found in other parts of Ethiopia. Although the stories I shall recount are self-revealing to some degree, I add notes to explain the background and context. One of the themes that emerges from my material is the striking difference in perspective between younger men (who have experience of travel and the world of government and politics) and older women (who reflect on the personal sadness and losses entailed in local struggles).

First, I will relate my own experience of fieldwork. When I started anthropological fieldwork in 1988, Gambela was a sensitive place, where many projects were concentrated which the Workers' Party apparently did not want publicized (Kurimoto 1996b). These included: a major resettlement programme, a state farm, and a dam construction and irrigation project in which numerous Russians were working. More than thirty Cubans were at the regional hospital and a group of North Koreans were working for a rice project. Moreover, there were two huge Sudanese refugee camps at Itang and Pinyudo (Fugnido), and the headquarters and training camps of the Sudan People's Liberation Army (SPLA), whose very

---

[2] According to the national population census of 1994, the population of Gambela region was 162,397, out of which 44,581 were Anywaa and 64,473 were Nuer (Central Statistical Authority 1995). Many Anywaa feel that their population was underestimated, and that of the Nuer overestimated. This has become a serious political issue under the ethnic-based regional autonomy system.

The Promise of 1991

existence on Ethiopian territory had been denied by the Ethiopian government. In reality, the SPLA enjoyed full support from the Workers' Party state, and SPLA men moved freely between refugee camps and training camps and between Sudan and Ethiopia. Finally, it was said that an Anywaa liberation movement organized by dissidents had been operating against the Ethiopian state and the SPLA. Because of this anti-Derg movement, originally called the Gambela Liberation Front (GLF) but by 1985 the Gambela People's Liberation Movement (GPLM), the Anywaa were generally considered as anti-government. The appointment by the central government of Nuer men to two key local posts, the head of administration and first secretary of the Party, was seen by most Anywaa as an attempt to control and oppress them. In May 1987, in retaliation for the attack by a GPLM platoon on a police post and settlers' villages, about eighty Anywaa were shot dead by police and militia in Gambela town. In 1988 there were twenty-seven Anywaa political detainees who would be released only after the fall of the Workers' Party state.

As I had previously spent some time in the southern Sudan conducting fieldwork among the Pari who are linguistically and historically closely related to the Anywaa, research in Gambela gave me opportunities to meet old friends who had become refugees or SPLA soldiers. This 'South Sudan connection' certainly helped me to secure myself and to stay in Gambela. At the same time, there was always a risk that the government would consider me as being engaged in activities incompatible with my research mission.[3] Soon I realized that many of the Anywaa had developed hostile feelings toward the SPLA because of the latter's arrogance and misbehaviour. Relations between the Anywaa and Sudanese refugees/SPLA became extremely tense after the Pinyudo massacre in September 1989 in which Pinyudo, the largest Anywaa village, was attacked and burnt down, and at least one hundred and twenty villagers were killed by a hostile party in which it was difficult to distinguish refugees from SPLA men. This was the village in which I had spent six weeks from mid-December 1988 during my first research trip.

Thus as a fieldworker I was caught in a very difficult and intricate position from the very beginning. There was also a fundamental obstacle in conducting fieldwork; after the Pinyudo massacre neither the regional administration, the state security office nor the Party office would give me permission to live in a village. As an alternative I was obliged to choose the small town of Abwobo (Abobo) as my base, and I started visiting neighbouring villages during the daytime. The village of Cwobo, across the Aluoro river from Abwobo town, was the place I had spent most of my time during the first stage. I had to be careful in choosing neutral and apolitical research topics. 'History' became one of them. This does not mean, of course, that I consider the subject of history neutral and

---

[3] Surveillance of me by the State Security office became loose as the time went by. But finally in February 1991 I was given an order by the office to leave Gambela within forty eight hours because I had visited an SPLA commander in Itang refugee camp.

apolitical. Narrating the past always reflects and is influenced by the present. What I thought was that by studying the remote past and the Anywaa language I might be seen as fairly harmless by Workers' Party officials.

History & the reflections of elders, 1988–90

In this way I started to look for elders and to talk them. Among the topics included were: the origin of *nyieya*ship and *kwaaro*ship, the two indigenous political offices connected with aristocratic or chiefly status respectively (see below), the process of migration and settlement, relations with other peoples, the encounter and interaction with Ethiopian highlanders and the British and so on. As I expected from my experience among the Pari, the Anywaa were highly conscious about their own past, and had a tradition of oratory. In most cases I had no difficulty in conducting interviews, and it was fascinating to listen to elders who talked tirelessly, eloquently, and powerfully, sometimes for more than an hour.

In Anywaa, history is called *lam*, while folk tales are called *leere*. *Lam* narrates what is true, what really happened in the past, while *leere* are just tales or fictions. What might seem to us to be of a mythical or legendary nature is what really happened for the Anywaa as long as it is narrated as *lam*. In Nilotic languages *lam* is a core religious and ritual notion. It means 'prayer', 'blessing', 'invocation', 'taking oath', 'sacrifice', and so on. What lies underneath is the power of words uttered from the mouth to determine the present and future state of being. Therefore, narrating history among the Anywaa is more than simply narrating events of the past. The narratives that follow are *lam*. What they tell is the 'truth' from their own point of view. If one says, 'I wish I could die', it becomes, according to the Anywaa, a self-fulfilling prophecy.

Several months after I started interviewing elders, when I had become confident that people knew who I was and what I was doing, I decided to start asking about the very recent past and present: the Derg era and after. In fact, these were common topics in everyday life, but I had not dared to record people's speech. Narratives about Derg times nearly always involved a nostalgic comparison with the previous period under Emperor Haile Selassie, a sort of 'good old days'. This does not necessarily mean that the empire was loved by the Anywaa. The period after the advent of empire in the 1880s up to the 1910s saw a number of violent clashes between Anywaa and imperial agents who tried to exploit local wealth such as ivory, cattle, and slaves. These events are still vividly told and are the basis for today's image of *gaala* (Ethiopian highlanders of 'red' skin). As the economic significance of Gambela to the empire declined, however, a mainly peaceful period ensued after the Ethio-Italian war.

Listening to these narratives, what struck me first was the predominant sense of powerlessness and speechlessness. The role of powerlessness is perhaps self-evident, but speechlessness requires some comment. I have argued elsewhere that in Anywaa narratives of the history of relations with outside powers such as the Ethiopian highlanders, British and Italians, idioms from the indigenous political system are used (Kurimoto 1992: 30-38). The Anywaa have a very pragmatic political ideology. Prior to the revolution they used to have two types of hereditary leaders: *nyieye* (nobles) and *kwaari* (village headmen). While all the *nyieye* belong to a single royal clan, *kwaari* belong to different clans, and people say, 'A *nyieya* is greater than a *kwaaro*.' The *nyieye* were the holders of the royal emblems, and without their consent no son of the royal clan could be considered as 'king'; however, the king still resides on the Sudanese side of the border. On the Ethiopian side, only the villages of Abwobo and Cwobo were under the *nyieye*, strictly speaking; though there were some others staying at their mothers' villages, more or less as guests. On the Ethiopian side of the border, there was not much difference in the extent of power a *nyieya* or a *kwaaro* exercised in practice, and this was mainly confined to the bounds of their local village. There were many *nyieye* and *kwaari* in the Ethiopian region of Anywaaland. In practice people supported their leaders, paying tribute and offering labour, but only as long as their material needs were met.[4] Therefore, a leader had to continually redistribute his wealth (partly collected from supporters and partly looted from outside) in order to maintain his power. Once he was unable to do so, he could not only lose their support, but be forcibly toppled by those who wanted a new leader. Such a 'rebellion' was called *agem*, and often took the form of an actual fight between two rival factions. (Later *agem* was adapted as the term for the 1974 revolution.) This ideology of proper leadership was extended to explain relations with outsiders. Typical expressions included: 'The Italians were good because they gave us [i.e., Anywaa soldiers] a lot of clothes and food,' 'Haile Selassie was a very good man. He gave us clothes.' Of course, not all Anywaa were given things by the Italians or by Haile Selassie, but they could at least maintain an image of generous leaders. The Derg, seen in this light, was quite different. Its style of rule did not fit the indigenous political ideology. It appears that the Anywaa became 'speechless' when they talked about the Derg not only because they feared it but also because it was beyond their conventional understanding.

Now let us listen to some narratives, more or less in chronological order, but moving from fairly confident statements of a kind that characterize public discourse, to more private, emotional and despairing accounts, particularly those of women. First I would like to quote from two relatively senior men discussing the course of recent history with reference to recognizable events, policies, and regimes. The first quotation is from Ugolli Ulwoc, an old man of Cwobo village speaking in June 1990, when the Workers' Party was still in power. He was approximately sixty-five years old when interviewed. He talked about taxation, poverty, hunger, and forced military recruitment.

[4] For the *nyieya*ship and *kwaaro*ship see Evans-Pritchard (1940, 1947), Lienhardt (1957, 1958) and also Kurimoto (1992).

Before, during the time of Haile Selassie who was great, as we stayed here, we paid three *birr* [as *gimira*, tax]. At first it was one *birr*, it was paid as help [to Haile Selassie]. After that the *birr* became three. When it became three *birr*, we stayed for a long long time .... When this government started [i.e. the Derg], *gimira* which was imposed by this government became fifty *birr*, forty *birr*, thirty *birr*. This amount of money, for us, Anywaa, is same as the money for marriage. For us, a wife of Anywaa was only three *dimui* [special beads used as bridewealth] .... This thing (*gimira*) started during our time, which makes us a little tired. When this government came, it did not give us money, it did not give us food, even the price of clothes became high for us. This is our life. Now Anywaa has nothing, no cars, they do nothing to satisfy them. People are caught just like chickens [for National Service, i.e., military service, for which many were recruited by force]. What can we say? It's just like that.

About food, our life has changed. It became bad. Our fields were taken, our children have nothing to eat. Some food was to be given to us saying, 'Now your hoes are taken,' but there is no such food. Later fields became only for sowing cotton. The fields by which we fed our children, became nothing. Now we eat hunger for a long time. 'Guests' [*welle*, settlers] were also brought by force, brought from over there [highlands]. It is like taking these people over there, to the war over there.

(Interview with Ugolli Ulwoc 1990)

Abwobo *wereda* including the village of Cwobo was the centre of various governmental development projects. These included the dam-irrigation project along the Aluoro river, a famine resettlement project for people from the north (cf. Alula Pankhurst on a similar project in the former province of Wellegga, in Chapter 7), an agricultural mechanization project, and a state farm. In fact the 2,500 ha. state farm, for cotton, was constructed where the fields and a couple of satellite settlements of Cwobo used to be.

In the second paragraph the narrator refers to settlers as 'guests' (*welle*). This word was also applied to refugees and to myself. In principle the Anywaa are very hospitable to guests, and food and accommodation are provided. This is a way of absorbing outsiders into their society. The use of the word implies that those settlers and refugees were not recognized as invaders or enemies initially. Another interesting point is that the narrator clearly sees the function of the state: just as settlers were brought by force from the highlands to Gambela, Anywaa men were taken from Gambela to the war fronts in the highlands.

Under the Derg, many of the indigenous Anywaa institutions and customs considered 'feudal' or 'anti-revolutionary' were banned and abolished. The most notable of these were political offices in which 'commoners' (*baangi*) were ruled by two types of hereditary leaders, nobles (*nyieye*, sing. *nyieya*) and village headmen (*kwaari*, sing. *kwaaro*). In addition, giving bridewealth in special old glass beads, *dimui*, was banned. The rule by aristocrats was replaced by peasant associations (*qebele*), and bridewealth in beads was replaced with cash. These two transformations were common themes when Anywaa talked about the changes induced by the Derg, as for example in the narratives of another man of Cwobo village. His name is Abagaala Ulok, and he was in his early fifties when interviewed in July 1990. While most of the Anywaa villages in Ethiopia had been ruled by *kwaari*, the villages of Cwobo and Abwobo were under descendants of Udiel Ngenynyo, a prominent *nyieya* who exercised great power as an agent of the Ethiopian empire and died in 1919 (Kurimoto 1992: 15–19).

> Our way of life before is, people stayed under *nyieya*. All the people were under *nyieya*. As people stayed under him, each clan (*tung*), their force was for *nyieya*. The work of all people [for *nyieya*] was cultivation. At that time, all things were obtained through cultivation. As to drinking beer at that time, they did not drink much. Elders drank beer. Young boys did not drink because they were small. The good way of life was later destroyed by beer.
>
> Now people have gone ahead and that way of life has changed, as the position of people became nearer to that of 'foreigners' [*jur*, or any kind of people other than Anywaa]. Foreigners came and brought education. That education brought a division among the people. People got thinking [*wic*, i.e. 'head']. Schools were opened and children went to school. According to the old way of life it was children who scared baboons and birds in the field. Now there are no children to scare them, as all of them go to school to get thinking. Elders are left and they cultivate by themselves and scare birds by themselves because their children are put in school.
>
> When the change came, this new government came, the revolution (*agem*) took place. ... As this government thought that all people should become as one, *nyieye* and *kwaari* were deposed. When they were deposed, the second thing (to be abolished) was *dimui*. ... Now a poor man may get money (and marry), but at that time a poor man could not get *dimui*. People started to marry by money.... Now those young people, why do they not multiply at all? It is because of beer. Those young boys who did not drink, now they have started to drink. Now those young boys, they do not raise around ten children as their fathers did.... Now people believe in the market (*gaba*; cf. Amharic *gabaya*, Oromo *gaba*). People go to the market and buy things and forget the work at home. Because of the maize which is brought for 'guests' (*welle*, settlers and refugees), people have left their work and think about the market.
>
> Now we peasants, we don't care about these things. There is nothing [that we can do]. In future we may get good work. A bad thing cannot stand alone. A good thing cannot stand alone either. The bad will go in future and the good will replace it.
>
> (Interview with Abagaala Ulok 1990)

Fear & Anger: Anywaa

When the offices of *nyieye* and *kwaari* were destroyed, everyone became commoners. Although the new political leaders such as administrators and party officials continued to be called *kwaari*, Anywaa generally understood that the revolution brought on the age of commoners. 'People became one' was a common saying to describe this change. Needless to say, 'oneness' implied 'equality'. Another common saying was, 'People began to stand up', as we shall see below. To stand or get up (*oo maal*) in contemporary Anywaa language has two important implications. One is that now people do not have to kneel down and crawl in front of *nyieye* or *kwaari*, and that women do not have to do the same in front of men. The other is that since the new notion of 'development' was translated into Anywaa as *oo maal*, 'People began to stand up' also means 'People became developed'. Thus the abolition of the old political system and the new revolutionary notions of equality and development were intermingled in Anywaa understandings.

In the narrative above, Abagaala also looks back with nostalgia to the pre-Derg era, the era of *nyieye* and *kwaari* as well as of Haile Selassie (the great *nyieya*). It was 'the good old days' when things were in order. We should not, however, take this widely used discourse at its face value. Although it appears that life before was much more secure and in order, there are certainly elements of idealization here. Indeed, some villages in Jor *wereda* did revolt against the abolition of *kwaari*, but in other areas it was accepted rather readily. Another point to be noted is that since the fall of the Derg there has not been much effort by 'traditionalists' to restore the *nyieya*ship and *kwaaro*ship. To my knowledge, only one *nyieya* and one *kwaaro* have been installed in the 1990s, and their status seem to be rather nominal and symbolic.

What the narrator sees as the most serious cause for demoralization is drinking. Here 'beer' refers to alchoholic drinks in general. People used to drink sorghum or maize beer brewed at home and occasionally some honey wine (according to Abagaala women and young men were not supposed to drink at all). After the revolution everyone started to drink, and new alcoholic drinks were introduced. These were locally distilled spirits (*areqi*), and industrially made beer, wine, brandy, gin and so on which were available at bars in Abwobo town. Local spirits were first introduced by settlers and were called *kambaatha areqi*, for its 'bitterness' or high alcoholic content. Afterwards, Anywaa women started to distil it at home.

At the end of the quotation, Abagaala expresses his sense of powerlessness: 'We do not care for these things. There is nothing.' Interestingly this 'we' is 'peasants'. In Anywaa peasants are called *jo-puro* which means literally 'those who cultivate'. This new 'we consciousness' of being peasants is certainly evidence of the internalization of revolutionary ideology. Although he deeply deplores the situation under the Derg regime, Abagaala still has some hope for the future. In fact, he acknowledges some positive aspects of the changes. One thing is the ban on the use of *dimui*. I hardly heard voices deploring the ban either from men and women. Since the quantity of this special bead circulating in Anywaa society was limited, control of the flow was a source of *nyieye*'s and *kwaari*'s power. Young men had difficulties in getting married, which, I believe, was a major motive that drove many to participate actively in the revolution. Abagaala also saw some advantage in the villagization that took place in Cwobo, because help was closer at hand if a person were sick, and meetings could be held more easily than when people were living far apart.

In the case of some other narratives, however, there is nothing to be appreciated. The sense of powerlessness and grievance can be very deep, as in the case of Uceri Akwer, son of a *nyieya*. In the 1950s he worked as an assistant for a Greek trader and lorry driver who used to transport goods between Gambela and Malakal in the Sudan. Then he became a guerrilla fighter in the Anya Nya movement that fought against the Sudanese government until 1972. Although he did not go to school at all, like the first two elders I have quoted, he was a man who knew the world outside Anywaa society. He was in his mid-sixties in 1989 and was known as an excellent narrator of oral history. The following is the concluding part of a long story in which he spoke quite eloquently, without faltering, starting from the origin of *nyieya*ship, migration to the present Anywaaland, encounters with Europeans and Ethiopian highlanders, up to the present.

> Now, after that this government [the Derg] came and said, 'You stand up from here. Your names are not here at all.' We said, 'Is it like that? Is this a new government? All right. We are also tired.... Here are the things.' Beads, drums and spears, all the things [royal emblems] with which we came from the river [the founder of *nyieya*'s clan is said to have come from the river] were destroyed, destroyed by commoners. They said, 'You are thieves. You are those who do things by force. You have spoilt the land. There is no *nyieya* at all. There is no *kwaaro* at all. There is no daughter of *nyieya*. Now the land belongs to us.'
>
> After that, Kambaathe [settlers] were brought to the village. Then we said, 'What are they?' It was said, 'They are Kambaathe.' 'Where are they from?' 'They are from over there [highlands].' 'They don't have their home?' It was said, 'They were dying of hunger.' 'Wai! Did all of them not die? You, they are going to take the land.' Kambaathe entered among the people. It was said, 'You, nobody can stay at his home. You mix together with these people [Kambaathe].' Some people refused and some went [to stay at 'integrated villages' together with settlers].
>
> People stayed. [Then a man said] 'Is this the way of living?' A man stood up, returned to his home and said, 'Cultivation [at home] is better [than that at the integrated village]. I have never eaten a tin [relief food]. I cultivated by hoe and had a very big granary. There was no hunger for me and my children. I could go to the river and get some fish. I could go to the forest and get some meat and eat it. Today, I have got a government that prohibits me from getting fish. I have got a

government that prohibits me from getting wild animals. I have got a government that prohibits me from getting food. Wuu ... Why? Was there any of my grandfathers who ate a tin?' ... Some villages fought [against the government] and people died. They were killed. When people stayed, some said, 'Were there Kambaathe here before?' And Kambaathe who were at Thatha were cursed [by Anywaa] and went away. Kambaathe who were at Pucala were cursed and went to the hinterland of the river.

Ajwil [Dinka, which also now means SPLA] then started to come along the Upeeno (Baro) river. It was said, 'Ajwil have come.' 'From where?' 'They came for their own government. This is Anyanya II.' 'Wuu ... Good. As they come, where are they going?' 'They come with their *kwaaro*, John Garang.' ... Anywaa said, 'You, we don't want to die without a reason. Do we have a *kwaaro* who will lead us among them [SPLA]? Is there an Anywaa who is a leader? Don't join them. Let us stay here.' First, when Ajwil came here, what they did was not good. As it was seen, it was found by Anywaa that they did a bad thing. Because Anywaa are black British, their mind is clear. They refused that. ... Then they [Ajwil] said, 'The food which is given [by UN], you Anywaa, your name is not there. This government is ours. .... There is no place for you here. This government is ours. Our *kwaaro* is John Garang.'

Our luck is bad. Even those in Ethiopia will get angry. They prohibit [Anywaa from getting] guns. They said, 'When Anywaa get guns, they will dominate us.' That policy was from there. They [Ajwil] combined themselves with Gaala (highlanders) and said, 'Anywaa have no place here.' The money which was given by the people of the world to all people was eaten by them. Anywaa were left outside. It was eaten by Kambaathe, Gaala, Ajwil and Nuer. It is these peoples that took guns. Now our problem is that the land is being invaded. Nuer had no land here. Ajwil had no land here. We did not know them at all. ... Ajwil, when they were satisfied and became fat, they came and excreted on Anywaa. They killed Anywaa [mentioning the Pinyudo massacre of September 1989]. [Among those killed] some were pregnant, some had small children and some were blind. They were thrown into the river and died. After that they did the same thing at Dippa [a village on the Thatha lake] and killed some people. People of down there [the lower Giilo river] came running and said, 'What to do, Jo-bat-Thatha? Let us go and fight against them. Why does Ajwil do this thing all the time? ... All those things, guns and cars, belong to them. Things given by the people of the world belong to them. Fuel belongs to them. But we buy them with our money. As they became rich, that money was it not from us?' ... But they are still doing the same thing. When he finds a woman, he rolls up her cloth of her back. 'Is this your vagina? Vagina of Anywaa is sweet. Vagina of Anywaa is sweet.' They insult us like this. This domination has been big and now we have nothing to do. We left that work in their hand. If they want to destroy us, they can do that.

Now we don't have our *kwaaro* over the land here. Really we have nobody, our relative or anybody. We are just staying here for *jwok* ['God', 'god']. Nobody has helped us with anything. We are dying like dogs. But no problem. [My story is] finished.

(Interview with Uceri Akwer 1989)

The abolition of indigenous political offices, the resettlement project, the coming of the SPLA and the Nuer are all narrated in one story. Significantly, he sees the situation in terms of different categories of people; the Anywaa were dominated by the Kambaathe (settlers), Gaala (highlanders), Dinka or Ajwil (SPLA), and Nuer. While those people had access to resources from outside, notably cars, guns and relief food, the Anywaa did not. This is the main reason for powerlessness. He concludes his long narrative with a deep sense of deprivation.

Personal experience & the reflections of women, 1993

The hostilities developed during the Derg era were to explode as 'ethnic' tensions in the next period when the all-powerful centralized state that sowed the seeds of the hostilities in the first place was gone. The new local administrative structure implemented by the EPRDF was initially welcomed by the majority of Anywaa. The oppressors (that is, mainly Nuer officials of the Derg and the SPLA) had gone, and dissident Anywaa who had organized the Gambela People's Liberation Movement (GPLM) had come back home and taken power at the regional level (Young 1999; Kurimoto 1997). For the first time in history, Anywaa had their 'own government'. When I visited Gambela in January 1993, two years after my last stay, a feeling of joy and relief was still prevalent among many people.

This 'honeymoon period' did not last, however. In fact, the period immediately after the fall of the Derg saw various violent clashes in the region among different interested parties: Anywaa-GPLM, Nuer-SPLA, EPRDF, and settlers. It seemed that hostilities accumulated during the Derg era suddenly burst out when the powerful central authority had gone. What made the situation more chaotic was a pattern of internal power struggles in the GPLM leadership in particular, and among the Anywaa elite in general. This is why the Anywaa call the change of power and the situation resulting from it *girrgirr*. This is an Amharic and not an Anywaa word; *girrgirr* means a rebellious outbreak or rioting. It is distinct from *abiot*, a revolution. In local Anywaa understanding, the *abiot* in which Mengistu Haile Mariam replaced Haile Selassie was straightforward – in the same way that an *agem* was a straightforward replacement of a sitting *nyieya* or *kwaaro*, usually by a close agnate. When Anywaa apply the term *girrgirr* to the taking of power by the EPRDF, they imply a state of extreme confusion, a sort of 'war against all'.

In early February 1993 I returned to Cwobo village where I had spent

Fear & Anger: Anywaa

Plate 12.3 'Linking arms.' Painting on the wall of the administrative office in Gambela town, 1993. The five persons represent five indigenous 'ethnic groups' with an Anywaa at the centre, a Nuer on the far left, a Komo and an Opuo on the right. Interestingly, only the Majangir is depicted as a woman. *[E. Kurimoto]*

a long time during the Derg period. Some women came to greet me and I asked them to tell me how they saw the present situation. In fact most of my previous informants had been men, and this was my first time to interview a group of women. They were more than willing to talk, but what impressed me was the dominant tone of despair and grievance in their narratives. The political development at the centre had nothing to do with them. Life as they saw it was in a process of continual deterioration. The common themes were death, poverty, and the demoralization and breakdown of the community. They did not see anything good in the present. This was striking to me because at the time the 'honeymoon' under the EPRDF still prevailed among many Anywaa men. Because of these interviews, I came to realize that women may have quite different views from men.

The first to talk was Aduk Medho, a very aged woman who already had had three children during the Italian occupation. She was born in Cwobo and had been living alone in her own small hut. She had three surviving children: two daughters in Gambela town and a son in Cwobo. Her deceased husband is from the royal (*nyieya*'s) clan.

> When there were *nyieya* and *kwaaro*, we were afraid of them. Now when this change [the fall of the Derg] came, commoners became *kwaari* (leaders) and *kwaari* became commoners. Now when people wanted to

231

settle down ... young people began to fear. There were problems of clothes and food. Everyone became a thief. When somebody was seen walking at night, he was considered a thief and killed at once. Old people became sad. Only old people remained. All young people went to the home of some other people, you [i.e., the author], the white people. People were afraid [to stay at] home here. They left their old mothers and fathers. They did not consider them. If he got married, he took his wife and went away leaving his mother. Old people became sad for the matter of young people. Like in this village here, no one can bear a child again. Those who can bear children, they went to you there. They ran away because of poverty and death. Death was not here at all [a long time ago]. There was no death. During the time of *nyieye* and *kwaari*, there was no death. People simply stayed at home. People simply danced to the drum. There was no problem. They cultivated sorghum. Food was quite enough. Anywaa were eating vegetables (*amaru*) and had only one cloth. There were not many clothes. If you got one bed sheet for wearing like this, you didn't need another. But today you, young people, have a lot of clothes in a hut for one person. When you see your friend wearing a cloth today and tomorrow, you want to dress like her also. You go to the place of death [i.e., the gold panning place] to find clothes. It is like that, my son. We are demoralized [*yiet-wa unyitai*, 'our stomach is unfolded, open'] and left out. We wish we could die.

(Interview with Aduk Medho 1993)

One of her main grievances is the exodus of young Anywaa, both men and women. They got out of Ethiopia, became refugees in neighbouring countries, namely the Sudan and Kenya. Some of them succeeded in resettling in the USA and Canada ('the land of white people'). The flow started during the Derg, and its fall did not stop the movement. On the contrary Anywaa took advantage of the chaotic political situation and continued to cross the borders. Now those settled in North America have started to telephone their families left behind. They transmit money and make long distance phone calls to their families in Gambela town, thanks to the development of remittance and communication systems – which drives more people to leave home. It seems Anywaa in North America may total at least several hundred, possibly more than a thousand, and I myself communicate with some of them by e-mail. This transnational movement of young Anywaa is a clear sign that they see their life at home as futureless.

The second woman was Dinguru Ugelo, a widow in her late forties, who was born in Cwobo. She was cultivating by herself. She was a jolly and talkative woman who gave me my Anywaa name, Ukunny Guula. She had two surviving children, a son and a daughter. The son had been away from home working at Dambala, a major gold panning place near Dimma and Gurafarda, seven days' walk from Cwobo. The daughter was married with five small children. Her husband was working as a manual labourer at the dam construction and irrigation project on the lower Aluoro river.

Fear & Anger: Anywaa

What this old woman talked about now is true. That is the way we lived before. During the time of Haile Selassie and *nyieya* we danced to the drum and got married there [found a husband at the dancing ground]. We went to our husband's home [after marriage] without any problem ... and bore children. We were good with our husbands. Like today and yesterday, Haile Selassie was destroyed, *nyieya* was destroyed, the land became the land of the government. The land of Derg. During the Derg time, there were also drums for dancing. After that when the land was changed [i.e., the fall of the Derg], we have had stealing, death, poverty and hunger. Something that destroyed all the people was spoken of by this old woman. The home was destroyed because of all this. Now your relatives to help you ... there is no help. As for cultivation to make a living, for people to settle down with food without hunger, there is nothing. A person has become all alone, husband and wife alike, this being the reason of destruction. This was caused by the *girrgirr* [Amharic, rioting and confusion]. It brought a big hunger. Now you can't differentiate a big person from a small person at home. Even this village, when you came here before [during the Derg era], you saw it as home. When you come now, it has been changed. What I tell you now, we can't do anything for this change. Before the old man [i.e., Haile Selassie] seemed to help us. *Nyieya* used to help us. Today, there is no *nyieya*. As there is no *nyieya* and no *kwaaro*, the home is destroyed. Today if somebody who wants to prepare his home, he may fear all those people because he may be killed. If he prepares like that, they may kill him. As things are like this, there is nobody to stand by one. People are afraid of that. That is right. People are afraid of only that. Because before, they knew drums [i.e., dancing], knew how to make sorghum beer and knew how to sing. Now there is no song, no sorghum beer and the food is never enough. When someone goes, she cannot find anything because she has no daughter, son, husband, and husband's brother. Because of this, destruction happened, and for you Ukunny Guula [author's Anywaa name], we have nothing to talk about.

(Interview with Dinguru Ugelo 1993)

The next narrator, Abang Uriet, was also a widow in her late forties. She was born in the next village. She had two sons and three daughters. The daughters were married: two in Cwobo, the other in a nearby village. The youngest daughter was still staying with Abang with a small baby girl. Two sons got married in Cwobo. The eldest son had been working at Dambala, and his two small daughters were with Abang. Like many others, Abang's household is female-headed. She was cultivating to sustain herself and other family members.

This life is not good. People are very much destroyed. You may have two or three children, they have nothing to eat. If you look around and go to another village for help, carrying a sack [to put something in it],

The Promise of 1991

you will simply get tired. We cry again. Before we had enough food. Now you may go to your husband's son to get some sorghum, you may go to your grandson to get some sorghum. That is how life is now, isn't it? We better die so that we never see the people again. Young people will be left by themselves with their own problem. Haile Selassie brought out a lot of things for people to eat. He disappeared like *nyieya*. *Nyieya* is absent. Haile Selassie is absent there. Now if somebody says, 'What is this? People are destroyed. Let me help them,' all the people in the world will look for him, catch him and kill him. This is because he is good and he wants people to be alive. Before when we stayed in this village here, was there grass like this? Was there a fence and a hut fallen down? [She is asking the author about the situation before.] People used to smear black mud [on the compound ground of a homestead]. Men were at *wi-mac* [gathering place of a clan]. Now men they have gone to cultivation. Women respected men. You could not drink water from your husband's calabash bowl. You could not eat the half of *kwon* [thick sorghum or maize porridge] of your husband. You could never do anything. You would crawl to approach your husband. You were afraid of your husband. Now people have stood up [i.e., 'no respect', 'developed']. There is nobody not standing. A small child stands up, an old man stands up. Because there is hunger, they have become thieves. That stealing is caused because of hunger. Before, he got some sorghum from his father, some *kwon* from his mother. Your father was there. Did you steal something? How could you steal something belonging to somebody, while you have things from your father and mother? Now they long for things. They are thieves, they are thieves. This is not because of money, but because of hunger. He steals because of hunger. He says, 'What shall I do? Shall I not just steal to eat and to be satisfied?'

Now whether the situation will return to normal, we the old people, don't know. We still don't know whether this place will be our home. Even men also, they don't know. Tomorrow if people try to cultivate, sorghum will not ripen. We stay like this. We are destroyed completely! The skin of people becomes whitish, and people move aimlessly about. People go there, go this way and go another way. If there was a sick old woman, people used to go and see her, and then go back to work. Now there is nothing like that. The smearing of mud, to make the home beautiful, nobody does that. The sorghum beer we used to make and then sat simply (to drink), and smearing oil (on the body), now there is nothing like that. We are longing (for food). We have nothing more to say. My talk is finished.

(Interview with Abang Uriet 1993)

As soon as Abang finished her narrative, Dinguru started to talk by herself in a very emotional way as follows.

Can't you see how we stay? Where are clothes? Where is happiness? Where is life? We have nothing and are in trouble. People want death only, only. We don't see something good. From where does something good come? We don't find meat and fish. For instance that old woman [Aduk Medho] is sick now. If she gets well-cooked meat or fish as soup, if she eats it, her body will become all right and she will get up and walk again. She sleeps and stays during the daytime. She sleeps and stays during the daytime. Even though death should be far away, she may die because of that hunger. She will die of hunger. An old woman eats something good and gets up with a stick. People laugh at her [because they are happy]. Now we are just staying. If you walk alone, you are still going poor. Your just walk and reach another village. [They ask] 'Who is this?' 'I am so and so.' 'Are you wives of people who have become as poor as this?' You sleep with hunger. You sleep with your body bent over [because of hunger]. You reach there and sleep on the ground with your body rolled up. In the next morning, if there is coffee, she will be called, 'Get up, come and take a cup of coffee.' If she drinks it, she gets satisfied. Then she sleeps. If you get a little thing, keep it to save a small child. You give a piece of *udeena* [injera] into the mouth of child. When you see him satisfied, you smile. When the child sleeps, you may go somewhere because the home is destroyed.... It is like that. I have told all stories. Clothes for the lower body cannot be found. Salt cannot be found. Coffee is expensive like oil. How can we get it? You can't find any money from the ground. Your daughter, you cannot ask her, 'My daughter give me money.' She goes with that money to buy something for herself. Now, Ukunny, we are not happy. There is nothing at all. We are just dead now. You can't express anything about yourself. If you go where shall you express yourself? You express yourself at your own home. We are dead.

(Dinguru Ugelo 1993)

Usually both before and after an interview I chatted with the informant and other people around. But to those women of Cwobo, I could not say anything. I was speechless, although apparently those women wanted to have a dialogue with me, addressing me and seeking an explanation of the present situation. Their sense of powerlessness and despair was so deep that no words of consolation seemed adequate.

Aftermath of the socialist state: looking back on the 1990s

When I stayed in Cwobo in February 1993 I was on my way back from Ukuuna where settlers had been massacred by Anywaa villagers during a chaotic period immediately after the fall of the Derg. According to the stories told to me, a notorious Anywaa outlaw robbed and killed a settler's family. Then a group of fleeing Derg soldiers arrived at Ukuuna, and settlers brought the case to them. A meeting was called and the soldiers killed the Anywaa peasant association chairman right there, and then fled

The Promise of 1991

to the east. After they left, Anywaa villagers started to attack settlers, burn huts and kill people indiscriminately. At the time there were about 3,000 settlers in Ukunna, and the rumours in Gambela town had it that many had died, but the exact number of casualties was not known. This was a horrible event in which victims of the Workers' Party regime killed other victims. Those whom I interviewed in Ukuuna were reluctant to talk about the incident at first, and when they eventually opened up, it was with so much pain.

Walking later through former settlers' villages between Ukuuna and Cwobo, a region where once hundreds of huts, schools, clinics and churches had existed but was now completely deserted with grasses and shrubs growing, I was caught by the horrible feeling that the land had been cursed and haunted by ghosts. I asked an old man accompanying me, a son of a *nyieya* in Abwobo, whether those murdered 'foreigners' (*jur*, i.e., settlers) could have made curses on the killers. He immediately said, 'No', but my question was obviously embarrassing for him. At a homestead of my friend where I took a rest on the way, I saw a herd of goats, looted from settlers. The man was a former *kwaaro* well respected for his dignity. He surprised me when he returned to me a neatly folded handkerchief that I had left there three years before. He was sincere enough to keep that tiny thing of mine, but for him, goats of settlers were just booty, which he happily showed me. Another man of Cwobo who kindly gave me accommodation and food had five rifles at his home collected during the *girrgirr*, evidence of the proliferation of firearms. I wondered at whom those rifles would be pointed next.

By the end of the 1990s, conditions had not improved in Gambela. Once again, international factors prompted a recurrence of violence. Relations between Anywaa and Nuer broke down again in January 1998 after a relatively peaceful period of five and a half years. On 5 January armed Nuer forces attacked Anywaa villages on the bank of the Upeeno (Baro) river in Itang *wereda*. Fourteen villages were burnt down and the villagers were displaced. That was the time of the second harvest, and grain was looted from granaries and fields. Primary schools and health centres were also destroyed. The total number of Nuer forces was said to be more than one thousand, and they were well armed with automatic rifles and some light machine-guns. When I stayed in Gambela in February, skirmishes were still continuing, but the regional government had done nothing to contain the fight or to help the 2,000 displaced villagers. The federal government was not taking any measure at the moment.

In order to understand this latest Anywaa-Nuer conflict, some notes on the background are needed. After the fall of the Derg in mid-1991, all Sudanese refugees, SPLA men, and Ethiopian Nuer officials and their families had fled to Sudanese territory across the border. Some of them regrouped themselves at Nasir on the Sobat river. This was the place where the three commanders of the SPLA led by Riek Machar revolted against John Garang, and declared the establishment of a new SPLA, the SPLA-Nasir faction. This faction repeatedly split, changing its name several times, and finally those remaining signed a 'Political Charter' in 1996 and then a 'Peace Accord' in 1997 with the government of Sudan. The Anywaa believe that Riek Machar, supported by numbers of both Sudanese and Ethiopian Nuer, is the architect of a conspiracy against them. Although he had joined the government of Sudan, Anywaa considered him as having his own 'government'. In the meantime, as the situation in Gambela quietened, numbers of Nuer (it is difficult to tell whether they were Sudanese or Ethiopian) crossed the border again to settle down in Ethiopia. Whatever their political motives might have been, they were apparently starving and acutely needed land for cultivation and grazing. Kong Diu who was said to be one of the commanders of Riek was their leader, and under the new political situation in Gambela, he claimed that six *qebeles* should be given to his people. This was a great threat to the Anywaa in Itang *wereda*, because if the claim were approved they would become a minority and lose their control of the *wereda* council.

One day in February 1998 I interviewed three men in order to try to get a clear picture of what had been happening on the ground. Two of them were peasants who had just come from the battlefields to Gambela town. They had been leaders of the peasant association in Itang *wereda* during the Derg era. The other was a former Derg administrator who is widely respected. He was one of the first Anywaa to complete twelfth grade (senior high school level) and then to obtain a diploma at an agricultural college. Like other Derg officials he was arrested after the fall of the regime and was detained again with other former Derg officials before the first elections under the new constitution in 1995. Among those who were detained were potential candidates for the election and so they could not stand.

In the course of interviews it became clear that these men had a clear view of the hierarchy of administrative levels in the new Ethiopia. There were echoes of the old revolutionary ideology in the words of the former official: '...who are the peasants? They are the government.' These men's statements were positive in their insistence that something had to be done. Responsibility for the violence in the frontier region near the Sudan border had to be accepted by somebody within the Ethiopian government or the situation would become like Rwanda.

The interviews indicated a strong sense of being autochthons of Gambela region; 'we were the first to settle here', 'the land belongs to us'. This sense of autochthony developed during the Derg era, and the present political and administrative framework provided a perfect hotbed for it to grow and bloom. What had once been accepted as a 'natural' process of migration and assimilation – the mixing of all kinds of peoples on the border – had become a burning political issue. In this regard, the results of the 1994 national population census which were published in 1995 were a shock to the Anywaa elite (see note 2). If the figures were correct, they had a total of some 45,000 against the Nuer's 64,000, and had become a

The Promise of 1991

minority (only 27.5 per cent of the total) in what was now defined as their homeland. In imperial times, this hardly would have mattered, but with elections and the ideology of democratic rights, it was profoundly disturbing. At the *wereda* level, the two western *wereda*s which were once dominated by the Anywaa were now completely swallowed by the Nuer. In Itang *wereda*, where the process of Nuer expansion was ongoing, the two groups shared almost equal proportions. In this context, the claim by the Nuer to six new *qebele*s became incendiary.

After talking to the men, I recalled the dejected and forlorn women who had spoken to me in 1993, at a time of relative political hope among Anywaa men. At that point, ethnic federalism had seemed a popular idea among men. It was perhaps women's particular experience – rooted as it is in crossing boundaries, in mixing and matching in order to survive in a changing world – that offered the clearest insights into the new state's dreams of ethnic purity. As the decade of the 1990s closed, the Anywaa no longer had to worry about the coercive power of the Workers' Party state, but they had everything to fear from their own nearest neighbours – neighbours just decades before with whom they had intermixed and sometimes intermarried.

Thirteen

Imperial Nostalgia
Christian Restoration & Civic Decay in Gondar

CRESSIDA MARCUS

In this chapter a depiction of the ethnographic landscape is offered as an attempt to capture the spirit of a people who have turned inwardly to focus on the spiritual maintenance of their community. The majority of Gondari are fervent Christians and their churchgoing and parochial activities are a focus for empowerment in their social and material environment. The Ethiopian Orthodox Church is presented as the site for community life in Gondar. Since 1991, post-war reconstruction in and around Gondar is manifested in the renovation and reconstruction of church buildings. Church construction exhibits the significance of the Ethiopian Orthodox Church in people's lives; and suggests empirically that it is the most important institution socio-culturally and possibly politically in Amhara Region. In response to social crisis, economic hardship and the political decline of Amhara, a muted form of resistance and protest is manifested in the people's attention to their churches. Church construction is a forum in which claims to legitimacy over the control of local space are real; and building projects are also symbolic statements about the possession of heritage and identity. Buildings past and present define civic and national identity; an identity that is also defined by collective representations, and historical consciousness. Nostalgia of past greatness is manifest in the built environment, and yet the cityscape is a symbolic structure onto which the concerns of contemporary community are projected. The observation of this patterning of old and new enables a consideration of the local dynamics within Amhara Region, as the region is challenged to define itself in the ethno-national landscape of today's Ethiopia (see Maps 1, 1.2).

Ethiopian Orthodox Christianity has been characterized as '*tabot* Christianity' (Teshale 1995: 3, 7–8); and in Amhara region, the church is the focus of national identity in the form of denominationalism. Each local parish (*debir*) is named after the *tabot*, which represents the saint or angel to whom the parish church is dedicated. The *tabot* in the parish church is the focus for ceremonial and the emblem of the denominational identity of Ethiopian Christians. At a parochial level the *tabot's* presence consecrates the church building and essentializes a nexus of local and national identity. The churches that house *tabot* sanctify space in Ethiopia and define the Christian landscape.

The sanctification of space

The churches concentrated in and around Gondar city are clustered around the castle compound, have prominent positions in surrounding urban neighbourhoods (*sefer*), or are dispersed in the sprawling fringes of the city, while a few are located in outlying peri-urban areas (see Figure 13.1). Places are often known by the name of the parish, in the name of which parish councils organize building projects, and the parish is also adopted by lay associations that provide the focus for the community. The churches themselves are monuments that command specific localities in the town (*akkababi*). Such a preponderance of churches defines the landscape and allows for a spirited ceremonial calendar, since each church has monthly and annual festivals to patron saints and angels. This urban vitality has its basis in the historic ecclesiastical importance of Gondar, the former capital of the Abyssinian empire and the Orthodox patriarchate. The presence of the church, especially in its capacity as the leading centre of church education in Ethiopia, links the city to the whole of Amhara Region and beyond.

In the national imaginary Gondar is unique and ennobled by a special nomenclature (*adbarat*).[1] This is a special title referring to the unique grouping of a great number of churches in and around the historic city of Gondar. Gondar is famed for its historic forty-four churches and amounts to a symbolic bastion of Christian fastness in the highlands of Ethiopia. There are a number of illusions that the appellation 'forty-four' maintains, the foremost being that there are forty-four actual churches in Gondar.[2] It is widely assumed that there are forty-four churches; indeed, even in the minds of those involved in the construction of new churches in and around Gondar, their number is immutable. The mythical number of churches, that is to say, the nominal forty-four, crystallizes Amharan historical identity by denoting a fixed and timeless place, a changeless sacerdotage, where ritualized life is undisturbed by the century one finds oneself in.

Gondar's combination of churches and castles is emblematic of the Amhara legacy of a grandiose ideology.[3] Gondar, a former Abyssinian

---

[1] *Adbarat* is a plurality of parishes. The forty-four churches are also referred to as *arba arat*. Forty-four is an hermetic number in Ethiopia. So, for example, there are forty-four Islamic shrines within the walled city of Harar.

[2] Two extant published lists of the forty-four churches differ. See *Hamar*, Tahsas and Tir issue, 1990 E.C. (in Amharic); and also an encomium for Lika Likawent Menkir, the former *aleqa* of Medhane Alem church, Gondar (in Amharic).

[3] Holcomb and Ibssa (1990: 143–4) list essential features of the modern national ideology; amounting to a vision of the nation state of Ethiopia being an ancient polity, with sacred

Imperial Nostalgia: Gondar

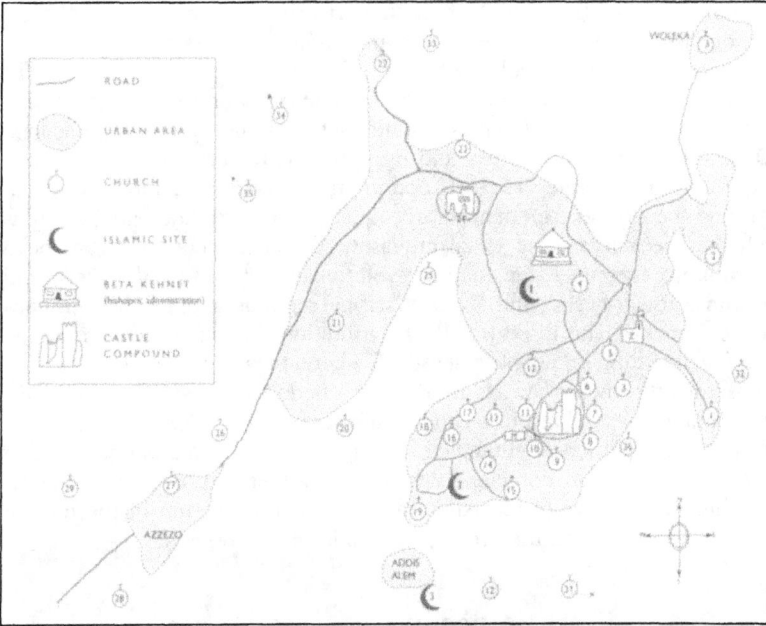

* indicates one of the 44 historic churches
1. *Debre Berhan Selassé: a national heritage site, typifies Gondarine style
2. Abune Aregawi: recently built
3. Weloqa Kidane Mehret: recently built
4. *Abun Bét Gebriel: a newly constructed basilica
5. *Medhané Alem: the pre-eminent church of recent times
6. *Aṭṭaṭami Mikaél: roof recently replaced
7. *Elfenny Giyorgis: circular church recently rebuilt
8. *Addebabay Tekle Haymanot: recently rebuilt in circular style
9. *Qeddus Rufaél (Abune Harrawi): modern and historic combined, recently rebuilt in basilica style
10. *Addebaby Iyyesus (Bale Egziabher): rebuilt in basilica style
11. *Gemja Bét Maryam: circular style
12. *Abba Jalé Tekle Haymanot: rebuilt in basilica style
13. *Yohannes Welde Negodgwad: circular style
14. *Qeddus Cherqos: recently rebuilt in basilica style
15. *Debre Met'maq Maryam: rebuilt in circular style
16. *Ab-Egzi (Kidane Mehret): circular style
17. *Fit Mikaél: circular style
18. *Beata Maryam (Uraél): grand circular style, rebuilt with expansion for a holy spring
19. *Fit Abbo: grand circular style, rebuilt
20. *Aberra Giyorgis: new basilica style church
21. *Ledeta Maryam: grand circular style, rebuilt within ruined arcades of the historic building
22. *Debre Tsehay Qwsqwam: a national heritage site, grand circular style within crenellated walls of palatial compound
23. *Qeha Iyyesus: circular style, rebuilt
24. *Qeddus Fasiledes: a national heritage site, not a functioning church
25. *Meṭmequ Yohannes: construction of circular style church under way near to historic crenellated church
26. *Abba Samuél (Qeddus Mikaél): ruined, recently constructed makeshift church nearby
27. Abba Gennetu Mikaél: built in modern times
28. *Azezo Tekle Haymanot: grand circular style, within remains of historic church
29. *Loza Maryam: circular style, rebuilt
30. *Ṭedda Egziabhér Ab: dilapidated but a formerly grand circular church
31. *Fanta Ledeta
32. *Defecha Kidane Mehret: grand circular style in ruins, separate square church built in modern times
33. *Abba Anṭonyos: small circular church, rebuilt
34. *Gonderoch Giyorgis: circular church, rebuilt
35. *Gonderoch Maryam: rebuilt
36. *Selestu Meet: ruined but still venerated
37. *Arbaetu Ensesa: ruined but still venerated

Islamic Landmarks
1. Islamic centre
2. Grand Mosque
3. Islamic shrines

M Municipality
Z Zonal administration

Figure 13.1 The sacred geography of Gondar [C. Marcus]

capital, is an emblem of Amhara imperialism, and today of Amharan nationalism. The great stone castles and crenellated fortress-like churches are monuments that evoke nostalgic pride in the Gondari, and wistful reminiscences of pre-revolutionary order. The Amhara no longer dominate the national scene, and the lack of national pre-eminence is a palpable loss. During the twentieth century and into the twenty-first, the Amhara have experienced extreme social disruption in the course of modernization. Among the stresses experienced are war, famine, migration, emigration as refugees, secularism and social criticism. Although there has been social disruption, especially due to the revolutionary reforms of the Derg, old attitudes towards the Amhara and self-perceptions about being Amhara are entrenched and at odds with the conditions in which they live today. Historicity, that is, people's sense of where they have come from and consequently where they are heading, is a bedrock of people's experience. Examining the Amhara in terms of their historicity evidences a conflict in their diminished status, entangled in national and personal pride. Situating the Amhara in their homeland and observing their deprivation challenges the historical image of the Amhara, which is the fulcrum of the tension between the past and the present in Gondarine historicity.

Historic decline & dispossession

To understand the degree of destitution found in Gondar today is to situate Gondar within Ethiopian history. The noble beginnings of Gondar presaged a far more impressive city than the provincial centre it is today. Gondar is a city in terms of its function as the centre for secular and church matters for the North Gondar Zone, but it does not measure up to the scale of a modern city.[4] Nevertheless urbanization continues apace, and the town is straining to accommodate the many impoverished newcomers who leave the surrounding farmlands (*getter*) in search of a brighter future. Destitution in and around Gondar is palpable. In living memory there once was a level of prosperity now barely recognizable in the shabby town centre. Since the demise of Haile Selassie I, a dramatic and tragic deterioration of Gondar has occurred. Lions, by imperial order, were once kept in captivity as a signature of the might of the Solomonic Empire. Under the Derg administration, the lions were not allocated food so that all but one starved to death in the castle compound; the last one, being in its death throes, broke out of captivity, ran amok and was shot. Their fate is a terrible synecdoche for the decline of an empire. Rancour resulting from the defeat of the Derg deepened a bitterness about an issue that is heartfelt and analytically fascinating – the decline of the Amhara as the national elite of Ethiopia.

[3 (cont.)] boundaries, which should be ruled by the Abyssinians (sic: Amhara), whose influence is a beneficial and civilizing force.

[4] Urbanization in Ethiopia was inhibited by various factors that are too complex to be developed here (see Crummey 1987).

Imperial Nostalgia: Gondar

At the beginning of the Derg a tragic drain on the human resources of Gondar began with the onslaught of the 'red terror', and then recruitment into the Derg army decimated the younger generation. Gondar was particularly affected by an exodus of those who fled Ethiopia in fear of persecution, especially because of its proximity to the Sudanese border. Even before the exodus, during Haile Selassie's reign, children, often the offspring of prominent Amhara families of Gondar, were selected to be educated in Addis Ababa and groomed for public service in order to advance the modernization of Ethiopia. This generation of Gondari never returned; thus a vacuum of trained professionals developed and was an unanticipated feature of the social trauma precipitated by the revolution. The attrition of human lives and resources was accompanied by the decline of the town's infrastructure and civic amenities.

Gondar's landscape reflects this, official administrative buildings are so deteriorated that some look derelict from the outside. As the war between the Derg's forces and Eritrea intensified, civic amenities in Gondar were overlooked in favour of the war effort. Civic construction was limited and included a few recreation centres, such as the Jan Tekle Werka tree recreation centre, formerly designated for party members. Today, the municipality allocates land on which very modest houses made from cement blocks or huts, generally in wattle and daub (*chika*), are erected. Residents complain that the lots are simply too small to accommodate them. The Gondari live in extremely squalid and desperately impoverished conditions. Houses frequently have no pit latrines. The roads are pitted, and human excrement litters even main thoroughfares. The city's supplies of water and electricity are erratic, and schemes underway to alleviate resource problems are undermined. Economic hardship has increased since the fall of the Derg, especially for those who relied upon governmental food subsidy. Poverty infuses all aspects of life and the destitute are everywhere.

The plight of the people of Gondar, the degree of land degradation around Gondar and the deteriorated state of the city, induce a terrible sense of loss in those who have been returning to Gondar for the first time since the fall of the Derg. They grieve the loss of image of the Happy Valley, the land immortalized in Samuel Johnson's *Rasselas, Prince of Abyssinia*. Once, the rivers flowing through Gondar were full, the landscape was clothed by trees, the people were content, honey was pure, and *teff* in plentiful supply. Today, bereft Gondari bemoan a situation in which honey is adulterated and the former staple cereal *teff* is now a luxury for some urban and rural households.

Emphasis on local conditions in Gondar dispels the mythology surrounding the Amhara of Ethiopia. Contemporary description fits a trajectory of decline, even though the role of the Amhara in the nineteenth-century consolidation of Ethiopia is undeniable. Because of the crucial influence that Amharan dominance had nationally, history in Amharan experience inevitably has particular resonance. Holcomb and Ibssa preface their disclaimer of the natural integrity of Amhara supremacy with the reflection that in any consideration of the composition of Ethiopia, there is recourse to the Amharan historical mythology and record (1990: xiv–xv). The key significance of this place in history for the Gondari today is that they no longer enjoy the spoils and trappings of empire. The importance of the imaginary and historical Abyssinia as composites in Amhara nationalism is poignant, since loss and desperation make conditions in Amhara Region today appear singularly bleak.

Parochial restoration & the construction of churches

Against this background, Christian activity can be understood to be a site of resistance, caught up in a spiral of historic decline. Gondar's deterioration and position as a backwater is defied by the faithful's estimation of the Gondarine churches' sacred importance. The central role of faith in the lives and attitudes of the people defines civic identity in formal and informal surroundings, and enriches the fabric of existence. The societal invigoration brought about by the popularity of the church is symbolized by the restoration and reconstruction of a number of old Gondarine churches, and the construction of new churches in and around the city, as well as throughout Christian communities in Ethiopia. The construction of religious buildings provides a focus of development for communities, and churches are symbols of achievement and prosperity. The maintenance and construction of, and communal investment in, religious buildings is a way of participating in the rejuvenation of the community. The development of church building is a movement that recaptures the built environment and materially represents the societal sentiments involved in the process of redefinition. By constructing buildings, Gondar is attempting to recapture its moral and ideological universe.

The reconstruction of historic buildings and churches is an idiom for the social reconstruction of Amhara Region. The construction of churches and quasi-ecclesiastical buildings is arguably more pertinent in terms of Amharan nationalism than the commanding castles or the administrative buildings. A restoration is underway in Gondar, which far outstrips the pace of reconstruction of civic amenities. The renovation of the castles has been initiated and funded by the central government. In contrast, the significance of church construction is that the impetus for it is generated by popular support. Related to the contemporary function of buildings is a temporal sensitivity in terms of the possession of the past. Landscape is a site of contestation: the conflict of interests of sacred and secular meet head on in churches that are designated heritage sites. Church buildings simultaneously function as meeting houses, at the centre of the parish; and as mnemonics of the historic image of Ethiopia. Certainly, the churches are not perceived by the parish councils that run them to be unchanging, historic buildings that must be preserved in a pristine state. Without the presence of the congregations, the churches would not be the very buildings they are. In fact, many of the churches would be in ruins if it were not for their valued position in the community.

The ethnography that follows here suggests that the manifold ways that churches are redefined, relocated, rebuilt or left to deteriorate indicate that although the Orthodox Church has an image of timelessness, it is a living church, vulnerable to the will of historical subjects. Churches evidence their embeddedness in Amharan history by their changing characters. The churches are monuments to historic periods, especially the Gondarine period in Ethiopian history, in which the 'forty-four' churches were supposedly erected; and to the contemporary period, in which reconstruction in modern materials such as concrete promises durability and symbolizes the endeavour to resist the onslaughts of fortune. Some churches are no longer standing. Even as new churches are built, churches on other sites are delapidated. Decline and absence of church buidings is commonly attributed to their destruction by foreign aggressors. The Dervish incursion and Italian occupation may well be responsible for razing and capturing churches; however, churches also decline because of local conditions. Many *debir* churches, whose wealth and prominence stemmed from tribute passed from satellite churches, are now impoverished and chastened by the need for voluntary contributions. Some struggle because they are unlucky in being located near a notable, possibly miraculous, neighbouring church. Some of the poorer churches were formerly notable, and now parade their deteriorated splendour on feast days, timeworn religious paraphernalia such as faded and streaked parasols, antique 'golden' crosses, and musty vestments.

Contemporary construction is a symbolic reference to the majesty of buildings in stone. The grandeur of Gondarine churches in the historical imagination should ideally never be forgotten, and at the very least must be commemorated by retaining the name of the parish. The Gondarine collective memory has forgotten many of the original 'forty-four' churches; but others, such as Arbaetu Ensesa, Abba Samuél and Selestu Meet, are commemorated, even though they no longer have a church building (see Fig. 13.1). New and antique buildings monumentalize the spiritual essence of the church, which is simultaneously the repository of local heritage. However, there is concern in parishes that their ambitions for construction are impeded by the prescriptions of government. Today the Ministry of Culture is in conflict with the churches because of Gondar's significance as a national icon. The church conforms to this evaluation of Gondar's heritage inasmuch as the church and municipality must formally request permission to organize the civic annual holiday of the religious celebration of *Ṭimqet* from the Ministry of Culture. But within the confines of the churches, the policy of conservation conflicts with the ambitions of the parish council and the autonomy of the individual church. Construction committees set up by parish councils must apply for building permission from the municipality.

The Promise of 1991

Parishes are constantly building additions and in some cases completely rebuilding churches. A remarkable level of activity is apparent: it seems at every turn in church compounds, and in their vicinity, some sort of construction work is under way. Many churches have been rebuilt, since replacing, not conserving buildings is culturally valued. Abun Bét Gebriel, Abba Jalé Tekle Haymanot and Qeddus Rufaél, among others, boast that they have been completely rebuilt. Ledeta Maryam was rebuilt over a decade ago within the ruins of the original arcades of the round Gondarine church. Some church buildings such as Qeddus Cherqos have been modified, expanded around the *meqdes*, the inner sanctum that contains the *tabot*. Certain churches, consecrated ones such as Debre Tsehay Qwsqwam and Debre Berhan Selassé, Metmequ Yohannes, and ruined ones preserved as archeological sites, such as Abba Samuél, are under the aegis of the Ministry of Culture, which acts as the guardian of Gondar's heritage.

Most parish councils have construction committees that produce plans and put them into effect. Planning and erecting churches is a phenomenon that in part responds to the changing needs of neighbourhoods (*sefer*). The growing need for a community focus and the construction of churches are not simply responses to the increase in population by the church authorities. Churches are often built by the newcomers. For example, the modern Abune Aregawi has been adopted by a Tigray congregation who came as refugees during civil war and then as returnees from Sudan. The parish has collected enough *birr* to build a larger church: I was proudly shown the pile of stones that will provide its foundations. The spread of the city towards Azzezo is producing church growth, as in the reconstruction of Abba Samuél, Aberra Giyorgis and Metmequ Yohannes, churches on the periphery of the town located in the area of greatest urban growth. Metmequ Yohannes parish recently began the construction of an entirely new building in the circular style, right next to the old crenellated church, with a huge portcullis-style gateway.

Church-building is the focus for the construction of community, whether by denizens, incoming rural people, or the displaced and dispossessed. Outlying communities spreading into peri-urban areas that are known by the name of a former *debir* church build on old sites, and reconsecrated ground. Strangely, despite the immutability of the number of churches (*adbarat*), church building is widespread. Gondarine heritage is a source of pride, even though no one but a handful of elderly priests can even approach giving a list of the forty-four churches. The historic and sacred value of the original churches still sanctifies Gondar.

The reconstruction of Abba Samuél church is important to the villagers who live in the peri-urban area. As the population of Azzezo swells, spreading outwards along the Gondar road, the importance of Abba Samuél increases. Yet Abba Samuél is in ruins, and enquiries concerning its location led to some confusion. When I enquired from locals where Abba Samuél is located, some could not say, while others would say it was near Mikaél. Alternatively, others suggested that it was Mikaél. In contrast

Imperial Nostalgia: Gondar

to other churches whose names are known, this church is an enigma. The church is known by two names; Abba Samuél is also referred to as Qeddus Mikaél. On inspection the church land can be seen as the site of two churches. The confusion is the result of the difficulties involved in reconsecrating Abba Samuél. Originally, Abba Samuél was one of the forty-four churches, but the *tabot* was removed in the nineteenth century during the Dervish invasion for safe-keeping. The church was razed by the Dervishes but the *tabot* was placed in a tiny chapel nearby, which functioned only as the *meqdes*. There it stayed until the time of the Italians, when the landholder *(balabbat)*, Qes Gebow Kassa, removed himself and the *tabot* to a distant piece of land that he possessed, a place called Giv Assara Maryam. Remarkably, though the Abba Samuél *tabot* was elsewhere, the *tabot* was revered by locals in Gondar, who continued a sodality *(senbettê)* in its name. People gave offerings *(silet)* to the ruins of the church and continued to hold a *senbetté* in the *tabot*'s name.

During the Derg period, the *senbetté* started to devise a project to rebuild the church, applying to Gondar's church administration to do so and requesting the return of the *tabot*, which Giv Assara Maryam parish refused to return. Frustrated attempts to reclaim the *tabot* and land dogged the community's progress. The request for the reconstruction of the church was rejected on health grounds. Derg officials reasoned that there was a need to keep the church's graveyard separate from the nearby abattoir, and burial of the dead would pose a health hazard so near to the plant. There was concern that the Derg would use the church's land to build a factory, because the area is a site for factory development. Members of the reconstruction committee, set up by the *senbetté* that had been meeting at the ruins for years, were unperturbed and managed to build a new church close by, which they dedicated to the archangel Michael *(Mikaél)*, so he could watch over and guard Abba Samuél church. Since they did not have the original *tabot* and were forced to build on another site nearby, the *senbetté* decided to dedicate the new church to Mikaél, which they did on its completion, on Genbot Mikaél 1988 E.C (Ethiopian calendar). So it is that the church is Abba Samuél in name alone.

For all purposes other than authentic realization, Mikaél church has replaced Abba Samuél. But the congregation has regrouped under the protection of Mikaél, and with support from people of Azzezo from another Mikaél church there, Abba Gennetu Mikaél, they are presently petitioning to rebuild the Abba Samuél ruins. The palimpsest of buildings figures as a landmark for the ambit of the city, and as a sacred site of historic value. The significance of the reconstruction cannot be explained solely by the need of the increasingly populous urban community for a church; the parish needs to be in possession of the primordial *(tantawi)* power of the land *(ager)*. The battle to re-establish Abba Samuél church is not simply the restoration of the *tabot:* the demands of the state intervene in terms of conservation and modernization, which limit the realization of a local renaissance. The application for building permission has not been granted.

The Promise of 1991

This time the main obstacle is concern for the conservation of the ruin by the Ministry of Culture.

That churches manifest local and group sentiment, and the enshrining of historicity, is evident for other historic churches. Sacred power is possessed locally and historic ruins are hallowed. All that remains of Selastu Meet's church is a mound, yet local people meet fortnightly in its name at a nearby church in which the *tabot* is kept. As in the case of Abba Samuél, Arbaetu Ensesa *senbetté* meets among the stones of the ruined church. Every *senbetté* that is founded is a local association of parishioners who identify with a particular parish. The reverence accorded to the church, *tabot* and dedicated cosmological figure enables a local *communitas* of sentiment. The parish is a primordial grouping, which in Amharic is expressed by the word *wegen*, meaning an interest group, and membership in terms of kinship, religion or politic alliance. The primordial affinity to the parish is not to the building *per se*; people believe in the original power of the *tabot* and site of the church. Churches evidently manifest local and group sentiment, and parishes are bound together by a spirituality that encompasses a passionate sense of belonging to that place.

Church 'planting' has a significance that is mythologized in the oral history of churches; each church invariably has a miraculous tale concerning its origin. This may be related to a historical event, or to a miraculous happening linked with a particular day in the religious calendar, or *synaxarium*. Church planting, or founding, has a legendary importance because the land is forever transformed and protected by a particular saint or angel. Church planting is emulated today in reconstruction and additions. Today church building is symbolic of the maintenance of community, which is achieved in social, temporal, ritual and material ways. For Gondari the public meaning of building churches should not be underestimated.

Churches are maintained by popular support. The financing of church buildings confers a certain sense of empowerment on those involved in their construction, because it is a means of directing the development and growth of Gondar. All patronage and co-operation is channelled into private construction, to be owned and managed by individual churches, yet has a public meaning and impact. In a very direct way, one's labour, efforts and money can be seen as being actualized in the cement and timber of the holy places. The economy of the sacred is inherent in the memorialization of community. An ingenious practice has grown in popularity: one builds a mausoleum for one's family within the church compound, but on the understanding that the first floor room will be kept clear and used by the church for teaching and for the meeting of the parish council. Only prominent parishioners may afford such spacious accommodation. Their influence is recognized during their lifetime, because they have constructed a parish council office in which, no doubt, they meet with other churchmen for parish council meetings, and under which they anticipate they will be entombed.

Imperial Nostalgia: Gondar

The heads of churches (*aleqa*) may remark *en passant* that they have completely rebuilt their church since their installation in office. Particular *aleqa* are complimented by Gondari on their ability to manage construction projects. Ability in this respect equals the importance of their scholastic achievements, or their popularity as preachers. I have interviewed many of the heads of churches (*aleqa*) who chair the parish council meetings and are responsible for the development of the parish and for the setting up of projects. I have also interviewed occasional members of parish councils who either once acted as committee members or are presently employed. I refer to this type of active parish member collectively as churchmen. These prominent men are involved in relations with the ecclesiastical authorities and act as assistants in construction projects, for which special construction committees are formed. There is therefore a group interest in the maintenance and modernization of the parish churches. In practice, in the myriad of concerns involved in parish affairs, there is one issue especially involved in the management of the church's land and future – the self-sufficiency of the church.

Contemporary conditions, such as the growth of the city leading to a serious shortage of space for the burial of the dead, and the lack of governmentally channelled income to the church, has led the church and parish councils to seek out new revenues and alternative use of space. Commercial sources of revenue have encouraged a new entrepreneurial style and led to shops selling religious paraphernalia and to storehouses for grain being built around church perimeters. These storehouses are lucrative and bring in thousands of *birr* a year. Beata Maryam church built a large church, a concrete construction for a spring in the name of the angel Uraél, and a large commercial store that is rumoured to bring in a monthly income of seven thousand *birr*. Church incomes rely upon the return on such investments.

The economy of the sacred is entombed in mausoleums, and monumentalized in church structures. Community efforts to aggrandize their church are efforts to enhance the spiritual reputation of the parish. Offerings (*silet*) are managed by the parish council. Donations may be presented publicly on the steps of churches, or may be negotiated within the context of the parish council. However, congregations and parishes are economically and socially mixed, and consequently churches are involved in a competitive drive. The focus of this parochial competition is the construction of church buildings. Even by the poorest, offerings to churches are valued as an investment and an expression of faith. For one Maryam *silet*, a neighbourhood communal offering and feast to Mary, one neighbour of mine could afford only ten cents as an offering. Cash offerings ranging from ten cents up to hundreds of *birr* are collected. Offerings on feast days range from a few handfuls of grain, to ten *injera*, to a sheep or an ox. The parish tax is so small as to be negligible, especially since people tend to overlook its payment. In contrast to the reluctance to pay their parish taxes, people give, if possible ostentatiously, in the name of patron saints.

The Promise of 1991

Huge investment is needed to build large basilica-style churches made from concrete and corrugated iron roofs. Parish council buildings are most often constructed out of concrete blocks and cement, usually being built into the walls of the perimeter of the church compound. The development of church peripheries for commercial and parochial use is being established as a normative use of space. The use of the wall for the extension of church property echoes the *enqullal gimb*-style castellations, a signature of Gondarine architecture. Since land was confiscated from the church, the only space that was left inviolate was the compound of the churches – hence the effort to control the space inside a church's compound by the development of church grounds and church buildings. The *Beta Kehnet* (bishopric administration) in Gondar spent much of the early nineties staking a claim to the perimeter of its administrative compound and the construction of buildings upon it. The church is constrained by the need for planning permission from the government, but it exhibited an anxious desire to formalize its property, as a result of the trauma of confiscation of church land during the Derg.

Land was confiscated from the largest monastery down to the humblest parish. The parish of Abba Antons did not even retain the field and spring by which the annual festival of Timqet was celebrated. The churches of Gondar were originally *debir* churches: they had lesser rural parishes under their jurisdiction, and churches also controlled land rights over the urban land in their vicinity. The confiscation of land and redistribution impoverished the church, but so did successive policies promoting municipal modernization (cf. Solomon 1994). The allocation of urban land in Gondar was once a power enjoyed by the *aleqa* of *debir* churches. The constraints imposed upon the church meant that it had to turn towards the generosity of its congregations. Annually in Ethiopia, millions of *birr* are donated to the church, and numerous *tabotat* are housed in new churches. For such an impoverished population, this fervid dedication to the reconstruction of churches is remarkable. Individuals may well give great sums towards the renovation, and preferably construction, of a church. In a sense, the heritage and rich culture of the church is owned by the people.

In this section the ethnography of landscape has underscored claims to authority over Gondar. Conflicting claims by church and state regarding heritage and land illustrate the dialectic of power between these two defining institutions, in an ethnographic setting where decline is palpable. The authority of the state does impinge upon the activities of the church; however, the church is relatively autonomous, especially in contrast to the drastic involvement of the state in church matters during the Derg period. The authority of the state over the church is particularly controversial when vetoing public assembly, and is usually limited to authorization (*fakad*) of civic ceremonial, and planning permission. This chapter so far has focussed on the construction and restoration of churches, which is complicated by the archaeological concerns of conservation that conflict with the need of the parishes to maintain the monumental power of the churches of Gondar.

Imperial Nostalgia: Gondar

Plate 13.1 *Celebrating a church festival at Gondar* [C. Marcus]

The symbolic action of church construction is the focus for the parish community, and it has been intimated that this identification with Christian community is extended to all Amhara involved in activities within the confines of the church. Though calculated during the Derg era, the church's demise did not transpire. In the face of adversity the church was forced by modern events to respond to the needs of contemporary society, and became a popular stronghold; a movement that confirmed its enduringly powerful position as an ancient polity. In the following section the organizational importance of the church is considered as vital in fostering social gatherings at a parish level. The ethnography suggests that parish churches are venues that contribute to the advancement of civil society.

The societal edifice of the church

Church attendance is a ritualized activity by which the traumatized Gondari have strengthened and maintained their identity. Civic pride is almost indistinct from personal social honour (*kibir*) which is expressed through stylized piety. Many people will pass through their church's compound to pray, and to kiss the church. The frequency and length of time spent inside the churches depends on individuals, who will be constrained by other responsibilities, especially if they have salaried positions in the administration and the like. Especially for the elderly, who have more time to devote to the church, the churchyard is a place to sit and meet friends and foster one's spirituality. There is constant activity and movement inside the compounds of the churches, and the number of people will swell when there is a *qedasse*, or liturgy for the saint, or on special holidays. The crowds that appear for special days are too large to fit inside church buildings, and congregations spill out into church compounds. They greet the *tabot* by a cacophany of praise; ullulating, singing, clapping, blowing trumpets, and bowing to the *tabot* as it makes its way around the church (*oudet*). The church is the most popular social movement, and also the only time that people congregate in public. Church attendance is marked by the frequency of saint's days celebrated monthly and annual holidays: this is especially so in Gondar, where the great number of churches affords an unusually rich parochial life. The display of grandeur and wealth provided by mass offerings at annual and monthly celebrations reinforces a widespread faith in providence. Attendance at church is now seen as a beneficial activity, and no longer a controversial partiality.

I attend church on a regular basis in order to carry out participant observation and conduct comparative analysis of the congregations and churches of Gondar.[3] I follow the church calendar and attend churches on their saint's day, an activity which amounts to a constant round of urban pilgrimage. Churchmen, pious individuals, and the destitute are on the Gondarine circuit; most Christians, however, are more moderate and attend their local parish church on Sundays, its annual holidays and the monthly celebration of the saint's day. Many will choose a few other churches to go to on its saint's day, because they have a special relationship with that particular cosmological figure. Very importantly, attendance, religious dedication to saints and angels, and one's social involvement in church activities is perceived by Orthodox Christians to be based on personal interest and motivation (*fellagot*).

The church has transformed itself into a broad-based popular movement, by contrast with the former obligatory membership of the youth, women's or peasant associations. Once people were fearful of the reprisals of the Derg, which discouraged many from attending church openly. The church offered resistance to the repressive regime in worship and in religious observance, such as fasting. In the post-Derg period, there has been an eagerness to practise Christianity and to proclaim one's affiliation ostentatiously. Christianity has been adopted by the younger generation, and young people are often zealous in their church-going. Adolescents enthusiastically participate in Sunday schools and related youth activities such as liturgical theatre and choirs. More widely, piety is now accepted as a necessary social attribute, and political personalities regularly attend church. Social pressure today has prompted those who would have shunned religious activity during the Derg to resume church-related practices.

Religious attendance is highly valued, and *fellagot* is the generic word for the reasoning and motivation behind affiliation to certain churches and

---

[3] I first carried out anthropological research in Gondar in 1998, and at the time of writing in late 1999 I was still involved in fieldwork.

Imperial Nostalgia: Gondar

associations. During the Derg period people were constrained and pressurized into attending *qebele* meetings that were scheduled for times when people would expect to attend church (cf. Helen Pankhurst 1992: 149). The Amhara did not uniformly disagree with all social reform: indeed, some had been educated so as to pursue modernization. Nevertheless, much of the Derg's impact was experienced by urban Gondari as an attack on religion, and by extension, custom, prosperity and freedom. The mode of the mass organization of women, young people, occupations through trade unions and at a local level in peasant associations and *qebeles* was one of control (ibid.: 29–30). Today, when people are informed in person of *qebele* meetings, they will respond with excuses, such as 'Sorry, but as you know my husband and I are elderly and frail. We are too ill to come'; but they will still find the energy to visit churches religiously. Rationalization of the church's institutions was a process begun before the Derg, but the style of the Derg influenced church organization. The church now issues identity cards styled on those of mass organizations. The parish council runs the parish church, its activities and properties. The parish council is a body that is ultimately accountable to the central church authorities; nevertheless relative autonomy is enjoyed by the parish priest (*aleqa*). The parish councils and *senbetté* are local organizations at which local people who see one another regularly in their locality (*sefer*) can make decisions about parochial affairs and discuss local concerns.

People join groups and support churches on the basis of *fellagot*, in contrast to the once forced attendance and membership of mass organizations. Mass organization has been characterized as an educational step (Tarekegn 1996: 78), a start towards the growth of voluntary associations; however, these were controlled by cadres and therefore had no local autonomy. Mass organization stymied the growth of civil society. Simultaneous with the physical reconstruction of Gondar's churches is the social reconstruction of the fabric of society. Having experienced so much social disruption under the Derg and such deterioration and disappointment in the public sphere, people are offered a sanctuary by the church, which illustrates its potential as a forum for political and social reform. What has come to replace the communist state has been established for centuries – the Ethiopian Orthodox Church. Upholding the Tewahedo Faith has been a conservative act; nonetheless, it too has witnessed transformations in accordance with the times. Contrary to the received view of the Church as a fossilized and dying faith, it is a popular, vibrant religion (cf. Gifford 1998: 51).

Simultaneous with the activities defining the built environment of Gondar's sacred space is a movement of social reconstruction, which finds meaning in the religious idiom of Christian brotherhood, and social security in the meeting houses of para-church associations. Based on comparisons with other voluntary associations gathered in the field, the most popular of all lay associations are the *senbetté*, which are set up by interested groups of parishioners as a lay, para-religious institution. *Senbetté* are co-ordinated by the parish council. The *senbetté* is a lay sodality that meets fortnightly after the Sunday service for breakfast. It is modelled on the Last Supper, where Christ and his disciples ate together and shared the body and blood of the Messiah. The bread shared is *injera* or *senbett qiṭṭa*, types of rough flat loaf, the latter being the left-overs from the Eucharist bread. The wine (*weyne ṭejj*) is invariably substituted by local beer (*ṭella*).

The goals of the *senbetté* are modest, and their activities are regarded by the authorities as a traditional folk custom. The *senbetté* is understood to be a democratically run association, where those who run it are fairly elected by members of the *senbetté*. Each member pays a small monthly membership fee, but if for some reason they can no longer afford it, then continuation of their membership is strictly forbidden. In contrast to the exclusion of those unable to pay, the conviviality between members is pronounced. Eating blessed food together is a ritualized consumption that unites them. The sharing of substance and extension of *sacra* from the ecclesiastical social strata to community level, through the sharing of bread, is experienced as beneficial. The egalitarian character of *senbetté*, however, is more idealistic than actual. Parish *senbetté*s are stratified into number one, two and three *senbetté*, and so on, representing economic superiority. The physical stratification of buildings one, two, three and so on expresses the inequality and hierarchy in daily life and the inevitability of cleavage and hegemony in the cultural imagination. Christian brotherhood, then, is an idiomatic ethos of friendship, but is not an absolute value because the *senbetté* is connected to the wider society, and influenced by the overriding principle of stratification.

The goal of each *senbetté* is to construct a building that will house the people for only a few hours every fortnight. The construction of *senbetté* houses marks the growth in the number of members and of *senbetté*, and thus the ascendancy of the parish and its members. The money collected and made from the *senbetté* provides enough to build quite substantial buildings of wattle and daub. The houses are symbols of a collective desire for church growth and attainable prosperity. Many churches now boast three to five *senbetté* when once there were two or possibly none. The Ethiopian Orthodox Church offers a release from grinding poverty, and unsurprisingly, more than spiritual salvation. The importance of the *senbetté* is as a nexus of locality and a primordial base which provides a sense of well-being and belonging. The reverence accorded to *tabotat* no longer located in a *sefer* by the local people suggests that local sentiments may be behind the tremendous growth of *senbetté* groups.

According to informants, the freedom to ask for assistance from *senbetté* members is an incentive for joining. At a meeting I attended, a man came needing help to pay off his son's fine, imposed by the *qebele* for having broken the teeth of another boy when fighting with a stick (*dulla*). He stood up and pleaded that the fine was harshly set at one thousand *birr*, and that any contribution would be appreciated. People request financial assistance

Imperial Nostalgia: Gondar

from other members in the form of voluntary contributions; and in some cases a rotary system for the payment of a lump sum in the region of one hundred *birr* is established as a form of funerary insurance. Variants of the self-help groups *iqub* and *iddir* are surfacing within the context of the *senbetté*, and this has the effect of exacerbating the tendency for exclusion and stratification latent in all *senbetté*. The Christian and quasi-democratic character and function of *senbetté* enables it to adapt to the needs of members.

Although membership may be denied to those who cannot make the necessary payments, it is a charitably oriented institution. Surplus food is handed out to the poor, although this is not simply altruistic: displaying a surplus increases the social standing of the people whose turn it is to provide and prepare the food. The primacy of the *senbetté* is partly economic, for most people will be able to support membership only in one *senbetté*, which may entail joint membership for a married couple. For economic and social reasons, a majority prefer *senbetté* membership over other associations such as *mahebar* and *iddir*.[6] *Mahebar* and *iqub* are the mark of success, and the social capital at stake means that they are completely exclusive. Regardless of the social capital involved in the exclusivity of smaller groups such as *mahebar* and *iqub*, *senbetté* is taken as an indicator of status and productive membership of the community. The place of the individual is respected, and this social capital (*kibir*) encourages active participation in the *senbetté*.

The drawing power of the *senbetté* is not primarily that it is a religious circle, but a meeting place for the community. The *senbetté* has increased in importance socially and taken root in the growing number of meeting houses in recent times since the relaxation of the Derg's regulations and especially during the post-Derg era. The peri-urban parish of Abba Antons had all its land confiscated during the Derg; the land was redistributed to households that became secularized. Farmers who had benefited from the confiscated land returned to the fold at the end of the Derg. In a *volte face* the farmers' affiliation changed from the peasant association to the parish. Significantly, these farmers refused to pay their parish tax but joined the *senbetté*. Their refusal to pay their dues has forfeited official entitlement to the services of the church, for they are not officially recognized parishioners. The *aleqa*, frustrated at their recalcitrance and disappointed by the response of the church administration to his admonition, takes solace in the bitter anticipation of their demise, for he is not duty bound to give them a Christian burial. The farmers responded to the contemporary

---

[6] *Mahebar, senbetté, iddir* and *iqub* are all forms of voluntary association. In brief, *mahebar* is a sumptuary party that meets monthly on its saint's day and is made up of friends and relatives (*zemed*) who eat together. *Iddir* is a funerary association that affords people the opportunity to provide for their own and their family's funerals, the trappings and food for the guests, and emotional support during a grieving period from the other members. *Iqub* is a fiscal rotary association dedicated to the accumulation of money, that is shared out among the members, according to set rules of conduct. All these associations are subject to interpretation and change; indeed, the boundaries between classified types are blurred, so that a *mahebar* may double up as a thrift club, and so on.

The Promise of 1991

focus of community by shifting from the peasant association to the *senbetté*; the latter was regarded as a civil association and not primarily as a church organization.

The Gondarine example affirms Gifford's generalization that 'religious groups are widely admitted to be the strongest form of associational life in contemporary Africa' (1998: 19–20). Due to the influence and popularity that the church enjoys in Gondar, social involvement in the church outstripped the mass-organization projects of the Derg. Local concerns, group and individual needs, are met by the pastoral and para-religious activities of the Orthodox Church. Membership in the church represents an opting into the denominational movement and allows participants the chance to be involved in the workings of an institution. Under the Derg the degree of state violence in Begemdir and Gondar was pronounced (Kiflu 1998: 136, 455–6), and consequently there is a perennial fear and mistrust of government, which produces a sense of dispossession from official political culture. Once the Derg lost control of Gondar, numbers at *qebele* meetings dwindled to almost nothing. Since attendance at mass organization meetings was no longer enforced, people took the opportunity to reject the imposition of totalitarian organization. In transitional Ethiopia, the much touted language of devolution, federalism and democracy did not persuade the Gondari that they were free (cf. Ottaway 1995; Vestal 1996). People use the vaunted concept of democratic freedom in protest, in the form of boycotting and silence.

After the Derg, freedom of religion engendered a freedom to join quasi-religious associations and social clubs. The movement of the church was in opposition to the ideological edifices of Marxism; and it is a denominational movement that is buoyant in the prevailing tide of national politics. Understanding the importance of the church in the lives of the Gondari is to discover a silent, monumental resistance to threats against autonomy. The key importance of Gondar as the symbolic heart of the Amhara Region has prompted an ethnography of landscape, to address the central concern of what happens to a nation in decline. An emblem of Ethiopian nationhood, the ruins and churches of Christian imperial Ethiopia endure as a monument to a bygone era; and are a melancholy reminder of what Gondar once was. At a local level people feel empowered, and a sense of belonging, meaning, and continuity is maintained by the constant improvement of church buildings. The church's place in society amounts to it being the only institution that is invested in by the Gondari and entrusted with the well-being of the people. Participation in church construction is a display of fortitude that does not require heads to be raised above the ramparts.

IV

'Ethiopia' from the Outside

Introduction

Donald L. Donham

The final two chapters by Wendy James and Alessandro Triulzi consider Ethiopia from without: in James's case, from across the border with the Sudan, and in Triulzi's case from the point of view of the educated and history-writing and reading diaspora. Like Douglas Johnson's essay in *Southern Marches*, these two papers give another perspective on Ethiopia, one from outside its formal borders.

Wendy James investigates the changing relationships among a series of towns on both sides of the Sudan/Ethiopia border. Well into the 1970s, these were largely trading centres functionally related to their hinterlands, living off a commerce that flowed relatively easily back and forth across the international frontier. In imperial days, after all, the cultural focus was on the centre. Frontiers did not matter all that much.

As the Derg began its project of state-building after 1974, all this changed as the boundary itself – especially its 'integrity' – became a cause for concern. Soldiers and citizens began to 'wave the flag', intent on defining, policing, and protecting the body politic. More than this political theatre, the Ethiopian state became interventionist across the boundary, providing essential support to Sudanese opposition movements, the SPLA in particular. After the rise of the EPRDF government in Ethiopia in 1991, this support ceased for a period, but it resumed in the mid-1990s. James shows the insecurity that has resulted for local peoples, as trading centres have been turned into garrisons interspersed with refugee camps. Consequently, rural population have had to choose between sheltering with soldiers and fleeing to refugee camps – often highly politicized spaces in themselves. By the end of the 1990s, local peoples had 'no place to hide'. They had to put themselves under the protection of one local faction or another. The net effect has been a splintering of cultural experience – now everyone has a different story – and a sense of repeated discontinuities in time and place.

These themes reverberate in the final essay in our volume. The Ethiopian intellectuals whose historical work Alessandro Triulzi discusses have not, for the most part, faced the kind of bodily threat that ordinary people on the western borders have. But most have lived much of their lives in the diaspora. As noted in the beginning, the creation of new identities typically requires a recasting of history, as present narratives are projected backward in time. The beguiling notion of that things have 'always been so' appears to be a part of the essentialism that makes action in the present both more confident and meaningful.

As with most aspects of Ethiopian history, the timing of 'ethnic' formation offers interesting contrasts with the rest of Africa. In colonial Africa, 'ethnic' identities were constructed early on as local elites responded to indirect rule by foreign powers, missionary translations of the Bible, and a variety of existing pre-colonial forms of difference. The writing of local histories by grassroots intellectuals, as J.D.Y. Peel has shown for the Yoruba (1989), was a crucial part of this process. When the academic history of Africa was finally organized after World War II, it inevitably built upon these earlier accounts – to such a degree that it took scholars some time to realize that 'the Yoruba' probably did not exist when their history was first constructed. Rather, such writings themselves did much to create such a coherent identity (out of the many less inclusive, regional and clan forms).

In Ethiopia, this sequence is reversed. The academic history of Ethiopia (though largely focused upon the old core) was established first, while 'ethnic' histories of specific peoples came later. Not only this, but the conditions of dissemination for Ethiopian intellectuals writing their 'own' histories have been different as well. There is today a flourishing range of independent publications, including newspapers, books, films, and internet sites on the Horn of Africa.

What we see, then, is that ethnic ideas and the search for clear definition have appealed most to intellectuals in the diaspora, perhaps even more than to local elites in Ethiopia, and the exchange of narratives between these two has been crucial. In contrast, people on the ground have often emphasized multiple identities and indeed have sometimes intentionally pursued mixed and hybrid strategies, reaching across boundaries to form ties as forms of social insurance. In many ways, the recent past in Ethiopia can be seen as a contradiction between these two tendencies – though the perspectives of ordinary people have less often found their way into representations of Ethiopia and its history.

Fourteen

No Place to Hide
Flag Waving on the Western Frontier

WENDY JAMES

Who does not imagine Ethiopia as a powerful and magnificent nation? Its power, however, has often been a remote and enchanting image (James 1990), rather than a fact of national wealth or military might imposing itself 'on the ground'. Frontiers used not to be marked on maps, and were scarcely more visible as one passed across them. Thus, even up to the late nineteenth century, while the kings and soldiers of old Abyssinia were certainly held in awe and sometimes feared, travellers could often negotiate fairly peaceful comings and goings into and from their territory. It is true that there have been periods of international conflict in the Sudanese border region, for example in the 1880s (Triulzi 1975) and later during the Second World War. But the borders themselves were not until recently a strategic factor shaping the building of nations or the pattern of international struggles. The disturbances of earlier times experienced in the remoter parts of western Ethiopia, for example those affecting the Gumuz of Gojjam and provoking the movement across the Blue Nile into Wellegga which I described in *Southern Marches* (James 1986), were produced by the ferment of national consolidation and political struggle at Ethiopia's centre. More recent displacements have been 'caused', not only by pressures from the centre, but from the presence of the border itself and what it represents in a very modern world.

I argue here that it was only after the socialist revolution in 1974 that the image of 'powerful Ethiopia' was carved out on the ground to become, for those who lived on or near its borders, a precisely defined, fearful and everyday reality. This chapter traces some of the ways in which the Ethiopian state has since then 'remapped' conditions in the western border regions. By this I do not mean any redrawing of the frontier as such, but rather the deliberate imposing of modern projects within the farthest border zones of the country, including strategic military projects which extended across the frontiers and had major international consequences.

'Ethiopia' from the Outside

As a result, people and places in the border zone have become sharply divided. Small towns on either side of the frontier itself have become military antagonists. The rural economy upon which they used to depend has been severely disrupted, and in some cases the rural population has been completely cleaned out of its home areas. 'Ethiopia' has become a tangible force shaping the lives of all those living on its western margins, people who used to dream and talk only of distant and magical emperors. Nearly everyone has been forced to 'take sides' in very pragmatic terms, to declare their allegiance and invite their fate, sometimes in shifting ways in successive times and places over the last generation.

The western Ethiopian region traversed by tributaries of the upper Blue and White Niles once possessed a productive network of trading relations, supporting a number of flourishing market villages and towns set among such local populations as the Gumuz, Bertha, Koma, as well as the better known Nuer and Anywaa or Anuak (see Map 1.4, p. 32). In the Blue Nile hills there was a string of such places, including the well known Metemma and Fazoghli. On a southern tributary of the river, leading to the famed market of Fadasi in Ethiopia, was the small town of Geissan, now right on the international border. It was probably engaged in trade even earlier, in the eighteenth century heyday of the old Kingdom of Sennar (for a general account of Sennar's southern trade see Spaulding 1985). The Turco-Egyptian occupation of the Sudan led to more vigorous, even exploitative gold-seeking, trading and slaving in the Ethiopian foothills, and while some of the ordinary people fled further into the hilly country, the rulers of Geissan attempted to charge tolls and retain control of the trade (see Schuver's account in James et al. 1996: 23–6). One of the places in the higher country with which Geissan maintained links was a small trading centre known in the late nineteenth century as Inzing, already inhabited by several Arab traders and notorious for slaving.[1] When the modern frontier between Menilek's Ethiopia and the Anglo-Egyptian Sudan was surveyed and fixed in 1902 (Gwynn 1901), the connection between these two places was severed: Inzing (later to become the town we know as Assosa) was included with Fadasi (now Bambeshi) in Ethiopia, while the gateway town of Geissan was cut off from them on the Sudan side. What had been seen as legitimate trade was reclassified as smuggling or worse (James 1979: 34–59).

To monitor the clandestine activities which nonetheless continued across the frontier, the Anglo-Egyptian government set up a number of police posts in formerly raided areas. One of these was at Surkum, some twenty miles from the border, set up in 1906; in 1910 it was transferred to the hill known as Jebel Kurmuk, right on the line of the frontier, in order to be more effective. This post eventually became the modern Sudanese

---

[1] Schuver mentions Inzing as 'a very prosperous village as several slave-merchants live here', with extensive cultivation and active Islamic teachers. It was sufficiently powerful in the early 1880s to have gone to war with its own local overlord in Gomasha (James et al. 1996: 108–9).

administrative centre of Kurmuk, built around a parade ground with a flagpost in the middle, a self-consciously 'political' town which nevertheless developed a certain amount of formal and informal trade across the adjoining stream bed to Ethiopia. A modest counterpart market town grew up on the Ethiopian side, known to those on the Sudan side as Dul (after the nearby gold-bearing mountain). The first taste of modern war came during 1940–42, when the Italian occupying forces in Ethiopia bombed Kurmuk and nearby localities, and British and allied troops moved from the Sudan into Ethiopia as a part of the liberation of the country and the restoration of the Emperor. Local villagers remember troops digging trenches and the Swahili greeting *Jambo!* as used by soldiers of the King's African Rifles from East Africa.

Trade of the modern kind was positively fostered in the early twentieth century between Ethiopia and Sudan along the Baro/Sobat valley, fostering the growth of several towns, especially Nasir in the Sudan and Gambela in Ethiopia. These were also valued as stages on an alternative route up to the important trade centre of Fadasi in the highlands. Local communities, especially Nuer and Anywaa, used to move up and down the Sobat valley with relative ease, and without much regard for their specific Ethiopian or Sudanese citizenship (Johnson 1986). The town of Gambela was firmly in place as a trading entrepot by 1904, and because of the importance of this trade, a political arrangement with the imperial Ethiopian government was made to give Gambela and the 'Baro Salient' a special status within Ethiopia. A British consulate was put in place, and access was facilitated for traders from both sides. A rusting steamer from the British days may still be seen near the town's bridge, reinforced during the time of the Derg. Ironically, this bridge itself bore a fading remnant of Mengistu's face as late as 1993.

Pre-Derg calm: Geissan, Kurmuk & Assosa in the 1960s and early 1970s

I stayed briefly with a merchant family in Geissan in late 1965. There were then few tensions on the border. They found it amusing, rather than seriously suspicious, that I was engaged in a study of the local 'Burun' peoples. On my first visit to Kurmuk shortly after, a couple of friendly policemen invited me for an informal afternoon's visit to the shops and bars on the other side, to the Ethiopian Kurmuk. A certain amount of normal trade was passing through to Assosa in Ethiopia. At that time the 'old' Sudanese civil war was felt to be very far away in the remote south, and no one seriously envisaged the possibility of it ever reaching so far north. Who were the people of Kurmuk? There were northern Sudanese merchants belonging to families, some long settled, from Khartoum and the central Blue Nile, and some visiting merchants from across the international frontier, many of whom had intermarried with those on the

Sudanese side. There were locals from the various indigenous tribal communities in the border region, such as Bertha, 'Burun', or Uduk, working as servants or labourers, and speaking Arabic as a second language to their own. The long-time residents also included a couple of Greek merchant families, who still acted as a channel of communication to the nearby Christian church at Chali (following the missionaries' deportation in 1964). There was a considerable transient population of the national Sudanese elite – civil servants, police, teachers, health workers and so on, nearly all of whom were on temporary assignment. Most of these professionals saw posting to Kurmuk as a kind of exile. There was no army presence there or in Geissan as far as I am aware. The town was everybody's town; and it was so in a fairly comfortable sort of way.

In the early 1970s, when Alessandro Triulzi carried out his initial historical research in western Ethiopia, Assosa was also a sleepy place. He writes that it had a few hundred people only, local Jabalawi and Bertha families, a few Sudanese traders and a few Ethiopian officials, traders and teachers. A malaria eradication campaign and a military garrison were the only signs of modernity. In the main square there was one shop run by a Tigrayan, where notables gathered, and a few coffee houses patronized by the local Ethiopians (Triulzi, personal communication). I had Sudanese friends who used to visit Assosa, informally and without difficulty, on foot from the Sudan side. During imperial times (and always excepting the years of the Italian occupation), it is clear that small border towns on the western fringes of Ethiopia, from Geissan, Kurmuk and Assosa down to Gambela, were (despite a few police) very remote and provincial in feel, their nationality understated, and their names virtually unknown to important circles in the capital cities of north east Africa. This relatively open framework made for great flexibility during the first Sudanese civil war (1956–72), when both displaced civilians and supporters of the southern-based guerrilla movement (Anyanya) could seek informal shelter on the Ethiopian side. The pre-1974 imperial Ethiopian regime was not hostile to the southern cause, but neither did it openly back the guerrillas. It played, in fact, a key role in securing the Sudanese peace agreement of 1972.

With the events of 1974 this relatively calm climate in the west began to change. I had the chance to visit Ethiopia for the first time and to embark on some field research in Wellegga in 1974 and 1975. I travelled first to the areas around Dembidollo and Mugi, Gidami and Begi. It was clear that the question of permission to visit places in the lowlands or near the Sudanese border was becoming more and more complicated. While in Addis Ababa I consulted and interviewed some people from Assosa about the history and ethnography of their region, and came to realize that my questions appeared more 'political' to them than I had intended. I never managed to get to Assosa. By 1975 even Begi (just inside Assosa district) was becoming very tense, and it was clear that as a foreign researcher I would have had a struggle to get permission to continue my work. I knew

No Place to Hide: the Western Frontier

Assosa was a rather 'Sudanese' sort of place; even then, people in highland Ethiopia used to talk about Assosa as 'the land of the Arabs'. My projected ethnographic study of the border zone north of the Blue Nile, mainly of the Gumuz-speaking communities in what was then western Gojjam, then had to be abandoned because of the general security situation in 1976.

Overview: the impact of socialist Ethiopia in the west

From 1974, and particularly from the early 1980s onwards, the western border regions of Ethiopia began to feel the heavy presence of government, and the border itself began to change in character. It became in fact one of the key frontiers of the Cold War. Armed resistance in Eritrea and north-western Ethiopia against the government in Addis Ababa was supported from the Sudanese side, and Ethiopia supported the Sudan People's Liberation Army, formed in 1983 to oppose the Sudanese government from its bases in the south. The SPLA was allowed to establish itself even in the Assosa district, while Gambela became a focal point in the Ethiopian political and logistical support of the SPLA. By the late 1980s there were substantial military training centres upstream at Bonga and downstream at Bilpam. There were also very large camps for Sudanese refugees at Itang and Fugnyido ('Pinyudo' in Anywaa), and a more modest one at Dimma, close to Mizan Tefere. These were run, in practice, mainly by the local leadership of the SPLA (perceived by most as Nuer/Dinka, though this was a simplification). Massive amounts of international aid were channelled via Addis Ababa to these camps, and had the effect of undermining the local productive economy of the Sobat-Baro Valley. The road to Gambela through Nekemte was reinforced, and a new strategic airport built near the town. The region also saw the establishment of several large 'settler' schemes for famine refugees from Wello and Tigray, in the heart of Nuer and Anywaa country, well away from the actual frontier and the temptation to use it for escape. This translation to the supposedly green and pleasant south ended in disaster for many of the settlements (see Eisei Kurimoto's account in Chapter 12).

As the towns of Assosa and Gambela on the Ethiopian side, and Kurmuk and Nasir on the Sudanese side were increasingly 'politicized', they lost their chief significance as centres for regional and international trade. During the late Mengistu period, border towns on both sides were profoundly shaped by the military priorities and patronage of Addis Ababa. Both Assosa and Gambela became centres for large state projects, including agricultural and famine resettlement schemes, military support, infrastructure and facilities for Ethiopian garrisons and the SPLA, and the channelling of relief aid to Sudanese refugee camps in the vicinity. Sightings of Russians and Cubans in these places were gleefully reported in the Western press. All the small and formerly 'innocent' border towns thus found themselves on the sharp edge of the 'Cold War' and became extremely sensitive hotspots in the course of the 1980s. In response to the increased Ethiopian and SPLA military presence in Assosa and Gambela, the Sudanese government established significant garrisons at Geissan and Kurmuk. Kurmuk for most of the 1980s was in Sudanese government hands, though it was close enough to the border to be taken twice by the SPLA; further south, Nasir, well inside the Sudan, was held by the SPLA from 1989 and benefited from the effects of military and economic aid from Ethiopia's socialist regime flowing through the Gambela district.

Thus, as the socialist regime entrenched its hold on the country, it 'remapped' conditions not only on its own side of the border but also beyond it to communities living in the Sudan, especially through its support for the SPLA. From the western regions, not surprisingly, came some of the rising impetus of the late 1980s against the Addis Ababa regime, especially on the part of the TPLF and the OLF with their tacit backing from the Sudan government. Indigenous border communities (like the Uduk, Koma, and Gumuz about whom I have written, or the Anywaa portrayed by Kurimoto, or the Bertha, whose history has been studied by Triulzi) were already used to having working relations with neighbours and kinsfolk on either side of the border. As the frontier sharpened, and nationalisms began to press their claims to strategic towns, local communities like these were inevitably caught up with both pro- and anti-Derg movements or southern and northern Sudanese antagonisms, and repeatedly forced to opt for safety with one side or another – whether moving from east to west to east, or north to south to north. In their later narratives, there are moments of pride in the capture of a town, or defining fears in the memory of a massacre, or the flight from an abandoned place of former safety. No narrative maintains a steady moral story. Most of the frontier zone's surviving rural inhabitants, like Kurimoto's women informants from the Anywaa villages, have had to flee and recuperate their lives more than once, and their bewildered narratives stand as a counterpoint to the rationality of those defining nationalisms which destroyed the integrity of their former existence. Despite the change of government and the new policies of regional democracy established in 1991, the Derg period has left a heavy legacy of disruption, displacements and intercommunal fear in this border region, from the hills well north of the Blue Nile right down to the plains well south of the Baro.

Detail: Assosa & the projects of the Derg

My notes on Assosa in the Derg period which follow in this paragraph are indebted to Alessandro Triulzi's own later interviews carried out in Addis Ababa.[2] The local effects of the arrival of the *zemecha* campaigning students

[2] Triulzi interviewed his former research assistant, Atieb Dafallah, in Addis Ababa in March 1999, shortly before Atieb became very ill. He died later in the year.

No Place to Hide: the Western Frontier

in 1975 were fairly disruptive and their motives not understood, as Assosa was a trading settlement and there was no landholding of a kind which could be described as feudal or oppressive (except in Begi). But the students were everywhere, appealing to Bertha emotions and creating difficulties between them and the local 'ruling families'. The first Derg cadre-appointed administrator of Assosa, Khedir Ahmed Zayd, became dissatisfied and escaped with armed followers to the Sudan, founding in Damazin the first anti-Derg movement in western Ethiopia (Jebhah al Wataniyya, which later became the Beni Shangul Peoples' Liberation Movement). Dissidents also left Begi for the forest, at one point attacking the town and later being among those recruited into the EDU (Ethiopian Democratic Union). The predicaments of the local population of the Assosa district were exacerbated by the presence, from about 1985, of major resettlement schemes for famine victims from Tigray and Wello, known locally as 'settlers of Mengistu, the Kafir [that is, in Arabic, the pagan or unbeliever], the Communist'. There were so many settlers that no village in the area remained the same (see Fig. 1.1, p. 18). Many Bertha were forced to leave their land. Some 55,000 families were settled, mainly in 49 zones with collective farms along the higher land from the Dabus to the hills near Assosa. Churches and schools were also built. Food was grown in plentiful quantities, which was good for the locals, but there was serious deforestation. The local Bertha, who were mainly Muslim, resented the fact that the Derg was 'against God', and while they appreciated some of the development projects that were brought in, as well as new schemes for roads, schools and electricity in the town, they suspected that these were just intended to buy them off. The greatest resentment was caused when the Derg brought in soldiers of the SPLA, who behaved very badly towards the Bertha – and they are remembered as helping the Derg dismember the Nimeiry regime in the Sudan, which was assisting the ELF and EPLF in the mid-1980s. The SPLA ('mainly Dinka and Uduk', regarded by the Bertha as 'just like beasts') had training centres at Yabus, Assosa, Geissan and Kurmuk (at a place called Agubela), where the Ethiopian army was also present. They had rockets and anti-aircraft guns. The SPLA moved here and there, carried out its own operations, and came back. The Sudanese army retaliated, destroying Kurmuk and Geissan. Kurmuk (on the Ethiopian side) was a big town under the Derg, but all the hospitals, schools and the water reservoir were demolished by artillery from Sudan. The *wereda* capital had to be moved eighteen kilometres back from the border.

These events occurred in late 1987, and I return to a different version of them below. However, it should be noted here that one of the projects of the Derg in the Assosa district was to allow the UNHCR to set up a refugee camp at Tsore, about half way between Assosa town and the Ethiopian Kurmuk on the border, in about May 1987. A large number of civilians had been encouraged by the SPLA to flee to Ethiopia after their villages had been burned by the Sudanese army in reprisal for SPLA activities. In

January 1990 they had to flee from Tsore. Around this time people were in fact fleeing in both directions: Oromo refugees sought shelter in the Yabus valley on the Sudan side, and northern Ethiopians were escaping from the resettlement camps. A report from Damazin, based on interviews with escapees from Assosa conducted by the Relief Society of Tigray in January 1986, describes the brutal behaviour of armed cadres who supervised work at the camps and delivered ideological training, with the power to imprison and punish. One escapee reported that most of their time was spent building the cadres' houses, making flags, banners and boards with slogans for the cadres' use and other construction work on the site.[3] Conditions for foreign missionaries on the Ethiopian side, meanwhile, had been deteriorating, as the government sought to suppress mission Christianity (Eide 2000).

Detail: Ethiopia's entanglement with the Sudanese civil war, & the fortunes of Kurmuk

The effects of modernization in socialist Ethiopia from 1974 extended in tangible form across the border. As explained above, Assosa town was experiencing something of a boom, in that state-sponsored agricultural production schemes were being set up, mining and other investment in the region was taking place and regional security was being strengthened. 'International' implications were few to begin with. Sudanese Kurmuk was also enjoying a period of relative economic growth and prosperity, because peace had been secured within the Sudan (as a result of the Addis Ababa Agreement of 1972), trade was increasing in and through the Blue Nile region and a certain amount of modern development was taking place. The town's population certainly increased, many of the newcomers from the north playing a part in the commercial revival of the region. The presence of the government in general was upgraded not only in Kurmuk but in the smaller market villages of the area. More schools and medical facilities, and also mosques and Islamic centres, were being set up. The Sudanese army was strengthened under Nimeiri, and there were recruitment drives across the country. Certainly by the late 1970s there were army garrisons stationed at a whole string of small market villages in the southern Blue Nile, including, for example, Chali (where I had carried out my original fieldwork in the 1960s, when even during the 'old' civil war, there had only been a police station). However, with the fresh outbreak of civil war on the Sudan side in 1983, Kurmuk's fate and that of its rural dependants began to be drawn firmly into the orbit of Addis Ababa and its deteriorating relationship with Khartoum.

Very soon after the Sudanese civil war broke out again in 1983, the

---

[3] Relief Society of Tigray, 'Report on interviews conducted in Damazin Camp in the Blue Nile Province of Sudan with Tigrean refugees who have escaped from resettlement camps in south-west Ethiopia', January 1986, pp. 13, 19.

No Place to Hide: the Western Frontier

Mengistu regime made plain its support for the resurgent southern movement, which was to become the Sudan People's Liberation Army/Movement (SPLA/M). This support included allowing the SPLA to set up military training and base-camp facilities inside Ethiopia, first in the Sobat/Baro valley and then further north, in the hilly area of the Assosa district, for example at Jebel Dul, more or less facing the town of Kurmuk. During this period of the mid-1980s, security became a major concern of the Ethiopian state in all its frontier regions. Internal opposition movements began to receive substantial assistance from outside, and even ordinary people living near the frontiers (on either side) became much more conscious of their relations with the Ethiopian centre. And even well beyond the frontiers, towns and rural areas which had formerly scarcely been aware of Addis Ababa, its government policies and its international position, became conscious that they were indirectly in its orbit. From being simply 'over the border' in some country quite separate from Ethiopia, they found themselves significantly *within the periphery* of Ethiopia and directly or indirectly affected by its central policies.

The SPLA had acquired a base in the Assosa area by about 1985 and began to infiltrate the Kurmuk district and to recruit from the rural population. The Sudanese army began counter-insurgency operations against local civilians in 1986, and by early 1987 large-scale burning of villages led to the flight of a large part of the population over to the Assosa district. It so happened that plans were already afoot (presumably among western NGOs) to set up a camp in that region to give shelter and aid to Oromo refugees fleeing persecution from Mengistu's forces (Barbara Harrell-Bond, personal communication). However, given the crisis developing on the Sudan side, this plan was changed to meet the demands of the SPLA, who put pressure on the Ethiopian government and the UNHCR, and the camp was opened to cater for Sudanese refugees from the Blue Nile. The site was at Tsore, in a valley just below the old trading centre of Gomasha, and close to Dul. The people who fled at this time were almost entirely rural. Many of the townspeople from Kurmuk itself, including many merchants, had already left for the northern cities. When the SPLA mounted an attack on Kurmuk in late 1987, partly from the Ethiopian side, the town virtually emptied; most of those with northern connections still left in the town went north, while those who were mainly locals of the district joined the Assosa camp. A similar pattern followed the simultaneous taking of Kurmuk's neighbouring town, Geissan.

The Khartoum government mounted a high profile campaign, across the Middle East, to raise support for the retaking of Kurmuk and Geissan. Radio propaganda was intense (see James 2000). These places, among the first in the northern Sudan to be taken by the SPLA, were represented as cities of the Arab homeland. The retaking took place just about a month later, after (as the SPLA claimed) the rebels had decided to leave with their captured equipment and loot. The military and government presence was re-established, but I do not believe that the former town population returned in any strength. Many remained for the next couple of years in the Tsore refugee camp across the border in the Assosa district. It seems to have been at this time that a number of Sudanese families of West African origin (commonly known as 'Fellata') began to move into Kurmuk town and take over vacant houses there. Many from this community apparently were being encouraged to extend their agricultural activities from the Gezira area southwards to this more remote part of the Blue Nile from which so many of its former occupants had fled. Tension meanwhile increased between Khartoum and Addis Ababa: the former stepped up its general support for the gathering forces opposed to the Mengistu regime. For instance, it opened a refugee camp inside the Kurmuk district (at Yabus) for Oromo refugees from Ethiopia and allowed at least the relief wing of the OLF to move southwards through Kurmuk to reach it. OLF aid workers were given hospitality and assistance by the Sudanese army on their way, by the garrison at Chali, for example. In mid-1989 the SPLA sacked this camp and later that year took Kurmuk and Geissan again. Many local recruits into the SPLA took part in these attacks.

The towns were again held only for a brief period; government rhetoric was stepped up once more. Sudanese attention was also being given at this time to the security incident at Jebelein on the White Nile, where (in brief) at Christmas Shilluk labourers had challenged the authority of their Arab employers on an agricultural scheme and many had been killed in reprisal. It is possible that, having sorted this out, the security forces simply moved on further south for the retaking of Kurmuk and Geissan. This movement of Sudanese forces in early January 1990 coincided with activity by anti-Mengistu forces on the other side of the border. Various accounts state that as Kurmuk was being retaken by the Sudan government forces, the OLF, EPLF and TPLF moved in a co-ordinated way towards Assosa, in order to capture the pro-Mengistu garrison there. The SPLA base and the refugee camp at Tsore were more or less en route, and the SPLA was advised to get the refugees out of the way. A few returned to seek safety with the advancing Sudanese forces, now in charge of Kurmuk again. Most fled southwards in front of the advancing guerrillas, shells flying over their heads towards Assosa. Together with the SPLA soldiers, they then retreated southwest and back into the Sudan at Yabus, facing fire from local supporters of the OLF on the way. Soon after their arrival back in the Sudan, they were bombed and fired on by Sudanese air and land forces. Eventually, after many months, the survivors ended up in Itang, then Nasir, then Itang again and Gambela, but all that is another story.

Aftermath of the Derg: reverberations on the Sudanese border, 1991–2000

From 1991, Gambela saw the ousting of the SPLA and refugee presence and the instatement of a locally-based Anywaa-led local government in

No Place to Hide: the Western Frontier

place of the Nuer and Dinka-led local government backed by the Mengistu regime. All events in Gambela since have been intimately connected with what has been happening on the Sudan side of the border. After the fall of Mengistu and the collapse of the Ethiopian lifeline in 1991, Nasir itself was the site of a early bid by regional commanders of the SPLA for leadership of the whole movement. The failure of this bid provoked a new era of turmoil in the southern Sudan, intensified by the action of the new Ethiopian government in 1992 in allowing Sudanese troops to pass through their territory, probably including both Beni Shangul and the Gambela region, on their way to attack areas of the far south still held by the SPLA. It was about this time that Sudanese consulates were allowed to open in Assosa and in Gambela, and their officials tried among other things to tempt refugees back across the border (where agricultural labour was needed). Most preferred the relative safety of their exile in Ethiopia to the explosive situation in their home areas straddling the front line of the Sudanese civil war.

The new dispensation in 1991 for regional devolution in Ethiopia included political recognition of the far western districts. Gambela itself had constitutional recognition of a kind it had never had before, as a new Region of Ethiopia, but as Kurimoto has described in detail, it lacked political stability and economic resources. Assosa became the capital of Beni Shangul, later amalgamated with the Metekel region of Gubba north of the Blue Nile to form Region 6 of Beni Shangul and Gumuz. According to the 1994 Ethiopian census, Assosa town had a fairly modest population at 11,749 (the wider Assosa zone, one of three zones in the region, had a total of over 208,000, while the region as a whole had some 600,000). A report on the new Region for the German Development Service in 1995[4] noted that there were still 50,000 Amhara settlers remaining from the mid-1980s scheme in Pawe, Metekel, who had been given the status of 'special *wereda*'. Near Assosa there were still some 60–80,000 settlers from Wello who had representation but no special status (compare Figure 1.1, p. 18) showing the plan of their settlements). These are substantial populations, who now live in a region formally named after the small and still very vulnerable Bertha, Koma, Mao, and Gumuz peoples, but which is also the home of many Arabic- and Oromo-speaking people who used to form a trading elite, and still retain family and other links on the Sudan side of the border.

After the change of government in Ethiopia in 1991 – and given the relatively good relations between the EPRDF and the Sudan government which then prevailed – Kurmuk and Geissan, now in Sudanese government hands, were no longer facing potential hostility across the border. They were able rather to concentrate on the question of the internal war zone of the Sudan, the front line still not far away to the south. They tried

---

[4] German Development Service (DED), 'Report on a Mission to identify co-operation possibilities in Region 6 Beni Shangul and Gumuz', compiled by Klaus Schmidt, April/May, 1995.

to attract the scattered population back from the war zone. Many different groups had been caught up in the course of the long, punctuated, scattered pattern of flight from 1990 further and further south and away from 'home'. In August 1991, I found not only a cross-section of southern Sudanese stranded at Nasir in upper Nile (which then flew a UN flag; see Plate 14.1) together with a few former Ethiopian soldiers who had left Gambela with the SPLA, but also numbers of Uduk, Meban and Koma from the Kurmuk district and Bertha-speaking 'Funj' from Geissan, plus a few Hausa-speaking women and children of a nomadic group who had been cut off by skirmishes. At various times between 1991 and 1995, scattered people from the Blue Nile tried to get back 'home', to the protection of the towns and small garrisons that were once again in Sudan government hands, even when this meant crossing the frontier from Ethiopia, and/or the front line of the Sudanese civil war. Some parties were ambushed on the way, and of those who did manage to get back to places such as Chali, or even Kurmuk, the vast majority were later to be dislodged yet again over the Ethiopian border by the definitive advances of the SPLA from late 1996 onwards.

It was when diplomatic relations between the Sudanese and Ethiopian governments were damaged by a series of events culminating in 1995 that Kurmuk had to orient itself again to the legalities and politics of the international frontier. The Sudanese consulates in Assosa and Gambela were closed. The SPLA was able to reconstruct its strength with the moral and political backing of both Ethiopia and the newly independent Eritrea. In late 1996 they took Kurmuk and Geissan again, for the third time. Their success was greatly helped by the realignment of opposition parties and forces in the Sudan as a whole, including the SPLA/M, who began collaborating under the National Democratic Alliance. The combined forces proved able to hold not only Kurmuk and Geissan but also other parts of the eastern Sudan. The situation was reflected in an article by Bona Malwal, who visited Kurmuk in mid-August 1997, as a guest specifically of Commander Malik Agar, originally from the Ingessana Hills just to the north of this region. Malwal reported that people were 'already speaking of the difference between being yourself; administering yourself; and being in control of your own future rather than being ruled by "foreign" officials from Khartoum with their superiority complexes'. He continued: 'I was pleasantly surprised to discover the determination of the people in and around Kurmuk to remain free of Khartoum's administrative stranglehold.... Kurmuk has been captured and overrun by the SPLA in the past.... Back in 1987 ... the government in Khartoum whipped up a racial hysteria claiming that black African hordes from the South had overrun a Northern Arab town. Observers could have been forgiven for accepting such hysteria until they found out the truth about the people of Kurmuk. The people of Southern Blue Nile are very much a black African people and are definitely non-Arab' (Malwal 1997: 8). He explained that while many were Muslims, they did not accept the policies

No Place to Hide: the Western Frontier

Plate 14.1   The UN flag flies over a damaged Nasir, 1991 [W. James]

of the National Islamic Front. The authorities had constructed a new and elegant regional headquarters at Kurmuk, where everything had been inscribed with Koranic verses. The SPLA decided to secularize everything in the building, which now flew the SPLA/M flag. Civil servants formerly employed by the NIF regime were reported as being very content to work for the new SPLA administration. The SPLA had also taken over a local gold mine. Malwal reported that the northern merchants (*jallaba*) who used to dominate the market had all gone, but that at the same time a steady stream of refugees was entering the town from the north, from NIF-controlled areas. He also referred to thousands of internally displaced, as well as to those who had crossed the frontier again to seek asylum in the Assosa district (ibid.: 9).

At about this time, a fresh initiative was taken in setting up a refugee camp in Assosa district, this time at Sherkole, even closer to the Sudanese border than Tsore. This camp gathered in northern and southern Sudanese from a wide variety of places, including some Uduk; though the UNHCR decided in 2000 to transfer several thousand Nuer in Sherkole to the more remote camp at Dimma, where most of the other Nuer refugees in Ethiopia had been located for some time. The foothills of Kurmuk are directly visible below Sherkole. An Ethiopian army camp sits cheek by jowl with the refugees, and SPLA trucks, according to eyewitnesses, openly patrol the nearby border. The refugees themselves appear to have no difficulty in visiting Kurmuk on the Sudan side, checking on the state of their home area and on those who have decided to resettle there. There are at the time of writing no barriers in the way of those who might wish to join or rejoin the armed struggle in their own country.

Conclusion

Wilson and Donnan's 1998 book devoted to 'border identities' draws attention to the substantive as well as symbolic significance of frontiers, the 'dialectical relationships between borders and their states' and the fact that an 'anthropology of borders ... reminds social scientists ... that nations and states, and their institutions, are composed of people who cannot or should not be reduced to the images which are constructed by the state, the media or of any other groups who wish to represent them' (Wilson and Donnan 1998: 3, 4). The western frontier region of Ethiopia illustrates themes of this kind very clearly, especially in relation to the local foci of power and its strategic definition, the towns.

The commonest approach of anthropologists, aid workers and so on to the changes affecting frontier regions has been to focus on displaced communities as such and the story of their memories, movements and adaptations. In a number of papers I have myself attempted to sketch the continuing story of the displaced Uduk, originally from the southern part of the Kurmuk district and now mostly upstream of Gambela, in a refugee scheme on the site of a former SPLA camp at Bonga (James 1994, 1996, 1997; see Plate 14.2). Like other displaced communities, they have found that assistance has been forthcoming only as a result of being securely under one form or another of local patronage, sometimes literally saluting one flag rather than another. Their experience and their memory is however set against a wider history, and a different sort of landscape.

An alternative analytic focus to that of displaced people is to look at the key strategic points of local power and significance within a contested or frontier zone, the small towns over which regimes, guerrillas, global forces and local elites strive to establish control and use to extend their wealth, patronage and symbolic influence. In the present discussion, I have adopted this focus. I have shown how a series of small border towns which remained almost unknown during former times of stability became quite notorious as a set of entangled military conflicts developed on the western edge of the socialist Ethiopian state, relating both to Cold War rivalries and to the rise of the militant Islamic regime in the Sudan. The names of the towns, perhaps especially that of Kurmuk, became glorified almost as flags in themselves, in the landscape of international politics – but what happened, in this process, to the people dwelling in those towns? I have shown how little continuity there has been: one set of inhabitants may flee, to be replaced by another, while a takeover may mean many different things to different people, and to the rhetoric of different warring parties. There is no normal continuity in the history of such a place, and no normal social reality. These towns were set against their rural hinterlands in new ways: people from the countryside were co-opted into armed movements, to help attack and defend garrisons, or they fled elsewhere, to rival garrisons or to the large agglomerations of the refugee camps, which as often as not were closely under the patronage of an embattled regime or opposition movement.

Memories of the same places are at odds. A set of quite different categories and cohorts of people will be able to tell variant stories of 'Kurmuk when I was young' or 'How we liberated Geissan' or 'Gambela belongs to us'. Those who have read the papers or heard the radio or television at specific points in recent years will retain quite clear but contradictory images of who the people of Kurmuk or Geissan or Assosa are, and to whom their allegiance should be recognized. Even some of those who stayed put through the most recent change-over in Kurmuk – at least according to Malwal's report – were happy to redefine themselves and their allegiance in seemingly paradoxical ways. I

Plate 14.2 Six hundred miles on foot and several years from home: a transit camp in Ethiopia, 1994. Uduk woman preparing wild roots [W. James]

noted earlier that Kurmuk was once a relatively comfortable place socially, though undeniably a bit of a backwater. But later came the armed forces of the Sudanese government and Sudanese rebels (not to mention possible backing from Ethiopian sources) to fight over whose town it was and to whose social and cultural heritage it belonged. This development came about primarily as a result of the changing international significance of events at the Ethiopian centre and their consequences for adjoining regions of the Sudan. The connected stories of Kurmuk and our other provincial towns began with the drawing of the international frontier and the sharpening of inter-state tensions almost a century ago. There have been quiet periods since – even in the 1960s the border had a fairly open and easygoing character. From 1974 onwards, however, the spatial history of socialist Ethiopia, its subsequent restructuring from 1991 and its fateful entanglements with the continuing Sudanese civil war threw these tiny and vulnerable towns into extraordinary strategic relief and a series of social convulsions.

Thus, since 1987 Kurmuk has been taken three times by the Sudan People's Liberation Army, and at the time of writing (late 2000) it remains in their hands. Geissan, long since having lost trade as its primary reason for existence, has had a similar recent politico-military history. Assosa, having been successfully taken by the combined forces of the OLF, EPLF and TPLF in early 1990, was then retaken by Mengistu's forces, only to be lost by them again in early 1991. The symbolic importance of Assosa and Gambela grew in the story of the Ethiopian peoples' struggles against Mengistu, just as the symbolic importance of Kurmuk had grown in the rhetoric of the Sudan government and the SPLA from its first capture and retaking in 1987. Today, Assosa and Gambela are brand new regional capitals; Geissan and Kurmuk are with the forces of the Sudanese opposition, the National Democratic Alliance (encompassing the SPLA), and the Sudanese civil war goes on.

The dramatic transformations which have affected Assosa, Gambela, Geissan and Kurmuk, and the local populations who used to live fairly peaceably in them or their vicinity, have been the direct result of the rise and fall of the Ethiopian socialist state. Especially when taken by occupying soldiery, when they can draw in local militia assistance, the towns have become fearful places for the rural people. At the same time, the dispersed and plural settlements of the countryside, as well as the regular migration routes of the pastoralists, have been disrupted. For security reasons, people have been forced to gather near the garrisons (whether of the government or the opposition movements), or, if they flee, into large and unaccustomed conglomerations in the refugee camps. There has been a polarization between the forces which concentrate the people in the government garrison towns and those which concentrate them under the care of the guerrilla movements or in refugee camps. Survival outside one kind of concentration or the other becomes very problematic.

The majority of the Uduk-speaking people (the group I know best from my first ethnographic research in the 1960s) were stranded in Itang in 1990 and 1991, and with all the other Sudanese had to leave on the fall of Mengistu. They spent a year stranded near Nasir between 1991 and 1992, and subsequently fled back across the border, where the authorities in Gambela had to decide what to do about them. Most of the survivors were still (as of late 2000) in a refugee scheme upstream of the town. Many of the narratives I have heard in the course of several visits to this displaced population echo those of Kurimoto's informants. They too have been not simply victims, but at various times victors of struggles at specific times and places; and like the various other border communities, include one-time supporters of several different armed forces in the circles of their neighbours and kin. One of the most down to earth, and yet eloquently

No Place to Hide: the Western Frontier

moving, occasions I experienced was in the swampy mud encampment of Nor Deng near Nasir in 1991. A few handfuls of relief grain, plus some tiny fresh-grown fingers of okra, had been collected by the neighbourhood women from several dozen shelters made of sticks and polythene. What was this feast for? It was a memorial meeting with prayers for the young men who had died over the previous year. Died, died how? Some were sons or nephews killed while fighting for the SPLA; some while fighting for the Sudan government; and some had just died for no particular reason. The 'feast' was to remember them all.

Fifteen

Battling with the Past
New Frameworks for Ethiopian Historiography

ALESSANDRO TRIULZI

To rethink the relationship between centre and periphery in today's Ethiopia is no easy task. It involves not only a complete 'new set of paradigms' (De Waal 1994: 39) for a fast-changing country which is in the process of reshaping its basic state structure and administration, but implies a revision of commonly held assumptions, including the old centre-periphery dichotomy envisioned in *Southern Marches* (Donham and James 1986). In fact, both the Western-based 'orientalist' model of interpretation and the current mode of rewriting Ethiopia's past along 'ethnonational' lines tend to portray an 'imagined' Ethiopia which mostly reflects the 'representational' mode of cultural and ideological constructs rather than the 'investigative' one of basic research (James 1990: 101). The view from the ground upwards, as it were, forces us to recognize the crucial importance of boundary zones and the qualitative nature of boundaries, internal as well as external. Far from being 'peripheral' or marginal, they have come to be pivotal in the way that mainstream historical narratives are now challenged.

Over the past decade Ethiopia's past has been fought over and constantly redefined by a number of revisions and reformulations which have attempted to modify the prevailing historical narrative of the country's past and to adapt it to changing political situations. I will try to analyse a few examples of such revisions because they reflect the extreme complexity of writing national history in an ethnically-bound climate such as that prevailing in the country today. There are some real difficulties in creating a multi-centred inclusive narrative of the nation's past until old and new stereotypes of self-assertion and exclusion are disbanded.

Up to the mid-1970s, parallel to other African post-independence experience, one can say that the writing of history in Ethiopia was mainly if not exclusively entrusted to Ethiopian university historians who had

Battling with the Past

been trained in the western academic tradition.[1] The foreign-trained Ethiopian scholars were expected, after their period of study abroad, to piece together the results of their archival searches in a series of carefully-researched statements about the country's past.[2] Such statements would eventually fill in the first modern assessment of the historical identity of the country.

As the modern writing of the country's history was conducted at a time of intense state-building and nationalist convictions throughout the African continent, the Ethiopian history which surfaced then was on the whole state-centred and institutional in character. It tended to extoll the centralizing and unitary role of Ethiopian monarchs, and concentrated on their innovative and modernizing role within Ethiopian society.[3] Methodologically, it combined archival data coming from European archives with mostly political and religious documents such as royal correspondence, chronicles, and land-grants deriving from the rich Ethiopian Ge'ez and Amharic tradition. The 'Church and State' tradition, in so far as it focused on the development and growth of the independent and literate Christian 'nation', was on the whole indifferent, or at best patronizing, towards the non-highland, non-Christian, non-literate, and non-cultivating components of the wider Ethiopian society.[4] It is perhaps because of this that the first modern reading of Ethiopian history was patterned according to the

[1] A rough list would include Ethiopian historians of Haile Selassie I University such as Sergew Hable Selassie, Tadesse Tamrat, Merid Wolde Aregay, Berhanou Abbebe and Fisseha Zewde at the Department of History, Aleme Eshete, Bairu Tafla, and Tsehai Berhane Selassie at the Institute of Ethiopian Studies, and a few expatriates respectively teaching at the Department (Sven Rubenson, Richard Caulk, Donald Crummey) or doing research at the IES (Richard Pankhurst, Stanislav Chojnacki). They were soon followed by younger scholars such as Bahru Zewde, Hussein Ahmed, Shiferaw Bekele, Tesema Ta'a, Abdussamed Ahmed, Tekalign Wolde Mariam, etc. who joined the Department of History which soon became 'the institutional home of Ethiopian historiography' (Crummey 1990: 105). Of the early Ethiopian historians, Sergew was trained in Germany, while Berhanou and Aleme studied in France. The rest were mostly trained in England and, since the mid-1980s, in the US. A few younger historians did not return from their study abroad (notably Mohammed Hassen, Gebru Tereke, Gulumma Gemeda, Shumet Sishagne, etc.); others quit the Department of History opting for a more politically-oriented career (Fisseha Zewde did so at the time of the Derg, Adhana Haile Adhana more recently). Only one historian, the young archaeologist Ayele Terekegn, has been purged from the University in 1993 by the present Government. I am limiting my observations only to few examples here. For more details see Crummey (1990).
[2] The initial plan included Sergew for ancient history, Tadesse for the medieval period up to 1600, Merid for the seventeenth and eighteenth centuries, Caulk, Rubenson for the nineteenth century, Bahru for the twentieth. Of these, only Sergew, Tadesse, and Bahru published formal assessments of the assigned period. Caulk could not finish his major study of Menilek's time (based on his PhD dissertation of 1966) before he died in 1983, Rubenson published an important monograph on the survival of Ethiopian independence (1976) and has since edited several volumes of textual documents (1987, 1994, 2000), Merid unfortunately never published his excellent PhD thesis (1971).
[3] See Merid Wolde Aregay (1971), Sergew Hable Selassie (1972), Tadesse Tamrat (1972), Zewde Gabre Selassie (1975), Bairu Tafla (1977).
[4] Exceptions include Bahru (1976), Quirin (1977), Braukamper (1980, 1983), Triulzi (1981), Donham and James (1986), McClellan (1988), and a score of Ethiopian MA students' theses. For details see Crummey (1990). A full bibliography is in Abbink (1991).

277

traditional unitary view of the country's past extolled by the Ethiopian highland *literati* and by the time-honoured western *éthiopisant* school of philologically-trained scholars studying the 'Christian Orient'.[5]

A growing number of non-Ethiopian historians, trained in the Africanist tradition of the time, accompanied and strengthened the modernizing and independentist reading of Ethiopian history. These scholars, some of whom were active teachers in the Department of History at Haile Selassie I University,[6] and many more in western universities, studied mainly nineteenth-century Ethiopia. They focused on the growth and expansion of the modern Ethiopian state and analysed its political, economic, and social foundations.[7] It is possibly because of their modernist outlook and historical training that only a handful of these historians came from a philologically-oriented tradition and were fully cognizant of its methodologies and achievements. Most of them did not question the Great Tradition of a centralizing, independent and unitary State rooted in an ancient past and led by an innovative monarchy.

The first break with this tradition – and the first gap between university historians and historical practitioners with a more alert eye to political and social change – came from within the Ethiopian student movement, who started analysing the class structure and the 'feudal foundations' of the Ethiopian Empire, focusing on the elitist and militaristic nature of the state and on the inequalities inherent in its social structure. The student movement readily acknowledged the inequalities between the different regional and ethnic components of the Empire, but after a heated and divisive debate consigned the national question to revolutionary change and regional autonomy to be achieved within the reformed State of the future.[8]

The centralized, authoritarian, and statist nature of the dictatorial regime that followed the fall of Haile Selassie's regime gave rise to a new wave of regional and ethnic demands. While opposition to the Derg's regimentation of the Ethiopian peoples and of their rights grew in the country, assumptions concerning national identity started being questioned more vigorously than ever particularly around the Eritrean question, soon to

---

[5] See, for instance, Tekle-Tsadik Mekouria (1960–1967), Ullendorff (1960).

[6] I refer here in particular to Sven Rubenson, Richard Caulk and Donald Crummey. Richard Pankhurst taught only intermittently in the Department of History. A British historian, David Chapple, has been teaching in the Department since the early 1970s but did not concern himself with Ethiopian history until quite recently. A few more expatriate historians, among whom one can list Patrick Gilkes, Peter Garretson, Mordecai Abir, Haggai Erlich, Harold Marcus, and myself, spent considerable time doing research in Ethiopia but never really taught there any significant amount of time.

[7] See, for instance, Kofi Darkwah 1975, Harold Marcus (1973), Mordechai Abir (1968), Haggai Erlich (1982), James McCann (1986).

[8] From the beginning, the debate over the national question appeared to be a divisive one for the student movement, particularly after Walleligne Makonnen article supporting self-determination appeared in Struggle (1969). See Ottaway (1978: 117–27); Triulzi (1983: 111–27); (Balsvik 1985: 277–89); Sorenson (1991: 301-17). For a denial of self-determination in Ethiopia, see the special issue dedicated to the problems of regionalism and religion in Ethiopia which appeared in *Challenge*, 1, 1970.

be followed by the Tigrayan revolt and the Oromo liberation movement.[9] Although the regime attempted to defuse the national question through the setting up of an Institute for the Study of Ethiopian Nationalities aimed at supporting Ethiopia's multi-ethnic society[10] and, towards the end, through various administrative reforms aimed at introducing last-minute regional autonomies (Gascon 1999: 185–209), the national identity issue remained unanswered throughout this period.

A new wave of historical revisionist writings accompanied this process. Not surprisingly, the bulk of the new literature came at first from ex-HSIU students who had been sent abroad and had decided not to return to Ethiopia due to conditions in the country which they considered oppressive to their nationality or to their studies, by involved scholar-politicians forced abroad by Derg policies, and increasingly by a growingly vocal diaspora which gave voice to its anger and anxiety through a highly passionate 'public history' aimed at redressing Ethiopia's past from official State history and pan-Ethiopian ideology.

Like most reactions to oppressive conditions and denials of identity, the new historians' narratives were intrinsically ideological and emotionally-bound. The Ethiopia they portrayed was the Ethiopia they imagined or hoped for the future. The fact that few authors were able to carry on archival research or go the field increased the ideological tone of their works. The last one, to my knowledge, who was able to carry on both archival and field research was Mohammed Hassen, the best of the new Oromo historians, but as he did not return to the country he could not complete his research among the western Oromo as planned. In fact, Mohammed's case is an interesting one: he was the last of the old historical school of HSIU students to be sent to the School of Oriental and African Studies by the Department of History in Addis Ababa, and the first not to return. His research on the Oromo of Ethiopia was supervised by the then doyen of African studies at SOAS, Roland Oliver, the same scholar who helped a score of post-independence African historians to write their dissertations on their country's past. The Ethiopian students of Roland Oliver included Tadesse Tamrat and Merid Wolde Aregay, the two most prominent historians of the old Haile Selassie I University. Indeed, one could argue that Mohammed Hassen's work, *The Oromo of Ethiopia*, was to Oromo studies what Tadesse Tamrat's *Church and State in Ethiopa* was for

---

[9] The Eritrean question was the cause of the first breakaway within the student movement in 1971 on the issue whether Eritrea was to find a solution within the national confines of a reformed Ethiopia, or outside it as an ex-colonial state wanting independence. In the end the 'national question' overcame the 'colonial' one argued strenuously by the Eritrean students. The Tigray and the Oromo questions followed argumentative patterns only partly similar to the Eritrean one, as the Tigrayans rebelled against repressive policies of the central state while the Oromo saw their plight within a form of internal colonialism. By the mid-1970s both the Tigray and the Oromo Fronts had taken to the field.

[10] See Ethiopian Government, Institute for the Study of Ethiopian Nationalities 1984. Donald Donham has investigated how the ambiguous mixture of Marxism, modernism, and repression which pervaded the Derg's revolution 'let loose a veritable deluge of new ethnic discourse' in the country (1999:129).

the *éthiopisants*: a modern statement of past glory and, inevitably, a by-product of cultural nationalism.[11]

In a way, Mohammed Hassen's case stands out as a turning point between 'university history', with its stated standards of objectivity and exacting archival and field research, and 'public history' with its forceful ethos aimed at giving a sense of identity to communities born and raised in the ethnically-uproooted diaspora world. From a historiographical point of view, this meant that, right from the beginning, the rewriting of one's group's past was embedded in a strenuously defended 'moral ethnicity' which tended to isolate each community within its own cultural and linguistic bounds, while at the same time lending ideological and political overtones to the general revision of Ethiopia's past. In particular the new historiography fostered a shared 'politics' of emotion and subjectivity typical of many diaspora situations and the reappropriation in exile of cultural homelands: 'It was only after becoming refugees,' writes Mekuria Bulcha, 'that many Oromos started to feel, "see", talk, write about and sing the natural beauty of their country' (1996: 57).[12]

It is unfortunate, though perhaps inevitable, that the rift between home conditions and diaspora expectations soon created a growing distance between university scholars and the new historians abroad, each group being suspicious of the other, each easily slanting the other's arguments as being biased and trapped within the respective State, 'ethnic', or community logic. The university historians' belief in a rich historiographic tradition as they daily face the burden of teaching and researching under stress in a country 'overburdened' by its past[13] is inevitably set against the experience and perspective of the diaspora historians. Thus opposed visions of the past have persisted, invalidating all attempts to construct or even search for alternative narratives based on reasoned statements and contextualized historical arguments.

To put it in another way, memory has won over history. Throughout the 1980s and early 1990s, the reappropriation of 'ethnic' memory and the self-representation of one's own past have been perceived as the only valid test of legitimacy by the self-asserting new historiographies. Thus renewed historical narratives based on each group's collective memory have surfaced more than ever both inside and outside Ethiopian society. Inevitably, most memories were constructed along nationalist lines and most 'positionings' about the past reflected political views. Perhaps this is

---

[11] See, for instance, Tadesse Tamrat's notion of 'sabeanization' in Ethiopia as 'the crucial process of the confrontation between the culturally superior Arabian (or sabeanized) groups and the natives of the interior' to explain the growth of the Christian kingdom or nation (1972: 5–13, 19–20, 302), or Mohammed's emphasis on the 'Oromo nation' having a 'national assembly' and a 'unitary national government' before the process of 'separation and migration' which is said to have started in the 14th century (1990: 5–14). For an interesting statement of the burgeoning Tigrayan historiography, see Adhana (1994: 17–20).
[12] See also Salman Rushdie, *Imaginary Homelands* 1991, and also Sorenson (1998: 245–49).
[13] The first author to claim that Ethiopia was a country 'burdened' by its past was J. Jesman (1963: 1). The concept has been taken up recently by Gebru (1991: xiii).

Battling with the Past

why the past has become such a 'contested terrain' in today's Ethiopia and why most narrative constructions of history in the country 'tend towards a process of retrospective projection that defines the national self' (Sorenson 1993: 38).

Thus to declare oneself Amhara, Tigrayan, or Oromo as distinct from Ethiopian implied the placing of 'our people' at the centre of the historical process, and inverted a long-held hegemonic narrative of the national identity. In an area so deeply overburdened with overlapping identities such as the Horn, to 'ethnicize' friends or foes was no longer a scholarly exercise but became a major instrument of 'exclusion and inclusion' within the body politic (Schlee and Shongolo 1995: 8). This is why 'ethnic' identities and definitions are today projected 'back into the past' and the writing of history, i.e. the self-centred reading of one's own past, tends to be 'constructed only out of the elements which fit the living and the elements which do not fit are reshaped (re-interpreted, re-evaluated) until they fit or they are rejected and simply excluded from what is handed on to the next generation' (ibid.: 16).

In a way, this is nothing new. All history is written or 'told' in the 'ethnographic present' of the researcher and his or her informant's moment of narrating or performing. Yet, our own historical 'present' – the past twenty years in Ethiopia – has witnessed so many complex and traumatic events that could not but affect the very making of historical writing and analysis. Thus, researching the country's past has been perceived, and is often practised, by several authors not as a mere advance in scholarly knowledge but as a direct intervention in the country's history, and a rightful step in the direction of a brighter future. In spite of John Lonsdale's warning that the future can be a particularly 'treacherous territory' for the historian who maintains it as his 'continued imagined ideal' (1989: 126), the reading of the past in Ethiopia appears to be coercively linked to the different expectations each group is advancing for its own imagined self in the country's future.

In recent times, as in other parts of the post-communist/post-colonial world, the re-writing of the country's history has therefore oscillated between Lowenthal's definition (1985) of the past as 'a foreign country', and the parallel pronouncement of 'the past is myself' which characterizes most ethnic history. I believe much of our work, regardless of our original intentions, has not been immune from this binary opposition which confronts us today – historians or not – in a most challenging way as academic classifications are easily manipulated in conflictual situations and 'there is no way to relegate such categories back into the ivory tower' (Schlee and Shongolo 1995: 9).

At the same time, must we consider alternative historical narratives, others' histories, so 'subversive' to the original communities as to endanger their very identity and their own representations of the past? Is it still possible in such a vast region to aim at one 'official' version of the past whether at state, regional or district level, and enforce it as it was done earlier through textbooks and state ideology? Is federal Ethiopia still in need of centre-oriented national or regional histories and, if not, what are the limits of a decentralized 'federal' history in the country? If Ethiopia is to be 're-written', whose past is to be written and who is to re-write it? To such large and weighty questions I believe there are no easy answers, yet historians must confront themselves with such issues if they are to meet the present challenge facing the country. The following are a few personal observations based on my own reading and researching at the 'periphery', so to speak, of Ethiopia's two main historical narratives, Greater Ethiopia and Oromiya.

Any attempt to address the complex issue of Ethiopia's past cannot ignore both the general pressures of the times and the specfic historical context within which Ethiopia is being re-written at present. I think we all agree that the 1990s have brought a peculiar 'wind of change' which is affecting not just east-west relations but new forms of unequal development across the African continent and in particular the Horn. The fall of Communist regimes in eastern Europe has spelled the end of similar regimes in Africa. We live in a world which is often described as being post-colonial – a term which is increasingly used to express both the countries which were subjected to external colonial regimes (e.g., European colonial rule) and those which were internally subjugated by some form of colonial-type internal domination (e.g., non-Russian republics under Soviet rule; African societies subdued by a domineering and exploiting centre, etc.).

At the same time, African societies, as opposed to African States, are finding new and often vocal ways to confront state power, assert new group identities distinct from that of the State, and take an active role in the new participatory state system which is to be established in the post-colonial world. After their 'disengagement' from the State which characterized the bleak 1980s, African societies are now said to express new forms of role reappropriation and identity: history is one of them. Thus a new flurry of local, regional and 'ethnic' memories confront one another in a frantic 'war of memories' (Brossat 1990) in which everybody makes claims on everybody else disputing their territory as well as their past, all broadly and fiercely opposing the repressive official state memory. In this way, African societies in transition reinterpret their own past as part of their reappropriation of new historical identities and political claims. Never before has the continent's future conditioned the reading of its past as it does at present.

One should not be surprised therefore if Ethiopia, in many ways a 'post-colonial' State where both history and society were traditionally monitored during both the 'feudal' and the 'socialist' rule of the centre, nowadays increasingly termed as 'colonial', has witnessed in recent years repeated attempts at re-writing its past. One wonders, however, whether the current rewritings of Ethiopian history have been able in fact to contribute a more balanced reappraisal of the region's past, or whether they have only reversed, not disbanded, a paradigm of historiographical

Battling with the Past

self-assurance and cultural 'centrism' which marred the old state-centred official memory (Sorenson 1998: 240). A few examples chosen from the new historiography of dissent will make this clear.

The first element in the current re-shaping of Ethiopian history is its stress on ethnicity and subjectivity as new forms of historiographical argument. These shifts in the historiographical debate, and their parallels in public discourse, are a sign that the new historians are looking for alternative forms of legitimacy in their work. The most evident mark is the self-declared 'ethnic' background of the historian which is nowadays openly revealed in most public history as part of the objective redressal of historical wrongs. Thus Mohammed Hassen contrasts the denial of Oromo history inscribed in the Ethiopianist 'historiography of the savants' by openly declaring that he is writing 'from an Oromo point of view' (1990: 3), and Asafa Jalata aims at recovering 'the history of the largest ethnonation in the Horn of Africa' from the standpoint of someone who 'was born and raised in Oromiya' and who 'experienced the brutality of Ethiopian colonialism' (1993: ix). In contrast, Teshale Tibebu refers to his own work as 'an Amharic social history of modern Ethiopia'. As to his cultural orientation, he claims to derive it 'from the Christian Amhara cultural universe in general, and its Gojjame details, in particular. I was raised in Dabra Marqos, in a clergy-ridden extended family, drinking from the fountain of Gojjam's Amhara genius, *qene*' (Teshale 1995: xxii). Of course, the departure from impassioned orthodoxy for the new public historians and their self-declared ethnic 'bias' should not blind us to the fact that the proclaimed neutrality of the scholar in actual practice is more ideal than real: many such statements, though disguised under the term 'Ethiopian', are no less biased regionally, ideologically, or culturally.

At the same time, the openly-expressed new subjectivity highlights the very process of becoming identity-conscious which is part of the public construction of group memory; in this sense it provides new light to the tormented background against which much public history is written. Hawani Debella, a young Oromo woman in exile, reported to Aneesa Kassam what it meant to be an Oromo under the Derg:

> The first thing that I do clearly remember, was the arrest of my mother's elder brother. He was a graduate from Addis Ababa University. He was arrested at his work place. His wife, who was then pregnant with their first child, came to my mother with the bad news. That same evening, she too was arrested. Both my aunt and uncle remained in jail for over nine years and their son was born and brought up in prison. This ten year period was a nightmare for the whole family. The whole family was persecuted by the government authorities, in particular by the *qabale*. They harassed everybody related to our family. They referred to us as *zernya Galla* ('racist Galla'). Even at school, the Amhara boys and girls who knew about the stand our family had taken continually provoked and taunted us. They called my sister and me bad names. They would form themselves into two groups. One group would

'Ethiopia' from the Outside

shout 'coca cola!' The other group would point to us and cry *farii Galla* ('cowardly Galla').

As I grew older... I slowly realised that we were being treated in this way because we were Oromo. It was a strange feeling, this feeling of not being wanted. You got this feeling in the neighbourhood in which you lived; you felt it at school, you got it when you went to the market place. You felt it when you went to church and when you went to work. It was everywhere. It made you afraid of everyone, it made you doubt everyone, hate everyone, including yourself.

(Hawani Debella and Aneesa Kassam 1996: 29-30)

This common story of discrimination and social rejection rising to the feeling of 'not being wanted' in your own surroundings is important, in my view, as it outlines the feeling of not-belonging which fuels today the historiographical debate and explains its categorical nature. The traumatic experience of exile which characterizes all diaspora groups, together with the sense of losing one's identity and the difficulties of forging a new one in the country of adoption, involves a major collective experience of 'fragmentation, multiple consciousness, nostalgia, resentment and myth-making' which is at the core of today's political and cultural fundamentalism (Sorenson 1998: 247). One should not be surprised therefore if the very name Ethiopia has come to signify different things to different people. For Teshale, Ethiopia is 'the second oldest Christian polity in Africa ... a country that was Christian when the Anglo-Saxons were prostrating themselves in front of pagan idols' (ibid.: xi-xii). To opponents of the Ethiopian state, the country is said to have increasingly identified with its basic feudal 'Abyssinian' nature (Crummey 1980) and, particularly after Menilek, its ruthless 'colonial' character (Sisay Ibssa 1990). Hence the name Ethiopia is rejected by Oromo nationalist historians today as the embodiment of a foreign 'colonial' imposition which subverted the canons of Oromo civilization and cut short its national history (Asafa Jalata 1993: 8). The term Ethiopia is thus interchangeably used in lieu of Abyssinia as the name of the abusive State where a (minority) Amhara-Tigrayan elite has imposed its rule over the (majority) Oromo people:

> In this study the terms Abyssinia and Ethiopia are used interchangeably to indicate the homeland of the Amharas and the Tigrayans, and the terms Abyssinians or Habashas or Ethiopians refer to these two peoples. Although the historical meaning of Ethiopia is applicable to all black peoples, its current meaning applies mainly to the Amharas and the Tigrayans. That is why Oromo nationalists say 'We are Oromiyans, not Ethiopians', recognizing the current meaning of Ethiopia.
>
> (Asafa Jalata 1993: 6)

Thus a culturally multifarious and 'poly-ethnic' country such as present-day Ethiopia is isolated in its 'Amhara-Tigrayan' components which, though culturally and politically dominant, cannot be enucleated as such to describe the complex nature of Ethiopian society. Yet, to Oromo nationalists Abyssinia is a 'foreign land' irretrievably moulded by Amhara-Tigrayan repressive policies. At the same time, Amhara and Tigrayan historians are today at odds to show how their two regions are not equally responsible for the country's historical burden and today compete with each other for a greater role in the country's development. To Gebru Tareke, Ethiopia is an Amhara-dominated 'empire-state' which 'is neither homogeneous, nor is it a historically constituted society. It is a conglomeration of diverse ethnic, linguistic, cultural and religious groups brought together and largely dominated by force.'[14] To Adhana Haile, Ethiopia is a 'state-nation-state', i.e., a state which was gradually 'denationalized' by the Shawan-Amhara ruling elites until EPRDF forces came to power and started forging 'a new unity in diversity' throughout the country (Adhana Haile 1994: 28).

Oromo and southern historians think differently: it is the 'Amhara-Tigrayan' complex which is responsible for the basic repressive nature of the Ethiopian state epitomized by the harsh 'colonial' rule of the gun-bearer *neftennya*. In the words of Baissa Lemmu:

> With opposing cultures and values, bitter historical events and memories between them, the Oromo and the Habasha have experienced difficult relations under an empire claiming self-transformation .... Millions of Oromos believe that their nation is unfairly treated, misgoverned, robbed, exploited and oppressed first by the Amhara and currently by the Tigrayans.
>
> (Baissa Lemmu 1998: 84, 90)

That the Oromo in the past have been 'essentialized as pure negativity' (Gadaa Melba 1988b: 7), and have been considered until recently as 'Ethiopia's Other' (Sorenson 1998: 229), I personally have no doubt. At the same time, Tigrayan historians lament today that it was 'Shawan monopoly of political power' (Adhana Haile 1994: 26) which depleted their region's wealth and marginalized its status in the country, while 'Amhara intellectuals' cry out against the ethnic politics of Weyane (i.e., Tigrayan) 'factionalism' and claim they have been reduced to the rank of an oppressed minority. Yet remarkably little is said of other minority groups, such as the Sidama, the pastoral Nilotic peoples or the lowland Omotic peoples of south-western Ethiopia, and of their cultural and political marginality in the country, or the role they played in the tormented making of the nation, except for their submission to external forms of cultural and linguistic dominance, and their enforced 'assimilation' within the prevailing political cultures of the time.

Similarly, the vexed question of origins of the two 'ethnonations', Ethiopia and Oromiya, is overly simplified by turning the old theory of 'we came first' on its head, and using an upside-down version of the old paradigm of exclusion. The Oromo nationalist case is significant here both for the historiographical simplification of a highly complex historical

---

[14] Gebru (1991: 37) derives the concept from Addis Hiwet (1975). See Adhana (1994: 16).

process which is still to be fully analysed, and for the danger of reiterating old anathemas of denial to reformulate new categories of otherness in the country. The reason why the Oromo historical example is particularly relevant is due to the fact that it is currently the most forcefully argued historiography of dissent, which is not surprising as the Oromo have indeed been denied historically a proper cultural and political space in the country. What is surprising, however, is the fact that their 'counter-hegemony strategy' appears to be based on no less exclusive attitudes and shows off the same 'psychology of distrust' of which the 'Abyssinians' are often accused (Adhana Haile 1994: 9–10).

The question of foreign origins is a case in point. Ethiopian discourse has argued traditionally that the Oromo were 'foreign invaders' to the country. In order to contrast this view, and to state that the Oromo were indigenous inhabitants, recent Oromo literature rather stresses their being 'original' and 'prior' to anybody else in the region. It is no longer 'the people of historic Abyssinia, the monophysite Christian majority' which embodies 'traditional Ethiopia' (Ullendorff 1960: 174), but the aboriginal 'Cushites' who are described in the new literature as the original inhabitants of the Horn:

> The Habasha people evolved through the children of Arab immigrants and Africans in the Horn of Africa, probably in the first half of the first millennium B.C. They later differentiated mainly into the Amhara and Tigrayan peoples.
> ... Even before Arab elements immigrated to the Horn of Africa and mixed with some Cushitic groups and developed into the Abyssinians or Habasha, the region was the home of the so-called Cushitic and negroid peoples.
> (Asafa Jalata 1993: 6, 17–18).

The insistence on the Arab or Semitic element in the making of the 'Abyssinians', and on the indigenous Oromo ancestors, appears to be used here not just to establish a historical fact (the undisputed belonging of the Oromo to the Ethiopian region), but to deny the alleged 'Arab' descendants the right to declare themselves part of the same 'belonging', a shared territory, a common if troubled history. To argue that the Oromo are indigenous inhabitants needs no labelling of any other group as 'foreigner'. One is not really deconstructing the traditional vision of the Oromo as 'foreign invaders' by constructing new images of racialized differences and essentialized traits in the region. Until the 'we came first' temptation which strengthened the old mythology of traditional Ethiopia is abandoned by all, together with all claims to constrain people to alleged 'original' homes, there will not be peace in the Horn.

At the core of the present reshaping or re-interpreting of the past in Ethiopia there is a dramatic surfacing of submerged and suppressed local identities that proclaim, often violently, their rights of existence. These are not mere ideological disputes about who came first or who has a better record in the past, but crucial claims over scarce land pasture and regional resources. The conflict between the pastoral Garri and their Gabbra allies against the Boran overlords over the Moyale region (Bassi 1997), the feud beteen the Bertha and the Gumuz of Beni Shangul with the Wello settlers (Young 1999), or the present wave of ethno-regional mobilization as a 'remote control' device to influence local politics in southern Ethiopia (Abbink 1995) are all examples of new tensions between old 'centres' and new 'peripheries', and of their political implications for the present.

What must concern us at this stage, in other words, is not so much the different interpretations of a common yet contested past, but the self-assurance shown or revealed by the new historical narratives in using old epistemological and interpretive models to 'redress' past wrongs: the wiping away of a historiographical tradition of exclusion and denial by applying to its old proponents its lopsided reversal. The fact that this literature is increasingly presented to redress a historiographical imbalance, and to reflect intentionally an internal point of view (be it Oromo, Amhara or Tigrayan), is simply not enough and may in fact be quite misleading: there is indeed a need in Ethiopian studies to redress past historical narratives, but this can be achieved only through fresh and original research, and comparative studies of different cultures at work, not political or counter-ideological statements.[15]

At the root of the current polemical literature there is, of course, a changed if fluid political situation in the Horn and the sudden explosion, and manipulation, of 'ethnic' claims thought to be the best arm – and perhaps the only antidote – to fight Ethiopian 'centralism', but one fears that an ethnic over-reaction risks being self-damaging to the very cause it is allegedly propounding. The current crisis in the Horn which includes both the recent border conflict with Eritrea and the present wave of instability in the wider region with its unsurpassed record of calamities, civil wars and violence (Abbink 1997, 1998a), has given rise to the spectre of what in Somali tradition is known as *Dad Cunkii* or 'the era of cannibalism' (Abdi Samatar 1992: 625). If Ethiopia is to avoid heading in the same direction, what is urgently required is a joint attempt by all concerned to understand the underlying causes, and the individual and collective responsibilities for the potential collapse of both State apparatus and civil order which yesterday ravaged Somali society and may tomorrow befall any other country.

Yet, attempts so far to explain and represent such an epochal crisis are altogether disappointing, encumbered as they are by opposing 'national mythologies', mutual intolerance and misleading calls for 'authenticity' which are used 'as a rhetorical tactic to dismiss inconvenient interpretations.' (Sorenson 1991: 314). The result is a common deficiency of well-balanced investigations, the 'nationalization of truth' taking a heavy toll even among its avowed worshippers, the scholars.

---

[15] See Tadesse Tamrat (1982, 1988); Mohammed Hassen (1991, 1994); Schlee and Shongolo 1995; Blackhurst in Baxter *et al.* (1996).

'Ethiopia' from the Outside

As a professional body we must claim the value of understanding more of our past at a collective level without having to be constrained by calls for the 'right' native, ethnic, or political lines of discourse. To attempt to see the past 'in the round' is a service to the 'moral economy' of our respective communities, and implies no submission to States or national movements, or to our feelings of belonging to any territory or group. The battle for Ethiopia's past, in this sense, is an urgent priority for us all, Ethiopians or not, and for the building of the country's future.

Bibliography

Abbink, Jon 1991. *Ethiopian Society and History: A Bibliography of Ethiopian Studies 1957–1990*. Leiden: African Studies Centre.
— 1993. 'Famine, gold and guns: the Suri of Southwestern Ethiopia, 1985–91'. *Disasters*. 17 (3): 218–25.
— 1994. 'Refractions of revolution in Ethiopian "Surmic" societies: an analysis of cultural response' in H.G. Marcus (ed.) *New Trends in Ethiopian Studies*. Lawrenceville, NJ: Red Sea Press, vol. 2, pp. 734–55.
— 1995. 'Transformations of violence in twentieth-century Ethiopia: cultural roots, political conjunctures'. *Focaal*. 25: 57–77.
— 1997. 'Ethnicity and constitutionalism in contemporary Ethiopia'. *Journal of African Law*.
— 1998a. 'The Eritrean-Ethiopian border dispute'. *African Affairs*. 97: 551–65.
— 1998b. 'Violence and political discourse among the Chai-Suri'. In *Surmic Languages and Cultures*. Gerrit J. Dimmendaal (ed.) Cologne: Rüdiger Köppe Verlag.
— 2000a. 'Tourism and its discontents. Suri–tourist encounters in southern Ethiopia', *Social Anthropology* 8(1): 1–17.
— 2000b. 'Violence and the crisis of conciliation: Suri, Dizi and the state in Southwest Ethiopia'. *Africa* 70(4): 527–50.
Abdi Ismail Samatar. 1992. 'Destruction of state and society in Somalia: beyond the tribal convention'. *Journal of Modern African Studies*. 30 (4): 625–41.
Abebe Zegeye and Siegfried Pausewang (eds). 1994. *Ethiopia in Change: Peasantry, Nationalism and Democracy*. London and New York: British Academy Press.
Abir, Mordechai. 1968. *Ethiopia: The Era of the Princes; The Challenge of Islam and Re-Unification of the Christian Empire, 1769–1855*. New York: Praeger.
Adams, W.M. 1992. *Green Development: Environment and Sustainability in the Third World*. London: Routledge.
Addis Hiwet. 1975. *Ethiopia: From Autocracy to Revolution*. London: Review of African Political Economy, Occasional Papers, 1.
Adhana Haile Adhana. 1988. 'Peasant response to famine in Ethiopia, 1975–1985'. *Journal of Ethiopian Studies*. 21: 1–56.
— 1994. 'Mutation of statehood and contemporary politics'. In A. Zegeye and S. Pausewang (eds) *Ethiopia In Change: Peasantry, Nationalism and Democracy*. London: British Academic Press.
Africa Watch. 1991. *Evil Days: 30 Years of War and Famine in Ethiopia*.
Ahmed Hassan Omer. 1987. 'Aspects of the history of Efrata Jille (Shoa region with particular reference to 20th Century)'. BA Thesis. Addis Ababa University.
Ahmed Hassan Omer. 1994. 'A historical survey of ethnic relations in Yefat and Temmuga Awrajja, Northeast Shawa, 1889–1974'. MA Thesis. Addis Ababa University.
Alemayehu Lirenso. 1989. Research Report No. 37. *Villagization and Agricultural Production in Ethiopia*. Addis Ababa: Institute of Development Research (IDR), Addis Ababa University.
Alemneh Dejene. 1990a. 'Peasants, environment, resettlement'. In S. Pausewang (ed.) *Ethiopia: Options for Rural Development*. London: Zed Books.
— 1990b. *Environment, Famine and Politics in Ethiopia: A View from the Village*. Boulder: Lynne Reinner.
Alemseged Abbay. 1998. *Identity jilted or re-imagining identity? The divergent paths of the Eritrean and Tigrayan nationalist struggles*. Lawrenceville, NJ: Red Sea Press.
Amborn, H. 1984. 'History of events and internal development: the example of the Burji-Konso cluster'. In Tadesse Bayene (ed.) *Proceedings of the Eighth International Conference of Ethiopian Studies*. Addis Ababa.
— 1989. 'Agricultural intensification in the Burji-Konso cluster of South West Ethiopia'. *Azania*. 24: 71–83.
Appadurai, A. 1996. *Modernity at Large: Cultural Dimensions of Globalization*. Minneapolis: University of

289

Bibliography

Minnesota Press.
Asafa Jalata. (ed.). 1998. *Oromo Nationalism and the Ethiopian Discourse: The Search for Freedom and Democracy*. Lawrenceville, NJ and Asmara: The Red Sea Press.
— 1993. *Oromia and Ethiopia: State Formation and Ethnonational Conflict, 1868-1992*. Boulder and London: Lynne Rienner.
Austin, H.H. 1902. 'A journey from Omdurman to Mombasa via Lake Rudolf'. *Geographical Journal*. 11: 372-96.
Bader, Christian. 2000. 'Notes sur les Balé du sud-ouest de l'Ethiopie'. *Annales d'Ethiopie*, XVI: 107-24.
Bahru Zewde. 1976. 'Relations between Ethiopia and the Sudan on the Western Ethiopian frontier (1898-1935)'. School of Oriental and African Studies, University of London: PhD Dissertation.
— 1991. *A History of Modern Ethiopia, 1855-1974*. London: James Currey.
— 1998. *A Short History of Ethiopia and the Horn*. Addis Ababa: Commercial Printing Enterprise.
Bairu Tafla. 1977. *A Chronicle of Emperor Yohannes IV (1872-1889)*. Wiesbaden: Steiner.
Bairu Tafla (ed.) 1987. *Asma Giyorgis and His Work: History of the Galla and the Kingdom of Shawa*. Stuttgart: Franz Steiner Verlag.
Baissa Lemmu. 1998. 'Contending nationalism in the Ethiopian Empire State and the Oromo struggle for self-determination'. In Asafa Jalata. (ed.). 1998. *Oromo Nationalism and the Ethiopian Discourse: The Search for Freedom and Democracy*. Lawrenceville, NJ, and Asmara: The Red Sea Press.
Balsvick, Randi R. 1985. *Haile Sellassie's Students: The Intellectual and Social Background to a Revolution, 1952-1977*. East Lansing: Michigan State University, African Studies Center.
Bassett, T.J. and D.E. Crummey (eds). 1993. *Land in African Agrarian Systems*. Wisconsin: University of Wisconsin Press.
Bassi, Marco. 1997. 'Returnees in Moyale district, Southern Ethiopia: new means for an old interethnic game'. In R. Hogg (ed.) *Pastoralists, Ethnicity and the State in Ethiopia*. London: Haan.
Baxter, P.T.W. 1991. 'Preface'. In Joseph van de Loo (ed.). *Guji Oromo Culture in Southern Ethiopia: Religious Capabilities in Ritual and Songs*. Berlin: Dietrich Reimer Verlag.
Baxter, P.T.W., and U. Almagor. 1978. *Age, Generation and Time: Some Features of East African Age Organization*. London: C. Hurst.
Baxter, Paul, Ian Hultin and Alessandro Triulzi (eds). 1996. *Being and Becoming Oromo: Historical and Anthropological Approaches*. Uppsala: Nordiska Afrikainstitutet.
Bayart, Jean-François. 1993. *The State in Africa: The Politics of the Belly*. London: Longman.
Bender, Marvin, *et al*. 1976. *Language in Ethiopia*. London: Oxford University Press.
Berihun Teferra. 1988. 'An economic study of the villagization program in Ethiopia: the case of a village in Selale Awraja of Shewa Region'. Addis Ababa University: BA Thesis.
Beyene Doilicho. 1993. Villagization in Selected Peasant Associations in southern Shewa: Implementational Strategies and some Consequences. Addis Ababa University: Institute of Development Research, Research Report no. 41.
Blackhurst, Hector. 1996. 'Adopting ambiguous position: Oromo relationships with strangers'. In Paul Baxter, Ian Hultin and Alessandro Triulzi (eds). 1996. *Being and Becoming Oromo: Historical and Anthropological Approaches*. Uppsala: Nordiska Afrikainstitutet.
Borelli, Jules. 1890. *Ethiopie meridionale: journal de mon voyage aux pays amhara. oromo et sidama, Septembre 1885 à Novembre 1888*. Paris: Ancienne Maison Quantin.
Braukämper, Ulrich. 1980. *Geschichte der Hadiya Süd-Äthiopiens: von den Anfangen bis zur Gegenwart*. Wiesbadern: Franz Steiner Verlag.
— 1983. *Die Kambata: Geschichte und Gesellschaft eines Süd-äthiopischen Bauernvolkes*. Wiesbaden: Franz Steiner Verlag.
Brooke, J.W. 1905. 'A journey west and north of Lake Rudolf'. *Geographical Journal*. 25: 525-31.
Brossat, A. *et al.* 1990. *A l'est la mémoire retrouvé*. Paris: Seuil.
Caulk, Richard. 1966. The origins and development of the foreign policy of Menelik II, 1865-1896. University of London: PhD Dissertation.
— 1972. 'Firearms and princely power in Ethiopia in the nineteenth century'. *Journal of African History*. 13 (4): 609-30.
Cavendish, H.S.H. 1898. 'Through Somaliland and around and south of Lake Rudolf'. *Geographical Journal*. 11: 372-96.
Cayla, Fabienne. 1997. 'Ethiopie: le nouveau modèle, un réalisme ethnique?' In *L'Afrique Politique 1996*. Paris: Karthala.

Bibliography

Central Statistical Authority. 1995. *The 1994 Population and Housing Census of Ethiopia: Results for Gambella Region. Volume 1. Statistical Report*. Addis Ababa: Central Statistical Authority.
— 1996. *The 1994 Population and Housing Census of Ethiopia: Results for Oromia Region*. Volume 1, part 1. Addis Ababa: Central Statistical Authority.
Chapple, David. 1990. 'Firearms again: the battle of Asem'. *Proceedings of the Fifth Seminar of the Department of History*. Addis Ababa: Addis Ababa University.
Clapham, Christopher. 1988. *Transformation and Continuity in Revolutionary Ethiopia*, Cambridge: CUP.
— 1992. 'The structure of regional conflict in Northern Ethiopia'. In M. Twaddle (ed.) *Imperialism, the State and the Third World*. London: British Academic Press.
— 1996a. *Africa and the International System: The Politics of State Survival*. Cambridge: Cambridge University Press.
— 1996b. 'Boundary and territory in the Horn of Africa'. In P. Nugent and A. I. Asiwaju (eds) *African Boundaries: Barriers, Conduits, and Opportunities*. London: Pinter.
Cohen, Gideon. 2000. 'Identity and Opportunity: The implications of using local languages in the primary education system of the Southern Nations, Nationalities and Peoples Region (SNNPR), Ethiopia.' PhD Thesis, University of London.
Cohen, John M., and Nils-Ivar Isaksson. 1987. *Villagization in the Arsi Region of Ethiopia*. Uppsala: Swedish University of Applied Sciences.
Conti Rossini, Carlo. 1929. *L'Abissinia*. Rome: Cremonese Editore.
Crummey, Donald. 1980. 'Abyssinian feudalism'. *Past and Present*. 89: 115–38.
— 1987. 'Some precursors of Addis Ababa: Towns in Christian Ethiopia in the eighteenth and nineteenth centuries.' *Proceedings of the International Symposium on the Centenary of Addis Ababa, November 24–25, 1986*, 9–31.
— 1990. 'Society, state and nationality in the recent historiography of Ethiopia'. *Journal of African History*. 31: 103–19.
Dakhlia, Jocelyn. 1992. 'L'historien pris au piège de la mémoire?'. In Bogumil Jewsievicki and Jean Létourneau (eds). *Constructions identitaires*. Quebec: Université Laval, Actes du Célat 6:73–83.
Darkwah, Kofi. 1975. *Shewa, Menelik and the Ethiopian Empire 1813–1889*. London: Heinemann.
Darley, H.A.C. 1969. *Slaves and ivory in Abyssinia: A Record of Adventure and Exploration Among the Ethiopian Slave-Raiders*. New York: Negro Universities Press.
Deguchi, Akira. 1996. 'Rainbow-like hierarchy: Dizi social organization'. In Sato, S. and Kurimoto, E. (eds) *Essays in Northeast African Studies*. Senri Ethnological Studies no.43. Osaka: National Museum of Ethnology.
DeMars, William. 1994. 'Tactics of protection: international human rights organizations in the Ethiopian conflict, 1980–1986'. In G. Shepherd *et al.* (eds) *Africa, Human Rights and the Global System*. Westport, CN: Greenwood Press.
Dessalegn Rahmato. 1985. *Agrarian Reform in Ethiopia*. Trenton, NJ: Red Sea Press.
— 1987. 'The political economy of development in Ethiopia'. In E. Keller and D. Rothchild (eds) *Afro-Marxist Regimes: Ideology and Public Policy*. Boulder: Lynne Rienner.
— 1988. 'Settlement and resettlement in Mettekel, Western Ethiopia'. *Africa* (Rome) 43: 14–43.
De Waal, A. 1994. 'Rethinking Ethiopia'. In C. Gurdon (ed.) *The Horn of Africa*. SOAS–GRC Geopolitical Series. London: UCL Press.
Dilebo, Getahun L. 1974. 'Emperor Menelik's Ethiopia, 1865–1916: National Unification or Amhara Communal Domination?' Howard University, Washington: PhD Dissertation.
Dimmendaal, Gerrit J. and Marco Last (eds) 1998. *Surmic Languages and Cultures*. Cologne: R. Köppe Verlag.
Donham, Donald L. 1986a. 'Old Abyssinia and the new Ethiopian empire: themes in social history'. In D. Donham and W. James (eds) *The Southern Marches of Imperial Ethiopia: Essays in History and Social Anthropology*. Cambridge: Cambridge University Press.
— 1986b.'From ritual kings to Ethiopian landlords in Maale'. In D. Donham and W. James (eds) *The Southern Marches of Imperial Ethiopia: Essays in History and Social Anthropology*. Cambridge: Cambridge University Press.
— 1999. *Marxist Modern: An Ethnographic History of the Ethiopian Revolution*. Berkeley, CA/Oxford: University of California Press/James Currey.
Donham, Donald L. and Wendy James (eds) 1986. *The Southern Marches of Imperial Ethiopia: Essays in History and Social Anthropology*. Cambridge: Cambridge University Press.
Duffield, Mark, and John Prendergast. 1994. *Without Troops and Tanks: Humanitarian Intervention in*

Bibliography

Ethiopia and Eritrea. Lawrenceville, NJ: Red Sea Press.
Eide, Øyvind M. 2000. *Revolution and Religion in Ethiopia: The Ethiopian Evangelical Church Mekane Yesus under the Dergue, 1974–85*. Oxford: James Currey.
Ergas, Z. 1980. 'Why did the Ujamaa village policy fail? Towards a global analysis'. *Journal of Modern African Studies*. 18 (3): 387–410.
Erlich, Haggai. 1982. *Ethiopia and Eritrea During the Scramble for Africa: A Political Biography of Ras Alula*. East Lansing.
Ethiopian Government, Institute for the Study of Ethiopian Nationalities. 1984. Excerpts from Documents of the Founding Congress of WPE on The People's Democratic State/The nationalities question. Mimeo.
— 1995. *Federal Negarit Gazeta*, 1, August 1995.
Evans-Pritchard, E.E. 1940. *The Political System of the Anuak of the Anglo-Egyptian Sudan*. New York: AMS Press (reprinted 1977).
— 1947. 'Further observations on the political system of the Anuak', *Sudan Notes and Records* 28: 62–97.
Fantahun Tiruneh. 1990. *The Ethiopian Students: Their Struggle to Articulate the Ethiopian Revolution*. Chicago: Nyala Type Publishers.
Fardon, Richard. 1996. 'Crossed Destinies: The Entangled Histories of West African Ethnic and National Identities'. In Louise de la Gorgendiere, Kenneth King, and Sarah Vaughan (eds) *Ethnicity in Africa: Roots, Meanings and Implications*. Edinburgh: Centre of African Studies, University of Edinburgh.
Fekade Azeze. 1998. *Unheard Voices*. Addis Ababa: Addis Ababa University Press.
Fortmann, L. 1980. *Peasants, Officials and Participation in Rural Tanzania: Experience of Villagization and Decentralization*. New York: Cornell University Press.
Freeman, D. 1999. 'Transforming traditions: the dynamics of cultural variation in the Gamo highlands, Southwest Ethiopia'. PhD Thesis, London School of Economics.
Fukui, Katsuyoshi. 1994. 'Conflict and ethnic interaction: the Mela and their neighbours'. In Fukui, K. and J. Markakis (eds) *Ethnicity and Conflict in the Horn of Africa*. London: James Currey.
Fukui, Katsuyoshi, and David Turton. 1979. *Warfare among East African Herders*. Senri Ethnological Studies no.3. Osaka: National Museum of Ethnology.
Fukui, Katsuyoshi, and John Markakis (eds) 1994. *Ethnicity and Conflict in the Horn of Africa*. London: James Currey.
Fukui, Katsuyoshi, Eisei Kurimoto, and Masayoshi Shigeta. 1997. *Ethiopia in Broader Perspective*. Papers of the XIIIth International Conference of Ethiopian Studies (3 vols). Kyoto: Shokado Book Sellers.
Futterknecht, C. 1997. 'Diary of a drought: the Borana of Southern Ethiopia, 1990–1993'. In R. Hogg (ed.) *Pastoralists, Ethnicity and the State in Ethiopia*. London: Haan.
Gabra Sellasé. 1959. *Tarika Zämän Zä Daggmawi Menelik*. Addis Ababa: Berhanenna Salam Printing Press.
Gadaa Melba. 1988a. *Ethiopia: An Introduction*. Khartoum.
Gadaa Melba. 1988b. *Oromia: An Introduction*. Khartoum.
Gamaledin, M. 1993. 'The decline of Afar pastoralism'. In J. Markakis (ed.) *Conflict and the Decline of Pastoralism in the Horn of Africa*. Basingstoke: Macmillan.
Garretson, Peter P. 1986. 'Vicious cycles: ivory, slaves and arms on the new Maji frontier'. In D. Donham and W. James (eds) *The Southern Marches of Imperial Ethiopia: Essays in History and Social Anthropology*. Cambridge: Cambridge University Press.
Gascon, Alain. 1999. 'Partager une terre sainte: Erythrée unitaire, Ethiopie fédérale'. In J. Bennemaison et al. (eds) *La Nation et le Territoire: Le Territoire, Lien ou Frontière, Tome 2*. Paris: L'Harmattan.
Gebru Tareke. 1984. 'Peasant resistance in Ethiopia: The case of Weyane'. *Journal of African History*. 25: 77–92.
— 1990. 'Continuity and discontinuity in peasant mobilization: the case of Bale and Tigray'. In M. Ottaway (ed.) *The Political Economy of Ethiopia*. New York: Praeger.
— 1991. *Ethiopia: Power and Protest. Peasant Revolts in the Twentieth Century*. Cambridge: Cambridge University Press.
Geertz, Clifford. 1973. *The Interpretation of Cultures*. New York: Basic Books.
Gerdesmeier, Volker. 1995. '"We are fighting each other now, but in the battlefield we were

Bibliography

friends.'" Gewalt, gewaltmonopol und ethnizität in Südwestäthiopien: gründe für eskalierende konflikte in der region'. In P. Bräunlein and A. Lauser (eds) *Krieg und Frieden: Ethnologische Perspektiven*. Bremen: Kea Edition.

Getu, M. 1997. 'Local versus outsider forms of natural resource management: the Tsamako experience in Southwest Ethiopia'. In Fukui, K., Kurimoto, E., and Shigeta, M. (eds) *Ethiopia in Broader Perspective: Papers of the 13th International Conference of Ethiopian Studies*. Kyoto: Shokado Booksellers.

Gezahegn, Petros. 1994. 'The Karo of the Lower Omo valley: subsistence, social organization and relation with neighbouring groups'. MA Thesis, Addis Ababa University.

Giddens, Anthony. 1990. *The Consequences of Modernity*. Cambridge: Polity Press.

Gifford, Paul (ed.) 1995. *The Christian Churches and The Democratisation of Africa*. Leiden, New York: E.J. Brill.

— 1998. *African Christianity: Its Public Role*. Bloomington and Indianapolis: Indiana University Press.

Gilkes, P. 1991. Review Article. 'Eritrea: historiography and mythology'. *African Affairs*. 90: 623-8.

Gufu Oba. 1998. 'Assessment of indigenous range management knowledge of the Booran pastoralists of Southern Ethiopia'. GTZ Borana Lowland Pastoral Development Programme in Co-operation with Oromiya Regional Bureau for Agricultural Development, Consultancy Report.

Gwynn, C.W. 1901. 'Surveys on the proposed Sudan-Abyssinian frontier'. *Geographical Journal*. 18: 562-73.

Haile Kiros Desta. 1992. 'Health activities of Save the Children Federation (USA) in Yefat and Temmug Awrajja'. Addis Ababa University: BA Thesis.

Hallpike, C.R. 1972. *The Konso of Ethiopia: The Study of the Values of a Cushitic People*. Oxford: Oxford University Press.

Hammond, Jenny. 1989. *Sweeter Than Honey: Testimonies of Tigrayan Women*. Oxford: Third World First.

— 1999. *Fire From the Ashes: A Chronicle of Revolution in Tigray, Ethiopia, 1975-1991*. New Jersey: Red Sea Press.

Harbeson, J.W. 1988. *The Ethiopian Transformation: The Quest for the Post Imperial State*. Boulder and London: Westview Press.

Harrison, J.J. 1901. 'A journey from Zeila to Lake Rudolf'. *Geographical Journal*. 18: 258-75.

Hawani Debella and Aneesa Kasam. 1996. 'Hawani's story'. In P.T.W. Baxter, J. Hultin and A. Triulzi (eds) *Being and Becoming Oromo: Historical and Anthropological Enquiries*. Trenton, NJ: Red Sea Press.

Helland, J. 1996. 'The political viability of Boorana pastoralism'. In P.T.W. Baxter, J. Hultin and A. Triulzi (eds) *Being and Becoming Oromo: Historical and Anthropological Enquiries*. Trenton, NJ: Red Sea Press.

Helland, J. 1998. 'Institutional erosion in the drylands: the case of the Borana pastoralists'. *EASSREA*. 14 (2).

Hendrie, Barbara. 1994. 'Relief behind the lines the cross-border operation in Tigray'. In J. Macrae and A. Zwi (eds) *War and Hunger: Rethinking International Responses to Complex Emergencies*. London: Zed Books.

Herbst, Jeffrey. 2000. *States and Power in Africa*.

Hill, G.B., and Powell, L.F. 1950. *Boswell's Life of Johnson*. Volume 5. (Tour of the Hebrides). Oxford: Clarendon Press.

Hinnant, J.T. 1977. 'The Gada system of the Guji of Southern Ethiopia'. University of Chicago: PhD Dissertation.

— 1985. 'The position of women in Guji Oromo society'. In Tadesse Bayene (ed.) *Proceedings of the Eighth International Conference on Ethiopian Studies*. Addis Ababa.

Hodson, A.W. 1927. *Seven Years in Southern Abyssinia*. London: T.F. Unwin.

— Undated. *Where Lions Reign: An Account of Lion Hunting and Exploration in South West Abyssinia*. London: Skeffington & Son Ltd.

Hoekstra, H.T. 1995. *Honey, We're Going to Africa!* Mukilteo: Wine Press Publishing.

Hogg, Richard (ed.) 1997. *Pastoralists, Ethnicity and the State in Ethiopia*. London: Haan.

Holcomb, Bonnie, and Sisay Ibbsa. 1990. *The Invention of Ethiopia*. Trenton, NJ: Red Sea Press.

Homi Bhabha. 1990. 'The third space: an interview with Homi Bhabha'. In Jonathan Rutherford (ed.) *Identity: Community, Culture, Difference*. London: Lawrence and Wishart.

Huntingford, G.W.B. (ed.), 1980. *The Periplus of the Erythrean Sea*. London: The Hakluyt Society.

Hyden, G. 1980. *Beyond Ujamma in Tanzania: Underdevelopment and Uncaptured Peasantry*. Berkeley:

Bibliography

University of California.
Iliffe, John. 1995. *Africans: The History of a Continent*. Cambridge: Cambridge University Press.
Inquai, Solomon. 1983–4. 'Revolution in Tigray: a historical background'. *Horn of Africa Journal*. 6 (4).
Jam Jam *Awraja* Villagization Co-ordinating Council. 1988. 'Villagization plan for 1988'. Kibre-Mengist.
James, Wendy. 1979. *'Kwanim Pa: the making of the Uduk People. An ethnographic study of survival in the Sudan-Ethiopian borderlands*. Oxford: Clarendon Press.
— 1990. 'Kings, commoners, and the ethnographic imagination in Sudan and Ethiopia'. In R. Fardon (ed.) *Localizing Strategies: Regional Traditions of Ethnographic Writing*. Edinburgh and Washington: Scottish Academic Press and Smithsonian Institution Press.
— 1994. 'Civil war and ethnic visibility: the Uduk of the Sudan-Ethiopia border'. In Fukui, K. and J. Markakis (eds) *Ethnicity and Conflict in the Horn of Africa*. London: James Currey.
— 1996. 'Uduk resettlement: dreams and realities'. In T. Allen (ed.) *In Search of Cool Ground: War, Flight and Homecoming in Northeast Africa*. London: James Currey.
— 1997. 'The names of fear: history, memory and the ethnography of feeling among Uduk refugees'. *Journal of the Royal Anthropological Institute* (n.s.) 3: 115–31.
— 2000. 'The multiple voices of Sudanese airspace'. In R. Fardon and G. Furniss (eds) *African Broadcast Cultures: Radio in Transition*. Oxford: James Currey.
James, Wendy, Gerd Baumann, and Douglas H. Johnson (eds) 1996. *Juan Maria Schuver's Travels in North East Africa, 1880–83*. London: The Hakluyt Society.
Jensen, Adolf (ed.). 1959. *Altvolken Süd-Äthiopien: Ergebnisse der Frobenius-Expeditionen 1950–52 und 1954–56*. Stuttgart: Kohlhammer Verlag.
Jesman, Czeslaw. 1963. *The Ethiopian Paradox*. London, New York and Nairobi: Oxford University Press.
Jewsiewicki, Bogumil and V.Y. Mudimbe. 1993. 'Africans' memories and contemporary history of Africa'. *History and Theory*. 32: 1–11.
Jewsiewicki, Bogumil, and David Newbury. 1986. *African Historiography: What History for Which Africa*. Beverly Hills: Sage.
Johnson, Douglas H. 1986. 'On the Nilotic frontier: Imperial Ethiopia in the Southern Sudan, 1898–1936'. In D. Donham and W. James (eds) *The Southern Marches of Imperial Ethiopia: Essays in History and Social Anthropology*. Cambridge: Cambridge University Press.
— Forthcoming. *The Root Causes of Sudan's Civil Wars*. Oxford: James Currey.
Key, G. 1967. *Social Aspects of Village Regrouping in Zambia*. Lusaka: University of Zambia, Institute for Social Research.
Kidane Mengisteab. 1997. 'New approaches to state building in Africa: the case of Ethiopia's ethnic-based federalism'. *African Studies Review*. 40 (3): 111–32.
Kiflu Taddesse. 1985. *Ya Tewled*. Volume 1. Addis Ababa: Reprinted by Bole Printing Organization.
— 1998. *The Generation, The History of the Ethiopian People's Revolutionary Party, Vol. II*. Lanhem, NY and Oxford: University Press of America.
Kurimoto, Eisei. 1992. 'Natives and outsiders: the historical experience of the Anywaa of Western Ethiopia'. *Journal of Asian and African Studies*. 43: 1–43.
— 1994. 'Inter-ethnic relations of the Anywaa (Anuak) in Western Ethiopia: with special reference to the Majangu'. In H. Marcus (ed.) *New Trends in Ethiopian Studies*. Volume 2. Lawrenceville, NJ: Red Sea Press.
— 1996a. *People Living Through Ethnic Conflicts*. Kyoto: Sekaishisosha. (Japanese).
— 1996b. 'People of the river: subsistence economy of the Anywaa (Anuak) of Western Ethiopia'. In S. Sato and E. Kurimoto (eds) *Essays in Northeast African Studies*. Senri Ethnological Studies no. 43. Osaka: National Museum of Ethnology.
— 1997. 'Politicization of ethnicity in Gambella'. In K. Fukui, E. Kurimoto and M. Shigeta (eds) *Ethiopia in Broader Perspective: Papers from the 13th International Conference of Ethiopian Studies, Vol. 2*. Kyoto: Shokado Booksellers.
Lamphear, John. 1992. *The Scattering Time: Turkana Responses to Colonial Rule*. Oxford: Clarendon Press.
Le Bon, Gustav. 1960. *The Crowd*. New York: Viking Press.
Levine, Donald. 1965. *Wax and Gold: Tradition and Innovation in Ethiopia*. Chicago: The University of Chicago Press.

Bibliography

Lienhardt, Godfrey. 1957. 'Anuak village headmen I'. *Africa*. 27: 341–55.
— 1958. 'Anuak village headmen II'. *Africa*. 28: 23–36.
— 1962. 'The situation of death: an aspect of Anuak philosophy'. *Anthropological Quarterly*. Vol. 35: 74–85.
Lonsdale, J. 1989. 'African pasts in Africa's future'. *Canadian Journal of African Studies*. 23 (1): 126–46.
Lowenthal, David. 1985. *The Past is a Foreign Country*. Cambridge: Cambridge University Press.
Lydall, J., and I. Strecker. 1979. *The Hamar of Southern Ethiopia: Baldambe Explains, Volume 2*. Hohenschäftlarn: Klaus Renner.
Mahtama Sellassie Wäldä Masqäl. 1965. *Hilqu Tıwlid Zä Negus Sahlä Sellasé*. Addis Ababa: Artistic Printers.
Malwal, Bona. 1997. 'In Kurmuk the thoughts are all about defeating the NIF regime'. *Sudan Democratic Gazette*. VIII, 88 (September): 8–9.
Marcus, Harold. 1973. *The Life and Times of Menelik II: Ethiopia 1844–1913*. Oxford: Clarendon Press.
— 1994. *A History of Ethiopia*. London and Berkeley: University of California Press.
Mathur, H.M. 1995. *Development, Displacement and Resettlement: Focus on Asian Experiences*. New Delhi: Vikas Publishing House.
Matsuda, Hiroshi. 1994. 'Annexation and assimilation: Koegu and their neighbours'. In Fukui, K. and J. Markakis (eds) *Ethnicity and Conflict in the Horn of Africa*. London: James Currey.
— 1997. 'How guns change the Muguji: ethnic identity and armament in a periphery'. In Fukui, K., Kurimoto, E., and Shigeta, M. (eds) *Ethiopia in Broader Perspective*. Volume 2. Kyoto: Shokado Book Sellers.
Maud, C.P. 1904. 'Exploration in the southern borderland of Abyssinia'. *Geographical Journal*. 23: 252–79.
Mbembe Achille. 1988. *Afriques indociles*. Paris, Karthala.
McCann, James. 1986. *From Poverty to Famine in Northeastern Ethiopia: A Rural History, 1900–1935*. Philadelphia: Philadelphia University Press.
McClellan, Charles. 1988. *State Transformations and National Integration: Gedeo and the Ethiopian Empire, 1895– 1935*. East Lansing: Michigan State University, African Studies Centre.
McHenry, D.E. 1976. 'The Ujamaa village in Tanzania: a comparison with Chinese, Soviet Union and Mexican experiences in collectivization'. *Comparative Studies in Society and History*. 18 (3): 347–70.
Medhane Tadesse. 1999. *The Eritrean-Ethiopian War: Prospect and Retrospect. Reflections on the Making of Conflicts in the Horn of Africa, 1991–1998*. Addis Ababa.
Mehtema Selassie Welde Masqel. 1965 E.C. 'Hilqu tewlid ze Negus Sahle Selassie'. Addis Ababa
Mekuria Bulcha. 1992. 'History and political culture versus Ethiopia's territorial integrity'. *Oromo Commentary*. 2 (1): 3–12.
— 1996. 'The survival and reconstruction of Oromo national identity'. In P.T.W. Baxter, J. Hultin and A. Triulzi (eds) *Being and Becoming Oromo: Historical and Anthropological Enquiries*. Trenton, NJ: Red Sea Press.
— 1997. 'Conquest and forced migration: An assessment of Oromo experience'. In Seyoum Y. Hameso, Trevor Trueman, Temesgen M. Erena (eds) 1998. *Ethiopia. Conquest and the Quest for Freedom and Democracy*, London: TSC Publications.
Mengistu H. Mariam. 1986. 'Report to the Central Committee of the Workers' Party of Ethiopia'. April 14, 1986.
Merid Wolde Aregai. 1971. 'Southern Ethiopia and the Christian Kingdom, 1508–1708, with special reference to the Galla migrations and their consequences'. University of London: PhD Dissertation.
— 1980. 'A reappraisal of the impact of firearms in the history of warfare in Ethiopia (c.1500–1800)'. *Journal of Ethiopian Studies*. 14: 98–121.
Messeret Lejebo. 1990. 'A study of the technological, social, socio-economic and environmental constraints affecting the performance of Ethiopian agriculture: with special reference to drought and famine crises in Konso'. *Giessener Schriften zur Wirtschafts und Regionalsoziologie*. Heft 18 Giessen.
Miers, S. 1997. 'Britain and the suppression of slavery in Ethiopia'. *Slavery and Abolition* 18: 257–88.
Ministry of Agriculture. 1986. *Villagization Guidelines*. Addis Ababa: Ministry of Agriculture. (in Amharic).
Miyawaki, Y. 1996. 'Maintaining continuity in a dualistic world: symbolism of the age grade succession rituals among the Hoor (Arbore) of Southwestern Ethiopia'. *Nilo-Ethiopian Studies*. 3

Bibliography

(4): 39–65.
— No date. 'Stop the development projects that destroy the life of local peoples'. Unpublished ms.
Mohammed Hassen. 1990. *The Oromo of Ethiopia: A History 1570–1860*. Cambridge: Cambridge University Press.
— 1991. 'Traditional methods of conflict resolution among the Oromo'. *Oromo Commentary*. 1 (1): 17–21.
— 1994. 'Some aspects of Oromo history that have been misunderstood'. *Journal of Oromo Studies*. 2: 77–90.
— 1996. 'The development of Oromo nationalism'. In P.T.W. Baxter, J. Hultin and A. Triulzi (eds) *Being and Becoming Oromo: Historical and Anthropological Enquiries*. Trenton, NJ: Red Sea Press.
National Villagization Co-ordinating Committee. 1987. *Rural Transformation*. Addis Ababa: NVCC.
Naty, Alexander. 1992. 'The culture of powerlessness and the spirit of rebellion among the Aari people of southwest Ethiopia'. Stanford University, Department of Anthropology: PhD Dissertation.
Naty, Alexander. 1994a. 'From independent chiefdoms to Abyssinian subjects: the Aari interpretation of conquest and colonization'. *Africa* (Rome) 49 (4): 498–515.
— 1994b. 'The thief-searching (*Leba Sha*) institution in Aariland, Southwest Ethiopia, 1890s–1930s'. *Ethnology* 33 (3): 261–72.
New York Times. January 6, 1985.
Nugent, Paul, and A.I. Asiwaju (eds) 1996. *African Boundaries: Barriers, Conduits and Opportunities*. London: Pinter.
Oberai, A.S. 1992. 'An overview of settlement policies and programs'. In P. Dieci and C. Viezzoli (eds) *Resettlement and Rural Development in Ethiopia*. Milan: Franco Angeli.
Omari, C.K. 1984. 'Ujamma policy and rural development in perspective'. In C.K. Omari (ed.) *Towards Rural Development in Tanzania*. Arusha: East African Publications.
Ottaway, Marina. 1995. 'The Ethiopian transition: democratization or new authoritarianism?' *Northeast African Studies* (n.s.) 3 (2): 67–84.
Ottaway, Marina. 1999. *Africa's New Leaders: Democracy or State Reconstruction?* Washington: Carnegie Endowment for International Peace.
Ottaway, Marina, and David Ottaway. 1978. *Ethiopia: Empire in Revolution*, New York: Africana Publishing Co.
Pankhurst, Alula. 1989. 'Settling for a new world: people and the state in an Ethiopian resettlement village'. University of Manchester: PhD Dissertation.
— 1991. 'People on the move: settlers leaving Ethiopian resettlement villages'. *Disasters*. 15 (1): 61–7.
— 1992. *Resettlement and Famine in Ethiopia: The Villagers' Experience*. Manchester: Manchester University Press.
— 1994. 'Responses to resettlement: household, marriage and divorce'. In C. Lepage (ed.) *Études Ethiopiennes: Actes de la Xe conférence internationale des Études Ethiopiennes, Paris 24–28 Août 1988*. Paris: Societé Française pour les Études Ethiopiennes.
— 1995. 'Negotiating tradition and change: the impact of the Constitution of the People's Democratic Republic of Ethiopia on marriage and divorce in a settler village in Western Ethiopia'. In E. Grande (ed.) *Transplants, Innovation and Legal Tradition in the Horn of Africa*. Turin: L'Harmattan.
— 1997. 'When the center relocates the periphery: resettlement during the Derg'. In Fukui, K., Kurimoto, E., and Shigeta, M. (eds) *Ethiopia in Broader Perspective: Papers of the 13th International Conference of Ethiopian Studies*. Kyoto: Shokado Book Sellers.
Pankhurst, Alula, and Tom Hockley. 1999. 'Resettlement revisited: field visit to the Qeto resettlement area in Western Wellega, January 25–29, 1999'. Report submitted to DFID, Addis Ababa.
Pankhurst, Helen. 1989. 'Villagization in a peasant association in Menz'. Addis Ababa: unpublished manuscript.
— 1992. *Gender, Development and Identity: An Ethiopian Study*. London: Atlantic Highland.
Pankhurst, Richard. 1990. *A Social History of Ethiopia: the Northern and Central Highlands From Early Medieval Times to the Rise of Emperor Tèwodros II*. Addis Ababa: Institute of Ethiopian Studies, Addis Ababa University.
— 1997. *The Ethiopian Borderlands: Essays in Regional History from Ancient Times to the End of the 18th Century*. Lawrenceville, NJ: Red Sea Press.
Parpart, J.L. 1995. 'Post-modernism, gender and development'. In J. Crush (ed.) *Power of Development*. London: Routledge.

Bibliography

Peel, J.D.Y. 1989. 'The Cultural Work of Yoruba Ethnogenesis'. In Elizabeth Tonkin, Maryon McDonald, and Malcolm Chapman (eds) *History and Ethnicity*. ASA Monography No. 27. London: Routledge.
Phillipson, David W. 1998. *Ancient Ethiopia: Aksum, its antecedents and successors*. London: British Museum Press.
Quirin, James. 1977. The Beta Israel (Felasha) in Ethiopian History: caste formation and culture change, 1270–1868. University of Minnesota: PhD Dissertation.
Ranger, T.O. 1976. 'Towards a usable African past'. In C. Fyfe (ed.) *African Studies Since 1945: A Tribute to Basil Davidson*. London: Longman.
— 1999. 'Concluding Comments'. In Paris Yeros (ed.) *Ethnicity and Nationalism in Africa*. London: Macmillan Press.
Ricci, M. 1945. 'Usanze matrimoniali, etica sessuale e credenze degli Arbore, degli Amar e dei Gheleba'. In *Studi Etiopici Raccolti da C. Conti Rossini*. Rome: Instituto per L'Oriente.
Richards, P. 1985. *Indigenous Agricultural Revolution: Ecology and Food Production in West Africa*. London: Unwin Hyman.
Rubenson, Sven. 1976. *The Survival of Ethiopian Independence*. London: Heineman.
— 1991. 'Conflict and environmental stress in Ethiopian history: looking for correlations'. *Journal of Ethiopian Studies*. 24: 74.
Rubenson, Sven et al. (eds). 1987. *Acta Aethiopica, Vol. I: Correspondence and Treaties 1800–1854*. Evanston, Ill: Northwestern University Press.
Rubenson, Sven et al. (eds).1994. *Acta Aethiopica, Vol. II: Tewodros and his Contemporaries 1855–1868*. Addis Ababa: Addis Ababa University Press/Lund: Lund University Press.
Rubenson, Sven et al. (eds). 2000. *Acta Aethiopica, Vol. III: Internal rivalries and Foreign Threats 1869–1879*. Addis Ababa: Addis Ababa University Press/Transaction Publishers: Rutgers University.
Rushdie, Salman. 1991. *Imaginary Homelands: Essays and Criticism, 1981–1991*. London: Granta Books.
Sahlins, Marshall. 1995. *How 'Natives' think: About Captain Cook, for example*. Chicago: The University of Chicago Press.
Sasse, H-J. 1986. 'A Southwest Ethiopian language area and its cultural background'. J.A. Fishman et al. (eds) *The Ferguson Impact. Volume 1: From Phonology to Society*. Berlin: Mouton de Gruyter.
Sato, R. 1997. 'Formation of the historical consciousness among the Majangir: a preliminary view with an analysis of the narratives on their own history'. *Swahili and African Studies*. 7: 78–104 (in Japanese).
Schlee, G. 1994 [1989]. *Identities on the Move: Clanship and Pastoralism in Northern Kenya*. Nairobi: Gideon S. Were Press.
Schlee, G., and A.A. Shongolo. 1995. 'Local war and its impact on ethnic and religious identification in Southern Ethiopia'. *Geo Journal*. 36 (1): 7–17.
Scott, James C. 1990. *Domination and the Arts of Resistance: Hidden Transcripts*. New Haven: Yale University Press.
— 1998. *Seeing Like a State: How Certain Schemes to Improve the Human Condition Have Failed*. New Haven: Yale University Press.
Sergrew Hable Sellassie. 1972. *Ancient and Medieval Ethiopian History to 1270*. Addis Ababa.
Seyoum Y. Hameso, Trevor Trueman and Temesgen M. Erena (eds). 1998. *Ethiopia: Conquest and the Quest for Freedom and Democracy*. London: TSC Publications.
Sisay Ibbsa. 1990. 'The place of Gada in independent Oromia'. *Waldhaansso*. 2: 3–8.
Skocpol, Theda. 1979. *States and Social Revolutions: A Comparative Analysis of France, Russia, and China*. Cambridge: Cambridge University Press.
Smith, D.A. 1897. *Through Unknown African Countries*. London: Arnold.
Solomon Addis. 1994. A history of the city of Gondar, 1934–1966 E.C. Addis Ababa University: MA thesis.
Sorenson, J. 1991. 'Discourse on Eritrean nationalism and identity'. *Journal of Modern African Studies*. 29 (2): 301–17.
— 1992. 'History and identity in the Horn of Africa'. *Dialectical Anthropology*. 17: 227–52.
— 1993. *Imagining Ethiopia: Struggles for History and Identity in the Horn of Africa*. New Brunswick, NJ: Rutgers University Press.
— 1998. 'Ethiopian discourse and Oromo nationalism'. In Asafa Jalata (ed.) *Oromo Nationalism and the Ethiopian Discourse: The Search for Freedom and Democracy*. Asmara: Red Sea Press.

Bibliography

Sorra Adi. 1998. *Proceedings of the meeting on Pastoral Oriented Development and Extension Concept (PODEC) for Boorana Zone.*

Sorrenson, M.P.K. 1976. *Land Reform in Kikuyu Country.* Nairobi: Oxford University Press.

Spaulding, Jay. 1985. *The Heroic Age in Sinnar.* East Lansing, MI: African Studies Centre.

Spitter, G. 1983. 'Administration in a peasant state'. *Sociologia Ruralis.* 22 (2): 130–43.

Stauder, J. 1970. 'Notes on the history of the Majangir and their relations with other ethnic groups of Southwest Ethiopia'. *Proceedings of the Third International Conference of Ethiopian Studies.* Addis Ababa: Haile Selassie I University.

— 1971. *The Majangir: Ecology and Society of a Southwest Ethiopian People.* London: Cambridge University Press.

— 1972. 'Anarchy and ecology: political society among the Majangir'. *Southwestern Journal of Anthropology.* 28: 153–68.

Strecker, Ivo. 1979. *The Hamar of Southern Ethiopia: Conversations in Dambatti, Volume 3.* Hohenschäftlarn: Klaus Renner.

— 1992. 'Berimba's resistance: the life and times of a great Hamar spokesman as told by his son Aike (Baldambe)'. Unpublished manuscript.

Taddesse Berisso. 1995. 'Agricultural and rural development policies in Ethiopia: a case study of villagization policy among the Guji–Oromo of Jam Jam Awraja'. Michigan State University: PhD Dissertation.

Tadesse Tamrat. 1972. *Church and State in Ethiopia, 1270–1527.* Oxford: Clarendon Press.

— 1982. 'Early trends of feudal superimposition on Gumuz society in Western Gojjam'. International Symposium on History and Ethnography in Ethiopian Studies. Addis Ababa University. Mimeograph.

— 1988. 'Processes of ethnic interaction and integration in Ethiopian history: the case of the Agaw'. *Journal of African History.* 29: 5–18.

Tadesse Wolde. 1999. 'Entering cattle gates: bond friendship between the Hor and their neighbours'. Paper presented to the workshop 'Regional Variations in the Cultural Systems of Southern Ethiopia'. Oxford, Harris-Manchester College. Unpublished.

Tarekegn Adebo. 1996. 'Democratic political development in reference to Ethiopia'. *Northeast African Studies* (N.S.) 3 (2): 53–96.

Tefera Fufa. 1988. 'Suffer the Oromo'. *Africa Events.* February 1988.

Teferra Haile Selassie. 1997. *The Ethiopian Revolution, 1974–1991: From a Monarchial Autocracy to a Military Oligarchy.* London: Kegan Paul International.

Tekeste Negash and K. Tronvoll. 2000. *Brothers at War: Making sense of the Eritrean-Ethiopian War.* Oxford & Athens, OH: James Currey & Ohio University Press.

Tekle Tsadik Mekouria. 1960–1967 [Eth. Cal. 1953–1960]. *Ya-Ityiopia Tarik* (2 vols; in Amharic). Addis Ababa.

Tesfaye Tafesse. 1994. *The Agricultural, Environmental and Social Impacts of the Villagization Programme in Northern Shewa, Ethiopia.* Addis Ababa University: IDR Research Report no. 44.

Teshale Tibebu. 1990. 'War culture and the quest for democracy in Ethiopia'. *Imbylta.* 1 (1).

— 1995. *The Making of Modern Ethiopia (1896–1974).* Lawrenceville, NJ: Red Sea Press.

Tilly, Charles. 1990. *Coercion, Capital, and European States, A.D. 990–1990.* Oxford: Blackwell.

Todd, D.M. 1976. 'Politics and change in Dimam, Southwest Ethiopia'. University of Kent: PhD Dissertation.

Togo, Takashi. 1982. 'Kalashinikov monogatari: AK totsugekiju no profile'. In Togo, T. (ed.) *Senjo wa bokurano omochabako.* Tokyo: Tokuma Shoten. (in Japanese).

Tokoi, Masami. 1992. *AK-47 and Kalashnikov Variation.* Tokyo: Dainippon Kaiga. (in Japanese).

Transitional Government of Ethiopia. 1991. *The 1984 Population and Housing Census of Ethiopia: Analytical Report at National Level.* Addis Ababa.

Triulzi, Alessandro. 1975. 'Trade, Islam and the Mahdia in Northwestern Wallagga, Ethiopia'. *Journal of African History* 16: 55–71.

— 1981. *Salt, Gold, and Legitimacy: Prelude to the History of a No-man's Land—Bela Shangul, Wallagga, Ethiopia (ca. 1800–1898).* Naples: Istituto Universitario Orientale, Seminario di Studi Africani.

— 1983. 'Competing views of national identity in Ethiopia'. In I.M. Lewis (ed.) *Nationalism and Self-Determination in the Horn of Africa.* London: Ithaca Press.

— 1996. 'United and divided: Borana and Gabaro among the Macha Oromo in Western Ethiopia'. In P.T.W. Baxter, J. Hultin and A. Triulzi (eds) *Being and Becoming Oromo: Historical and*

Bibliography

Anthropological Enquiries. Trenton, NJ: Red Sea Press.

Tronvoll, Kjetil. 2000. *Ethiopia: A New Start?* London: Minority Rights Group International.

Turton, David. 1973. The social organisation of the Mursi, a pastoral tribe of the lower Omo valley, Southwest Ethiopia. London School of Economics and Political Science: Ph.D Dissertation.

— 1978. 'Territorial organisation and age among the Mursi'. In P.T.W. Baxter and U. Almagor (eds) *Age, Generation and Time: Features of East African Age Group Systems*. London: C. Hurst.

— 1986. 'A problem of domination at the periphery: the Kwegu and the Mursi'. In D. Donham and W. James (eds) *The Southern Marches of Imperial Ethiopia: Essays in History and Social Anthropology*. Cambridge: Cambridge University Press.

— 1992. 'How to make a speech in Mursi'. In P. Crawford and J. Simonsen (eds) *Ethnographic Film Aesthetics and Narrative Traditions*. Åarhus: Intervention Press.

— 1995. 'History, age and the anthropologists'. In G. Ausenda (ed.) *After Empire: Towards an Ethnology of Europe's New Barbarians*. London and San Marino: Boydell Press and Center for Interdisciplinary Research on Social Stress.

— 1996. 'Migrants and refugees'. In T. Allen (ed.) *In Search of Cool Ground: War, Flight, and Homecoming in Northeast Africa*. London: James Currey.

Turton, D., and C. Ruggles. 1978. 'Agreeing to disagree: the measurement of duration in a Southwestern Ethiopian community'. *Current Anthropology*. 19 (3): 585–600.

Ullendorff, E. 1960. *The Ethiopians*. Oxford: Oxford University Press.

Uusitalo, M. 1989. 'Some particles and conjunctions in Konso discourse'. Paper presented to workshop on discourse analysis at the East Africa Bible Translation Centre in Nairobi, 1989.

Vannutelli, L., and C. Citerni. 1899. *L'Omo: Viaggio D'esplorazione Nell'Africa Orientale*. Milan: Ulrico Hoepli.

Vestal, M. 1996. 'An analysis of the new constitution of Ethiopia and the process of its adoption'. *Northeast African Studies* (n.s.) 3 (2): 21–38.

von Höhnel, L. 1894. *Discovery of Lake Rudolf and Stefanie*. London: Longmans Green.

Wallelign Makonnen. 1969. 'On the question of nationalities in Ethiopia'. *Struggle*. 17 January 1969.

Watson, E.E. 1998. 'Ground truths: land and power in Konso'. Cambridge University: PhD Dissertation.

Welby, M.S. 1900. 'King Menelik's dominions and the country between Lake Gallop (Rudolf) and the Nile Valley'. *Geographical Journal*. 16: 292–306.

— 1901. *Twixt Sirdar and Menelik: An Account of a Year's Expedition from Zeila to Cairo Through Unknown Abyssinia*. London: Harper.

Wilson, Thomas M. and H. Donnan. 1998. *Border Identities: Nation and State at International Frontiers*. Cambridge: Cambridge University Press.

Wolde Selassie Abbute. 1997. 'The dynamics of socio-economic differentiation and change in the Beles-Valley / Pawe resettlement area, North Western Ethiopia'. Addis Ababa University: MA Thesis.

Yasin Mohammed. 1990. 'The peopling of highland Illbabor'. *Proceedings of the Fifth Seminar of the Department of History*. Addis Ababa: Addis Ababa University.

Yimer Yesuf. 1996. 'A history of Qeto resettlement area (1985–1995)'. Addis Ababa University: BA Senior Essay.

Yohannes Hadaya and Gemechu Gedeno. 1996. *Konso–English–Amhara Agricultural Dictionary*. Farm Africa FRP Technical Pamphlet no. 10.

Young, Crawford. 1993. *The Rising Tide of Cultural Pluralism: The Nation-State at Bay?* Madison, WI: University of Wisconsin Press.

Young, John. 1998. *Peasant Revolution in Ethiopia: The Tigray People's Liberation Front, 1975–1991*. Cambridge: Cambridge University Press.

— 1999. 'Along Ethiopia's western frontier: Gambella and Benishangul in transition'. *Journal of Modern African Studies*. 38 (2): 321–46.

Zewde Gabre Sellasie. 1975. *Yohannes IV of Ethiopia: A Political Biography*. Oxford: Clarendon Press.

Index

Aari (land and people) 34-5, 59-73, 182-3; and state 59-63, 68-73
Abba Samuél church 245-8
Abi Adi 96
Abwobo **220**, 222, 224, 226, 236
Abyssinia *see* Ethiopia
Addis Ababa 1, 11, 16, 24-5, 27, 35, 55, 74, 85, 89, 90-3, 106, 108, 114, 159, 242, 263, 267; University 145
Addis Ababa Agreement (1972) 266
Adhana Haile 285-6
Adigrat 109, 111
Adirman, Muse 188, 190-1, 195
Adwa 11, 103, 107
Aduk, Medho 231-2, 235
Afa Xonso language 153, 198, 200, 202, 204-8, 218
Afar 27
Afar Liberation Front 85
Afar people 22, 30, 82
Afewerki, Isaias 5
Afqera 80-2
Africa 258, 277, 282; *see also* Horn of Africa *and name of country*
Agar, Commander Malik 270
age 136; *see also* elders
age systems 37-8, 42, 46-8, 50, 57, 119-20, 127-8, 152, 157, 161, 164, 166, 168, 170-1; *see also gada*, generation-grade systems
Agricultural Marketing Corporation 19, 132, 138
agriculture 10, 12, 20, 22, 117-18, 269; Anywaa 221, 225; Arsi 13; Assosa 266; Guji 123-4, 131-2; Hor 41; Jebelein 268; Konso 199, 201, 208; Majangir 192, 194; Muguji 175; Qeto 133, 135, 138-9, 143, 147, 149; Suri 160; Tigray 91, 95, 97, 114; surplus expropriation 15, 17, 19, 23, 30; *see also* cash crops; gardens *and name of crop*
aid, international 7, 13, 17,

23-4, 88, 263-4, 267; *see also* non-governmental organizations
air attacks *see* warfare, aerial
Akwer, Uceri 228
alcohol *see* drinking
Ali Mirah, Sultan 82
Aliyi family 85-6
All Ethiopia Socialist Movement *see* MEISON
Aluoro, River 225, 233
American Presbyterian Mission 187-9
Amhara 27, 239-40, 243-4, 256; and state 29, 244, 247, 251, 255-6; people 6, 11, 29, 63, 108, 154,192, 201-2, 220, 240-2, 244, 251, 252, 269, 283-6
Amhara National Democratic Movement 26, 108
Amharic language 11, 29, 41, 104, 145, 150, 152, 154, 160, 167, 194, 200, 205, 248, 277
ancestors 68, 128
Andom, General Aman 16
angels 239-40, 248, 252
Anger Gutin 141
animal husbandry *see* livestock
Anuak people *see* Anywaa people
Anya Nya movement 228, 262
Anywaa language 227
Anywaa people 21, 153-4, 166, 171, 186, 191, 193, 219-38, **220-1**, 260-1, 264, 268; and state 154, 221-2, 225-6, 228-30, 236-9
Arabic language 262, 269
Arabs 260, 263, 268, 286
Arba Minch 47, 51, 55, 200
Arbore people *see* Hor people
Aregash **95**
Aregay, Merid Wolde 279
Arero 210
Armar, Dalle 58
armed forces 23, 30, 257; and Aari 60, 68-71; and Anywaa 235-6; and Guji 130; and Hor 40-3, 53, 56-7; and Konso 202; and Majangir 185, 193; and Muguji 182; and Suri 158, **162**; Gondar 243;

Northern Shewa 81-3, 85, 88; Somali war 17, 86; Tigray 96-7, 102, 106-7; western border region 261, 264-71, 273-4
arms 11, 17, 41, 43, 55, 106, 152, 163-4, 166, 171, 173-84, 185, 192, 217, **220**, 229, 236, 265
Arsi 13
Asafa, Jalata 283
Assab 22
assemblies *see nab*s
Assile Hamar people 41-2
associations 256; *see also* herding; *luuba*s; peasants; *senbettés*; women; youths
Assosa 141, 260-71, 273-4
Audio Scripture International 189n
Awasa 199
Awash 13
*awraja*s 13, 91, 95-6, 118, 121-3, 131; *see also place name*
Awsa 82
Axum 10, 28, 105n, 107; Bank 105-6
Azzezo 246-7

Baaka chiefdom 60
Baale people 158
Bacha people 175
Baissa, Lemmu 285
*baito*s 23-4, 95, 111-14
Bakko 61
Bale 13, 19, 130
Bambeshi 260-1
Bank of Ethiopia 107-8
Banna people 176, 178-83
Baptist Mission of Ethiopia 77
barley 119
Baro, River *see* Upeeno, River; *see also* Sobat
Barre, Siad 82
beads *see* bridewealth
Bebeka 192-3
beer *see* drinking
bees *see* honey
Begemdir 256
Begi 262, 265
Beni Shangul 27, 269, 287
Beni Shangul People's Liberation Front 265
Beqawle *see* Karate

300

Index

Bertha language 270
Bertha people 260, 262, 264-5, 269, 287
Beta Kehnet 250
Bible 169, 194
Bilpam 263
Bio-Barka 65-6
Birale cotton plantation 58
Birru brothers 80-3
Blue Nile region 259-60, 266, 268, 270
bond-partners 175-6, 178-9, 181, 183
Bonga 263, 272
Borana 118, 198, 209-10, 212, 216; language 57; people 22, 57, 180, 209-10, 287
borders 259-75
Bore 118
bracelets 42
bridewealth 53, 178, 225-8
bridges 77-8, 261
brigandage 39
Britain 14, 116-17, 224, 260-1
British South Africa Company 116
Bruké Demissé 80-3
Bryant, Mike **169**
Bule Uraga *wereda* 118
burial 98n, 128, 189, 196, 247, 255
'Burun' people 261-2
Buso village 213

calendars 207-8, 240, 248, 252
Canada 232
cash crops 143, 149
caste 64, 68; *see also* class, social
castles 240-2, 244
Catholic Relief Agency 88
Catholics 51, 127
cattle 42, 54, 64-5, 75, 119, 156, 160, 162, 173, 176, 178-9, 181, 184; gates 37, 43, **43**, 45-6; see also oxen
Central Intelligence Department 101
cereals 12, 212, 243
Chali 262, 266, 268, 270
charities 255
chernan ritual 38
chiefs and chiefdoms 60, 156, 160, 162-3, 179
children 41-2, 44-6, 51-2, 63n, 120, 127, 136, 144-5, 150, 154, 198, 201, 211-13, 226, 243; *see also* youths
chisennya system 62
Christian Relief Development Association 88
Christianity 113, 136, 145-6, 157, 159, 262, 266, 277, 283-4, 286; *see also* Ethio-

pian Orthodox Church *and names of other sects*
civil service 56, 73, 109, 271
clans 30, 42, 47, 50, 119-20, 158, 175, 186, 191, 194, 196, 212, 224, 226, 228; *see also* name
class, social 15, 34, 63, 109, 278; *see also* caste
coffee 12, 53, 61, 65, 68, 119, 142, 148
collectivization 133, 135, 137-8, 149
colonialism 4, 10-12, 14, 28, 116, 258, 282-5
commerce *see* trade
Commission for Organizing the Party of the Working Class of Ethiopia *see* ISEP-AKO
commoditization 183
commoners 225, 227-8, 231
conscription *see* armed forces
construction 240, 244-50, 253-4, 256
cooperatives 19, 23, 34, 79, 131, 133, 135
cotton 58, 225
courts of law 48, 50, 205
crime 7, 127, 132, 183, 186, 229, 234
Cuba 17, 106, 221
culture 2, 6, 21, 153, 276, 280; Aari 62, 67, 73; Amhara 34, 201, 283; Gondar 239, 250; Guji 119, 125, 127; Konso 199, 201-2, 207, 210, 212-17; Maale 33; Majingir 196; Oromo 286; Sidam 285; Suri 156-9, 161, 163, 165, 168, 170-2; Tigray 104, 106; western border region 257-8, 273, 283-7; *see also* customs; tradition
Cushitic languages 205
Cushitic peoples 286
customs 151, 156, 162, 165, 172, 188-91, 195-6, 225, 254; *see also* culture; tradition
Cwobo 222, 224-6, 228, 230-3, 235-6

dabare system 124-5
Damazin 266
dams 221, 225, 233
dancing **220**
Dassanetch people 178, 181
death *see* burial
Dembidolo 191
democracy 113, 154, 209, 217, 238, 253, 256

Derg 3-5, 7, 11, 14-26, 28, 30, 34-5, 257, 261, 278-9; and Anywaa 219-20, 222-8, 230-3, 235-8; and Hor 37-58; and Konso 200, 210, 213-14; and Majangir 151, 185, 187-9, 191-2; and Muguji 174, 179, 184; and Oromo 283; and Suri 152, 156-8, 161, 163-4, 166-7; Assosa 264-6; Gondar 154, 242-3, 247, 250-3, 255-6; Northern Shewa 76-7, 79-89; Qeto 133-5, 137-9, 142, 149; Tigray 91-4, 96-100, 102-4, 106-15; western border region 268
Dervishes 245, 247
development 13, 16, 20, 23, 86, 112-15, 117, 134, 161, 193-5, 201-2, 209-10, 225, 227, 249, 265-6
Development Through Cooperation Campaign *see zemecha*
diaspora 2-3, 5, 25, 258, 279-80, 284
Dimma 263, 271
Dinka people 263, 265, 269
Disciples of Change 40n
discrimination 95, 113, 197, 284
diseases *see* health
districts *see weredas*
divorce 113-14, 135
Dizi people 155, 159, 162-6, 171, 175
Dollote IV (Wolekorro) 160
Donham, Donald L. 12, 62
Dordora 69
drinking 52, 127-8, 132, 196, 226-7
drought 71, 74-5, 141-2
drums 211, 213
Duba Fora 45, 47
Dul *see* Kurmuk

economy 12, and Aari 61, 68; and Anywaa 223; and Guji 123-5, 130-2; and Muguji 181, 183; and Suri 156, 159; global 10-13, 15, 17(*see also* globalization); Gondar 243, 248-9, 254-5; Qeto 138; reforms 23; Tigray 91, 99-100; western border region 260
education; and Aari 72-3; and Anywaa 226, 237; and Guji 125; and Hor 44, 51-2; and Konso 153, 200-1, 205, 208, 214, 216; and Majangir 194; and Suri

301

Index

158, 163; Gondar 240, 243, 253; Qeto 145, 150; Tigray 91, 94-5, 98, 111-14; *see also* literacy
Educational Sector Review 75
egalitarianism *see* equality
elders 42-3, **47**, 47-8, 55, 58, 66-7, 136, 162-6, 168, 208, 223-30
elections 41, 46-7, 95, 113-14, 151, 237-8
elites 6-7, 12, 17, 154, 278; Amhara 94, 242, 284-5; Anywaa 230, 237; Konso 28, 214, 217-18; northern Ethiopia 153; Northern Shewa 75-7, 80, 83; Qeto 135, 138; Sudanese 262; Suri 158, 160-1, 168, 171-2; Tigray and Tigrayans 4, 94, 114-15, 284; trading 269; western border region 272
Emergency Relief Desk 7, 24
encadrement 14, 17, 20, 22-3, 28, 33-4, 130-1, 151, 154, 219
ensete 12, 19
environment 129, 134, 146-7, 150, 209
equality 63-4, 159, 161, 195, 197, 209, 227
Eritrea 2, 4, 7, 14, 16-17, 21-2, 25-6, 28, 30, 69, 71, 88, 117, 243, 263, 270, 278, 279n, 287
Eritrean Liberation Front 265
Eritrean People's Liberation Front 3-4, 16-17, 23-5, 86, 265, 268, 274
Ethiopia; Civil Code 62; Constitution (1994) 201; Federal Democratic Republic of Ethiopia 27; House of People's Representatives 167; Ministry of Agriculture 51, 123, 138; Ministry of Culture 245, 247; Ministry of Education 51; Ministry of Health 51; Ministry of Land Reform 79; National Military Service and Civil Defence Proclamation 88; northern highlands 10, 12, 15; People's Democratic Republic of Ethiopia 21; Population and Housing Census (1994) 118, 221n, 269; Transitional Government 177, 182, 184; (western) and state 259-64, 266-7, 269-70,

272; western border region 259-75
Ethiopian Democratic Officers' Revolutionary Movement 109n
Ethiopian Democratic Union 35, 85-6, 265
Ethiopian Evangelical Church *see* Mekane Yesus Church
Ethiopian Orthodox Church 1-3, 10-13, 29, 33-5, 83, 154, 170, 201, 214, 239-56
Ethiopian People's Democratic Movement *see* Amhara National Democratic Movement
Ethiopian People's Revolutionary Democratic Front 3-6, 25-30, 257, 269, 285; and Anywaa 153, 220, 230-1; and Hor 34, 37, 49, 55-8; and Konso 198; and Majangir 185, 192-3; and Muguji 173n; and nationalities 21; and Suri 151, 156-8, **162**, 165-7; Council of Representatives 26; Gondar 154; Northern Shewa 89; Tigray 108, 114-15; western border region 269
Ethiopian People's Revolutionary Party 16, 35, 84-6, 98, 108
ethnicity 6, 16, 26-8, 30, 130, 156, 161, 230, 258, 276, 279-81, 283-4, 287; Anywaa 154, 238; Hor 40; Konso 153, 199; Majangir 191, 195, 197; Muguji 152, 174, 177, 184; Oromiya 36; Soviet Union 21; Suri 157, 165-8, 170-2; Tigrayans 4, 285; Wellegga-Qeto 35, 145, 147
ethnography 207, 220, 245, 251, 256, 262-3, 281
Europe 11, 282; *see also name of country*
Evangelical Christianity 151-3, 168-70, 185-97; *see also* Mekane Yesus Church
Evans-Pritchard, E. E. 186

Fadasi *see* Bambeshi
Falashas 112
families 45-6, 85, 101, 119-20, 125-7, 132, 134, 181, 211, 243, 265; *see also* households *and under name*
famines 19, 23, 28, 71, 75, 87-

8, 106, 130, 134, 141, 225, 265
Farm Africa 205
Farmers' Training School 53
farming *see* agriculture
Fazoghli 260
federalization 198-9, 202, 205, 210, 238, 282
Fellata 268
fertility 38, 43
festivals **251**
firearms *see* arms
fishing 175, 228
forests 129, 147, 193, 265
frontiers *see* borders
Fugnido *see* Pinyudo
fund'o system 34, 47-50
'Funj' people 270

gaala people (Anywaar term) 223, 229-30, 283-4
Gabbra people 287
gada 119-20, 127-8
Galla people *see* Oromo people
Gambela 19, 21, 27, 30, 153, 191-2, 220-3, 232, 236-7, 261, 263-4, 268-70, 272, 273, 274
Gambela People's Liberation Movement 222, 230
Gamo-Gofa 40, 59, 61
Gamo highlands 6
Garang, John 229, 237
gardens 124, 131, 140, 142
Garri people 287
gebbar system 12-15, 34, 60-2, 73, 156, 159
Gedeo people 29
Ge'ez language 11, 145, 277
Geissan 260-2, 264-5, 267-70, 273-4
generation-grade systems 209-14; *see also* age systems
German Development Service 269
Germany, East 106
gifts 38, 48, 74, 135, 141, 175; *see also* offerings
Gishé 80-2
globalization 25, 157, 165, 170-2; *see also* economy; multinationals
goats 50, 57-8, 173, 175, 236
Godare 185, 187-9, 191, 193-5
Gofa 175-6; people 60
Gojjam 13, 19, 259, 263, 283
gold 232, 271
Gondar 16, 29, 112, 154, 239-56
government, central *see* state
grain *see* cereals
Great Britain *see* Britain

Index

Green Campaign 86
groups 65-6, 142, 197, 199, 283, 285
GTZ 209-10, 212
Guji people 29, 35, 116-32, **120-1**; and state 116-32; *see also* Oromo people
Gumuz language 263
Gumuz people 259-60, 264, 269, 287
Gurafarda 192
Gurageland 19

Habasha *see* Ethiopia
Haile Selassie I, Emperor 13, 33, 39, 61, 76, 90, 92, 160, 223-5, 227, 230, 233-4, 242, 278
Haile Selassie I University 76, 278-9; *see also* Addis Ababa University
Hamar Banna 57
Hamar people 41-2, 54, 58, 176-7, 181
Harar 27
Hararghe 130
Hassen, Mohammed 279-80, 283
Hausa language 270
health 68, 129, 247; services 41, 91, 95, 109, 111, 114, 125, 135, 138, 158, 170, 187, 196, 221
herding associations 163
heritage *see* history
hierarchy 19, 64, 159, 237, 254
historiography 220, 258, 276-87
history 1, 4, 153, 206, 222-30, 239-48, 258, 262, 272-3, 276-7
Hoekstra, Harvey *see* Odola
homesteads *see* housing
homicide *see* crime
honey 58, 119, 148, 175, 181, 184, 243
Hor language 40-1
Hor people 22, 34, 37-58, 152, 183; and state 34, 42, 45, 49-50, 56-8
Horn of Africa 281-3, 286-7
hospitals see health services
households 136, 141, 144, 212, 255; *see also* families
housing 119, 123-4, 129, 131, 144, 147-8, 150, 189, 243, 254
human rights 5, 24, 132, 150
hunting 175, 178, 183

identity 2, 25-6, 153, 156, 258, 278-82, 286; Aari 67; Afar 30; Amhara 29, 63; Gamo

highlands 6; Gondar 239-40, 244; Konso 204, 206, 216, 218; Kunama 30; Muguji 177, 179, 182; Oromo 283-4; Qeto 134, 146, 149; Suri 157, 161, 168-72; Tigray 13, 98; Wellegga 13
Ideological School 22
ideologies 5, 21, 155, 160, 196, 214, 224, 227, 237-8, 240, 256, 266, 276, 279-80, 286-7
Idinitt people 175
Ilemi Triangle 174
IMALEDH 87
imperialism 241
industry 91, 114
Ingessana Hills 270
initiation 38, 165-7
Institute for the Study of Ethiopian Nationalities 21, 161, 279
insurgency 130
Inzing *see* Assosa
irrigation *see* water
ISEPAKO 87, 102
Islam 2, 10, 12-13, 22, 27, 29, 91, 113, 121, 127, 136, 146, 265, 270-2
Islamic Front for the Liberation of Oromiya 27, 29
Israel 106, 112, 116
Italy 14, 28, 39, 92, 224, 231; and Aari 62; and Sidam 54; and Suri 159-60; Battle of Adwa 11; cooperative villages 116; Eritreans 25; Gondar 245, 247; Kurmuk 261; rifles 177, 179; World War II 67, 160, 261
Itang 221, 236-8, 263, 268, 274
ivory 174, 178, 183-4

jald'abas see leaders
JamJam 118, 121-4, 127, 129-32
Jaman 185, 193, 195
Jebelein 268
Jimma 13, 159
Jinka 52-3, 57, 68-9, 73, 177, 179
Johnson, Samuel 204, 243
Juba 179

Kara language 176
Kara people 152, 173, 175-9, 181-2
Karate (formerly Beqawle) 55, 200, 205, 215: region 207
Karmet people 41

Karo 173, 176-7; people 176
Kefa 13
Kenya 55, 116-17, 174, 232
kernat ritual 48, 56
Khartoum 270
Kibish 161
Kibo, Daniel **169**
kings, ritual 60, 154, 224
kinship 34, 153
Koegu people *see* Muguji people
Koma people 260, 264, 269-70
Kombolcha 74
komorus **160**, 160-3, 166-8, 172
Kong Diu 237
Konso land and people 27-8, 44, 47, 49-50, 153, 198-218, **203-4**; and state 198-9, 201-2, 209, 214-18
Konso language *see* Afa Xonso
Kumi 186, 191, 193
Kunama people 30
Kurmuk 260-74
kwaari see leaders
Kwegu people 152, 175-6; *see also* Idinitt, Nyidi, Koegu

labour 123-4, 128, 133, 135, 137, 139-43, 149, 176, 196, 269
lam concept 223
land 4, 220, 277, 287; Aari 62-3, 65, 72; Anywaa 228, 237; Bertha 265; Borana 209; Gondar 243, 247-50, 255; Guji 119; Hor 38, 42, 58; Kara 176; Konso 212, 214; Majangir 152, 193; Muguji 176; Northern Shewa 74, 78-80; Nuer 237; Qeto 134, 139-42, 146, 148-50; reform 14-17, 23, 35, 63, 78-80; Suri 161; Tigray 94-5, 114
Land to the Tiller decree 77-8, 92
landlords 13, 15, 34, 38, 40, 62, 64-7, 78-80, 162, 214, 265
languages 6, 10, 27, 40, 134, 154, 169, 280, 285; *see also name*
Lasta 142, 146
law 46, 49, 62, 113-14, 156, 160, 211; *see also* courts of law
leaders 152, 157; *jald'abas* 38, 42, 44, 46, 48-51, 56; *kwaari* 224, 226-8, 230-1, 233, 236; *mura* 38, 42, 44, 48; *nyieye* 224, 226-8, 230-1, 233, 236; *poqallas* 44,

303

Index

212-16; *qawot*s 37-8, 42-3, 45, 48, 50, 54; *tapa* 186-91, 196
leba sha 60
Lemma **47**, 47-9
Leninism 20-1, 87
Liberia 13
lineage 119-20, 158
literacy 46, 77, 87, 128, 135; *see also* education
livestock 37, 39, 50, 64, 72, 123-4, 127, 129, 131-2, 143-4; *see also name of animal*
Local Community Development Office 77
local government 15, 160, 168, 201, 209, 212, 216, 218, 268-9; see also *awrajas*; *baito*s; *qebele*s; *t'eklay gizat*; *wereda*s
Lonsdale, John 281
Lower Omo *see* Omo
Lutheran World Federation 169
*luuba*s 44

Maale 213; people 33-4, 60, 151, 154
Mago National Park 178, 183
Mago, River 175
maize 119, 138, 143, 149
Majangir people 29, 151-3, 175, 185-97; and state 185, 187, 193, 196-7
Maji 159, 163-4, 168, 174, 186
Malwal, Bona 270-1, 273
Mao people 269
markets and marketing 4, 23, 39, 55, 78, 91, 111, 138, 148, 159, 207, 226, 260, 266, 271
marriage 38, 44, 95, 113, 119, 135, 141, 144, 148, 176, 228, 233
Marriage and Divorce Committee 136
Marxism 14-15, 20-1, 40, 87, 151, 256
masculinity cult 59-60, 62, 66-7
Mau Mau 117
Meban people 270
Mebrat **95**
Mechara 143-4, 148
Meelanir clan 186, 191, 194, 196
Me'en people 155, 163
meetings 41-2, 46, 56, 110, 130, 132, 162, 166, 192-3, 205-8, 210, 235
MEISON 16, 85-6, 98

Mekane Yesus 51-2, 169, 214
Mekelle 91, 95, 100-1, 103, 106-14
Mekuria, Bulcha 280
men 134-7, 151-3, 164, 167, 189, 209-10, 220-1, 224, 227, 231-2, 234, 238, 275
Mengistu Haile-Mariam **5**, 16, 87, 117, 230, 261, 267-9, 274
Menilek II, Emperor 25, 39, 90, 118, 156, 177, 202, 284
Menz 80, 83
metal products 112
Metekel 269
Metemma 260
Meti 185-6, 191-3, 195
migration 28, 237, 274; *see also* refugees; settlers; villagization
military training *see* armed forces
missionaries 152-3, 157, 168-71, **169**, 262, 266; *see also name*
Mitchell Cotts 13
modernism and modernity 16-17, 19-20, 22, 33, 35, 61, 151, 155, 163, 170, 214-15, 242, 247, 253, 266
moieties 119-20
money 159, 181
Moyale 287
Muguji language 176
Muguji people 27, 152, 164, 173-84; and state 174, 184
multinationals 13; *see also* economy; globalization
mura see leaders
murder *see* crime
Mursi people 152, 158, 161, 175-6, 178-9
Muslims *see* Islam
myths and mythologizing 10-11, 105-6, 223, 240, 243, 248, 286-7

Naita *see* Shulugui, Mount
Nasir 236, 261, 263-4, 268-70, **271**, 274
National Democratic Alliance 270, 274
National Islamic Front 270-1
National Shengo 21
National Villagization Coordinating Committee 117
nationalism 13, 25, 30, 130, 242, 244, 264, 277, 280, 284-5
nationalities and nationality 4-6, 11, 21, 23-9, 156, 161-5, 168, 184, 220; *see also*

name of people
nationalization 14
neftennya see settlers
nger ritual 38, 57
Nile *see* Blue Nile region
Nilotic languages 175, 223
Nilotic peoples 285
Nimeiry, Jaafar Mohammed 265-6
non-governmental organizations 24, 51-4, 153, 201, 267; *see also name*
Nor Deng camp 275
North America 25; *see also* Canada; United States
North Korea 106, 221
North Gondar Zone 242
Northern Shewa Region 74-89; and state 74
Norway 7
Norwegian Church Aid 51
Norwegian Save the Children 42, 52
Nuer people 21, 153, 191, 221n, 222, 229-30, 236-8, 260-1, 263, 269, 271
Nyangatom people 152, 164, 171, 175-83
Nyidi people 175
nyieye see leaders

Odola 187-9, 191, 195-6
offerings 249-51; *see also* gifts
Ogaden 53
Oliver, Roland 279
Omo, Lower 173-6, 178, 182-3
Omo National Park 178
Omo, River 175, 178-9
Omo-rate 177
Omotic languages 173
Omotic region 285
Organization of African Unity 17
Oromifa *see* Oromo language
Oromiya 27, 29, 36, 118, 153, 192, 199, 282-3, 285; Southern 118
Oromo language 29, 34, 118, 145-6, 150, 269
Oromo Liberation Front 16, 24, 26-7, 29-30, 147, 264, 268, 274, 279
Oromo people 13, 26, 29, 56, 108, 130, 148, 206, 266-8, 279-80; *see also* Guji people
Oromo People's Democratic Organization 26, 29, 109
Orthodox Church *see* Christianity
oxen 133, 138, 140, 143, 147, 149; *see also* cattle

304

Index

Pari people 222
parishes 244-51, 253-5
pastoralism 12, 22, 37-8, 41, 45, 49, 54, 175, 220, 274, 285, 287; *see also* herding associations
Pawe 269
peasant associations 15, 19-20, 23-4, 50; Aari 64, 69-72; Anywaa 226, 235; Gondar 256; Guji 118, 121-3, 130; Hor 47; Konso 210, 213; Maale 33; Majangir 189; Northern Shewa 79; Qeto 135-6, 138-9, 144; Suri 163; Tigray 91
peasants 15-17, 19-20, 35-6, 153, 220; Aari 61-6, 71-2; Amhara 34; Anywaa 227, 237; Gamo highlands 6; Guji 117-18, 122-5, 130-2; Majangir 185; Northern Shewa 74, 76-9, 81, 83-4; Qeto 149; Tigray 4, 92-4, 99, 110
Pentecostalism 127
Pinyudo **221**, 221-2, 229, 263
police 39-40, 48, 50-1, 55, 57, 67, 187, 222, 260
politics 3, 7, 11-15, 17, 20, 22-5, 27-30, 36, 257, 259, 270, 276-7, 280, 282, 284-5, 287; Aari 59, 62, 65, 73; Anywaa 154, 224-5, 227, 230-2, 238; Benishangul-Gumuz 30; Gambela 30, 261, 263; Gondar 239, 256; Guji 119, 123, 125, 130-2; Hor 34, 40-2, 48, 56-7; Konso 200-2, **203**, 205, 216-17; Kurmuk 272; Maale 33; Majangir 185-6, 191-2, 196-7; Muguji 173, 177, 179, 184; Northern Shewa 75, 83; Nuer 237; Oromo 286; Qeto 138; Suri 156-61, 163-5, 167-8, 170-2; Tigray 92-4, 96, 99, 102, 108-9, 115
polygyny 125
populations 118-19, 141, 144, 220, 221n, 237, 269
poqallas see leaders
poverty 94, 100, 132, 242-3, 255
Presbyterianism 169; *see also* American Presbyterian Mission
prisons 100-2, 106, 109, 112
privacy 127-8, 132
propaganda 105, 134, 267
prostitution 91, 96, 101-2

Protestantism 63-4, 121, 127, 213-16; *see also name of sect or organization*
provincial government see *awrajas*
Provisional Military Administrative Council *see* Derg
Provisional Office for Mass Organizational Affairs 86

Qale-Hiywot Church 51, 169
*qallu*s 119-21
qawot see leaders
*qebele*s 27, 41, 45-6, 48, 50, 91, 112, 152, 177, 189, 195, 237-8, 253, 254, 256, 283; *see also place name*
Qeto 133-50; and state 149
Qey Afer 48, 52

Radda Barna (Swedish Save the Children) 51-2
radio 39-40, 55, 78, 267
rainfall 207-8
refugees 221-2, 225, 232, 236, 246, 257, 263, 265-9, 271-4
regional government *see t'eklay gizat*
regionalization 114, 134, 145, 147, 150, 165, 264, 269, 279
Relief and Rehabilitation Commission 51, 56, 88, 177
Relief Society of Tigray 7, 103, 266
religion 120-1, 156, 201, 277; *see also name*
resettlement *see* settlers; villagization
Rhodesia 116
rice 221
Riek Machar 236-7
rituals 38, 64, 121, 158, 240; see also *chernan*; drinking; *gada*; *kernat*; *nger*
roads 52-3, 77-8, 92, 97, 99
Roman Catholics *see* Catholics
Rossini, Conti 27
Rufo Ali **44**, 45

saints 239-40, 248, 252
Save the Children Fund (USA) 88
School of Oriental and African Studies 279
schools *see* education
Scott, James C. 2-3
Seleshi, Ras Mesfin 13
senbetté associations 253-6
Sennar 260

settlers 15, 22, 59-60, 62-4, 67, 72, 88, 133-50, 159, 168, 220-1, 225, 228-30, 235-6, 265, 269, 285, 287; *see also* refugees; villagization
Shabo people 193
Shakacho people 192
sharecropping 139-40, 143, 148-9
Shawans *see* Amhara people
sheep 66, 173, 175
Sheraro 96-7
Sherkole 271
Shewa 13, 19-20, 28, 34-5, 128; *see also* Northern Shewa Region
Shulugui, Mount 164, 171
Sidam people 44, 52-4, 56, 285
slavery 10, 39, 60, 159, 260
Sobat, River 261; *see also* Baro
social services 68, 125, 131-2
socialism 117, 161, 214
society and social factors 277-8, 282, 284-5; Gondar 239, 242, 244, 251-6; Guji 125-9, 132; Majangir 186-7, 196; Muguji 182-3; Qeto 137, 150; Suri 156-7, 159, 168, 171; towns 272-4
Somali (region) 27, 30
Somali Republic 22
Somalia and Somalis 13, 16-17, 20, 22, 53, 82, 86, 130, 179-80, 287
sorghum 37-8, 41-2, 52, 58, 137, 143, 149, 175, 181, 184, 234
South Omo People's Democratic Organization 57
South Omo People's Development Association 57
South Omo Zone 199
Southern Nations, Nationalities and Peoples Region 27, 29, 153, 199
Southern Oromiya *see* Oromiya
Southern Region 192
Soviet Union 6-7, 17, 21, 23, 25, 97, 106, 114, 282
spirits 190, 195-6, 212, 215
Stalin, Joseph 4-5, 21, 25, 27n
state 1, 3, 9-15, 19, 22-5, 33, 35, 257, 276-8, 284-5; 287; Africa 7, 277, 282; and nationalities 27; and Soviet Union 17; Marxist-Leninist 20; *see also names of areas and peoples*

305

Index

Staton, C. 77
Stauder, J. 186-7
Stephanie, Lake 37
stick-duelling 168, 172
students 15-16, 34, 56, 64, 67, 75-9, 85, 92-3, 98-9, 103-4, 162-3, 264-5, 278-9
Sudan 30, 92, 103, 111, 154, 163-4, 174-5, 181-2, 184, 224, 228, 232, 243, 246, 259-75
Sudan People's Liberation Army 152, 178-80, 192-3, 221-2, 229-30, 236-7, 257, 263-5, 267-72, 274-5
Sudan Interior Mission 63
sugar 13
Suri language 160
Suri people 27, 151-2, 155-72, **162, 169**, 178; and state 155-6, 159-61, 165-8, 170-1
Surkum 260
Surma Council 151-2, 167-8, 171
Surma people see Suri people
Surmic languages 173, 175, 192
surplus expropriation see agriculture
Swedish Philadelphia Church Mission 177
symbolism 239, 244-5, 248, 251, 256, 272, 274

taboos 45, 211, 213
tabotat 239-40, 246-8, 250-2, 254
Tabya 39-42, 44, 46-7, 49-51, 53, 56-8
Tadesse Tamrat 279-80
Tanzania 35, 117
tapa see leaders
Tareke, Gebru 285
Tatek Military Training Centre 53
taxation; Aari 61-2, 68, 71-2; Anywaa 225; Dizi 249, 255; Guji 130, 132; Hor 39, 46; Konso 214; Muguji 184; Qeto 138-9, 148, 150; Suri 156, 158, 160, 163; Tigray 100, 114
taxis 106
Tepi 153, 185, 192-3
Teshale Tibebu 283-4
Tewodros, Emperor 11, 23
Tigray 4, 7, 13, 19, 21, 26-8, 35, 69, 71, 88, 90-115, 117, 263, 265, 279; and state 94-6
Tigray National Organization 93

Tigrayan People's Liberation Front 3-5, 7, 21, 23-6, 35, 86, 92-115, 151-2, 156, 264, 268, 274; Agriculture Department 97; People's Department 94; underground 102-4
Tigrayans 29, 246, 284-6
Tigrinya language 104
Timqet 245, 250
Tirmaga people see Suri people
Toposa people 152, 164, 178
torture 100, 102, 109
tourism 152, 157, 168-9, 172
towns 61, 90-102, 105, 107, 109-15, 148, 215, 243, 257, 260, 272, 274; see also name
trade 12, 257, 269; Aari 61; Africa 10; Anywaa **221**; Hor 38, 43, 53; Muguji 182, 184; Qeto 139, 143, 148-9; Suri 159; Tigray 91; towns 260-1, 263, 266, 274 (see also name of town); see also *fund'o*
trade unions 253
tradition 152-4, 201, 213-14, 216-17, 227; see also culture; customs
transport 53
travel 53, 139
trees 147; see also forests
Tsamako people 41, 58
Tsore 265-6, 267-8, 271
Turkana people 165n

Uduk language 274
Uduk people 262, 264-5, 270-2, **273**
Uganda 178-9
Ukuuna 235-6
Union of Ethiopian Marxist-Leninist Organizations see IMALEDH
Unionist Party 14
United Nations 14, 28; High Commission for Refugees 265, 267, 271
United States 5, 7, 232
Upeeno, River 236
urbanization see towns
USAID 7
USSR see Soviet Union

veterinary services 41-2, 53, 158, 163
villagization 19-21, 35, 41, 68-9, 116-18, **120-1**, 121-32, 151, 185-91, 193-6, 228; Co-ordinating Committees 121, 131; see also settlers

warfare 38, 60, 92, 106, 225, 247; aerial 82, 96-7, 107, 114; Ethiopian-Eritrean War 7, 30, 88; Ethiopian-Somali War 17, 53, 86; Northern Shewa 85; Sudan 164, 174, 180, 184, 261-2, 266, 269-70, 274; Tigray 88; Uganda 178; see also World War II
water 38, 51-2, 58, 77, 82, 125, 146, 170, 221, 225, 233
weapons see arms
Weezar, River 77
Wellegga 13, 19, 35, 133-50, 259, 262
Wello 19, 80, 82, 133, 135, 137, 139-42, 144-5, 149-50, 263, 265, 269, 287
wells see water
weredas 27, 48, 50, 57, 91, 95-6, 113, 151, 153, 157, 162, 167, 198-204; see also place name
Werre, Ilu 80, 82-3
Weyane rebellion (1940s) 4
Woito, River see Limo, River
women; Aari 64, 66, 69-70; Anywaa 153, 221, 224, 227, 230-5, 238, 264; associations 46-7, 71, 102-3, 113, 136, 252; Gondar 252-3; Guji 125-8; Hor 44-5; Konso 218; Majangir 185; Nor Deng camp 275; Qeto 135-7; Tigray 35, 91, **94-5**, 96, 100-2, 110, 113; Uduk **273**
wori see families
work see labour
Workers' Party of Ethiopia see Derg
World War II 159, 179, 258, 259
Wungabaino Hamar people 41-2
Wycliffe Bible School 169, 205

Yabus River 265-6, 268
Yekatit Silsa Siddist Political School 53
Yoruba people 258
youths 57, 189-91, 194-6, 252; associations 46, 71, **220**, 252; see also children

zemecha 15-17, 34, 40, 64-5, 77-9, 162-3, 213, 264-5
Zenawi, Meles 5

www.ingramcontent.com/pod-product-compliance
Lightning Source LLC
Chambersburg PA
CBHW051419290426
44109CB00016B/1358